Avoiding Attack

Avoiding Attack

The Evolutionary Ecology of Crypsis, Aposematism, and Mimicry

Second edition

Graeme D. Ruxton
William L. Allen
Thomas N. Sherratt
Michael P. Speed

OXFORD
UNIVERSITY PRESS

Great Clarendon Street, Oxford, OX2 6DP,
United Kingdom

Oxford University Press is a department of the University of Oxford.
It furthers the University's objective of excellence in research, scholarship,
and education by publishing worldwide. Oxford is a registered trade mark of
Oxford University Press in the UK and in certain other countries

First Edition published in 2004
Second Edition published in 2018

Impression: 1

Published in the United States of America by Oxford University Press
198 Madison Avenue, New York, NY 10016, United States of America

British Library Cataloguing in Publication Data
Data available

Library of Congress Control Number: 2018944296

ISBN 978–0–19–968867–8 (hbk.)
ISBN 978–0–19–968868–5 (pbk.)

DOI: 10.1093/oso/9780199688678.001.0001

Printed in Great Britain by
Bell & Bain Ltd., Glasgow

Dedicated to
Katherine
Jioh Allen
Frances, Finlay, Stuart, Kaelan, Nathan, Kyah, & Jakob
Yvonne

Acknowledgements

Many colleagues helped with reading text, suggesting references, and providing figures. In no particular order, we would like to acknowledge our thanks to Nicholas Scott-Samuel, Anna Hughes, Willem Bekkers, Jolyon Troscianko, Roger Hanlon, Sami Merilaita, Jennifer Kelley, Olivier Penacchio, Innes Cuthill, Sonke Johnsen, Malcolm Edmunds, Jayne Yack, Takeo Kuriyama, Ullasa Kodandaramaiah, Carlos Cordero, Bree Putman, Dinesh Rao, Pierre-Paul Bitton, Lynne Isbell, Chris Hassall, John Skelhorn, Francis Gilbert, Thomas Reader, Krushnamegh Kunte, Karl Loeffler-Henry, Thomas Aubier, Mathieu Chouteau, Kyle Summers, David Kikuchi, Mikhail Kozlov, Kevin Arbuckle, Butch Brodie, Johanna Mappes, Hannah Rowland, Bibi Rojas, Janne Valkonen, & Elena Zvereva. We apologize to those that should be on this list but slipped our mind. None of those named bear any responsibility for our errors and omissions.

The smartest thing the original three authors of *Avoiding Attack* did was bring in Will Allen as a co-author—his drive, energy, and scholarship have been transformative—and every chapter of this book is better for his contributions; some simply would not exist at all without him. The joint smartest thing we did was hire Rosalind Humphries as research assistant to coordinate the four authors. Her friendly, unflappable organization and very gentle chivvying were essential in completing this project and we all owe her a huge debt. Our final smart call was to stay with OUP—Bethany Kershaw was exceptionally patient, always wise in her guidance, and everything we could hope for in an Editor; the production and copyediting side of the book (by Suryajeet Mullick and Julian Thomas particularly) was managed with collegiate efficiency and good grace. Julian in particular combines a very keen eye for detail with exceptional intellectual engagement, professionalism, and wit.

Contents

Introduction

In 2004, the first edition of '*Avoiding Attack: The Evolutionary Ecology of Crypsis, Warning Signals, and Mimicry*' by Ruxton et al. was published. The book aimed to provide a systematic and up-to-date review and synthesis of widespread anti-predator defences. In it, we focussed on sensorially mediated defences and the many factors that underpin these adaptations, aiming to set out the state-of-understanding in the fascinating world of anti-predator adaptations, and highlight which topics within the field seem most ripe for further investigation.

Since the publication of the first edition, many research opportunities have been realized, and our understanding of the diverse and captivating strategies that have evolved in nature has developed significantly. In light of this, we here present a second edition of *Avoiding Attack*. We have strived to update and develop our previous work, particularly in areas where scientific advance has been most radical in the past 13 years, including: mechanisms of crypsis, understanding among-species variation in anti-predator defences through evolutionary genomics and phylogenetics, and the causes and consequences of variation in secondary defences. Since the first book, emphasis in the field of anti-predatory adaptations has also shifted from a very mechanistic perspective to an integrated understanding including broad evolutionary and ecological consequences of adaptations; we hope that this new book broadens our previous focus to consider these consequences more systematically. To do so, most chapters now follow a similar structure where appropriate, first discussing the natural history and key examples of a defence, before shifting to explain the mechanism via which it works to promote survival, followed by the evolutionary history and costs and benefits of the trait. We then review the ecological factors that the trait responds to, as well as co-evolutionary considerations. Each chapter then finishes with an open section where we outline what we see as key unresolved challenges and propose some more-or-less easily achievable potential avenues for resolving these. In addition to our subject focus, we have broadened our author-base too, bringing Will Allen on board to compensate for senescence in the original three.

As with the previous edition, our own interest in predator–prey interactions remains as strong as ever, but we would also argue that general scientific interest in anti-predatory defences and signals has never been greater, as evidenced by recent high-profile reviews and special editions of well-known journals (Cuthill et al., 2017; Caro et al., 2017). There are very few species for which perception by their predators or prey is not a major influence on fitness. Hence the phenomenon of predation and the many sensory adaptations surrounding it remain of great ecological and evolutionary significance.

Interactions between predators and their prey can often usefully be broken down into a sequence of stages, beginning with 1) encounter (spatial and temporal proximity), and leading through: 2) detection, 3) identification, 4) approach, 5) subjugation and, ultimately, 6) consumption (see Endler, 1991; Caro, 2005). In the literature, anti-predatory defences employed by prey individuals ahead of subjugation (stages 1–4) are typically referred to as 'primary defences', affecting the likelihood of the predator physically contacting the prey. While much of this book focusses on primary defences, we also consider 'secondary defences'—those adaptations which act once subjugation (or contact) has begun (stages 5 and 6). We think this view of a predator–prey interaction as a sequence of stages is a powerful

Avoiding Attack: The Evolutionary Ecology of Crypsis, Aposematism, and Mimicry. Second Edition. Graeme D. Ruxton, William L. Allen, Thomas N. Sherratt, & Michael P. Speed, Oxford University Press (2018). © Graeme D. Ruxton, William L. Allen, Thomas N. Sherratt, & Michael P. Speed 2018. DOI: 10.1093/oso/9780199688678.001.0001

conceptual tool for understanding co-evolution of investment by the parties involved, and so we devote the second section of this chapter to a further exploration of this sequence.

There are no perfect defences. No animal has evolved a defence that gives 100 per cent protection against all possible predatory threats. This is not surprising. Predators will often evolve to counteract defensive developments in their prey, and defences can be expensive to maintain. A suit of armour might offer you protection from surprise attack from a medieval swordsman; but you can easily imagine how exhausting walking about in that armour might be, and how it might limit your ability to go about all sorts of daily activities (not to mention how it might limit opportunities for reproduction). Thus we would expect organisms to invest judiciously in anti-predator defences. In the context of the sequence of a predation event, we might expect that if prey invest strongly in being able to flee effectively from predators then they do not also invest in costly secondary defences too. Aside from simultaneous heavy investment in two defences being unattractive from a fitness perspective, it may be impossible from a design perspective. Our notional suit of armour likely makes fleeing from an onrushing swordsman impossible. Similarly, design tradeoffs exist between trying to hide from predators using camouflage and signal to predators using conspicuous coloration.

In terms of general trends across defences acting at different stages of the predation process, we can make some predictions. A predation event can end at different points in the process, so for a prey type that has several defences that act at different stages of the process, the earlier-acting defence will be employed more often. This does not necessarily mean that the earlier-acting defence will be more honed by evolution; it probably does mean, however, that if the two defences have costs that are imposed every time they are engaged, then high costs can be maintained for the late-acting defence rather than the early-acting one. The stakes get higher for the prey the deeper into the sequence a predation event gets as it becomes less likely that the prey will be bailed-out by a later-acting defence. Similarly, the deeper into a predation event we move the more time and effort the predator will have invested in that attack and the less easily it should be convinced to give up on this particular prey item. Thus, we can see obviously very expensive defences—such as autotomy of body parts—occurring as very late-acting defences.

Naturally occurring predation is unpredictable in space and time, in large part because we would expect predators to exploit any predictability in prey to increase predation rates (and for prey to exploit predator predictability to reduce predation). Further, naturally occurring predation can often be adversely affected by the presence of human observers. This creates real challenges studying predation in the field and, as a generality, these challenges increase as we advance through the sequence of a predation event. If we switch from the field to the laboratory

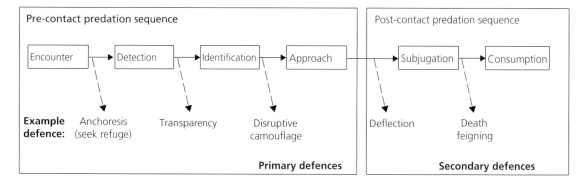

Figure 0.1 The classical predation sequence from search to subjugation (continuous lines) and examples of anti-predator behaviours that have evolved to inhibit specific stages of the sequence (dashed lines). These behaviours can arise pre-contact (as primary defences) or post-contact (as secondary defences) with the predator. Behavioural responses range from inhibiting search (by fighting back or seeking refuge, for instance) to inhibiting consumption (by death feigning, for instance).

then ethical issues often arise, especially when studying anti-predator adaptations in vertebrates. Again, these challenges generally increase as we move deeper into the predation sequence. So this book is in large part a celebration of the fantastic ingenuity of generations of researchers in inventing an extraordinary diversity of imaginative solutions to the challenges of our field. It greatly helps that the researchers you will find in our reference list are exceptionally diverse in their background: you will find individuals that would be comfortable describing themselves as one or more of the following: naturalist, ethologist, animal behaviourist, animal ecologist, evolutionary biologist, vision scientist, psychologist, physicist, biochemist, or mathematical biologist.

Chapter summary

One widespread primary defence is crypsis, an umbrella term for the various strategies organisms use to prevent detection. These are covered in Section 1 of the book: Chapter 1 discusses background matching, an adaptation that remains relatively understudied, perhaps because as 'standard' camouflage it is unfairly considered less interesting than other defences. Among other things, we aim to show how understanding background matching addresses important issues in evolutionary biology, such as the maintenance of polymorphisms. As achieving highly effective background matching in a complex environment is very hard to achieve, there is room for other types of crypsis. In Chapter 2 we explore disruptive camouflage, a topic that has perhaps seen the largest advances since the first edition, which contained a four-page review that concluded there was little empirical evidence for disruption. While background matching aims to minimize the signal sent to the receiver, disruptive camouflage aims also to increase the sensory noise sent to receivers by creating false edges and features that make detection or recognition of an animal's outline and shape difficult. Detecting an object and recognizing what it is are conceptually distinct processes but they are functionally related, and while a predator can detect that something is present without recognizing what it is, the reverse cannot occur. Disruptive camouflage exploits features of its viewer's perceptual processes such as edge detection and figure-ground segregation to achieve camouflage.

Chapter 3 goes on to explore countershading, a term which describes the common colour phenotype where the surfaces of an animal that orient towards the sun are darker than surfaces that face away from the sun. The number of camouflage-related mechanisms that might select for countershading is considerable. We list six, including three that fall under the banner of self-shadow concealment, the hypothesis that animals gain protection from predation by hiding their self-cast shadows. Thus a major task of this chapter is in clearly explaining and distinguishing these alternative mechanisms. Our interest in countershading is about light and vision. Countershading occurs because ambient light is generally highly directional in nature—there is no other sensory system that is likely to be subverted by some equivalent of countershading because there is nothing analogous to the directional nature of ambient light in other modalities—although there is strong directionality in extrinsic thermal radiation and the earth's magnetic field, these are not used in predator–prey interactions.

Concluding Section 1, we discuss the cryptic nature of transparency and silvering in Chapter 4. Transparency adaptations are only common in pelagic habitats and it is important to note that they are not the same as invisibility—reflection and refraction still occur, potentially creating visible features. Interestingly, transparency and polarization vision might be linked phenomena, but recently there has been more scepticism of this idea. Silvering, by contrast to transparency, is only common in mesopelagic habitats. Silvering presents a similar solution to transparency, but allows organs to be opaque without reducing crypsis; we cover some of the downsides of this adaptation.

In Section 2 of the book, we go on to consider adaptations through which prey can avoid attack after they have been detected. Prey species may evolve traits that render themselves unprofitable to predators and ways of advertising these defensive qualities. Chapter 5 first covers the nature of secondary defences, such as armour, spines, and toxins. These are the adaptations that underpin and co-evolve with the defensive advertising described in Chapters 6 and 7 on aposematism and mimicry.

We thus need to understand secondary defences to understand defensive signalling fully. Further, the existence of secondary defences will affect selection on other defences in camouflage. Secondary defences can protect the bearer but also may offer protection to other individuals that the predator subsequently encounters. Plants are discussed more in this chapter than any other. Camouflage is hard for plants because of the demands of photosynthesis and limits on mobility. Their lack of mobility also robs them of fleeing as a defence, or seeking shelter in protective environments like burrows, factors that favour heavy investment in secondary defences. In Chapter 6, we look at aposematism, where a secondary defence combines with a warning signal that advertises possession of the defence to predators. Aposematic signals normally seem to have a visual component. Aposematism has evolved many times and there has been convergent evolution towards some signal phenotypes such as repeating bands of dark colour interspersed with yellow, orange, or red. In recent years fascinating new findings have emerged and the chapter focusses on these discoveries, for example evidence that selection for aposematism changes spatio-temporally—particularly in non-tropical regions aposematism is less effective at the time of year when many predators are young of that year that have not yet learned to avoid aposematic signals.

Chapter 7 turns to look at the occurrence, evolution, and maintenance of defence through Müllerian mimicry. Müller's original theory made a lot of simplifying assumptions about the way in which predators behave when faced with unfamiliar prey, but recent experiments and theory have helped identify how and why predators decide whether to sample these prey. In so doing, this work has helped reveal a much greater range of situations in which we might find Müllerian mimicry. Most examples we find of Müllerian mimicry are invertebrates, although it is also well known in vertebrates such as poison dart frogs and catfish. We discuss how mimicry rings and spatial mosaics are two particularly important and interesting features of the biogeography of coloration; hybrid zones between morphs are an excellent place to study selection and speciation processes. Previous work has been heavily focussed on avian predators, but they are not the only visual predators and we dis-

cuss exciting prospects for non-visual and multi-modal mimicry that are beginning to emerge. Finally, Section 2 concludes with a discussion of elusiveness in Chapter 8, and how prey signal this to predators through pursuit deterrence and perceptual advertisement. These strategies are distinguished from aposematism on the basis that prey are not signalling their difficulty of capture rather than the presence of a post-capture defence. Elusiveness signalling is an area where theory is strong and there are now many empirical examples of elusiveness signalling, especially from terrestrial vertebrates.

The final section of the book, Section 3, concerns anti-predatory defences that are focussed around misleading predators in one way or another. We first look at the evolution and maintenance of Batesian mimicry and masquerade in Chapter 9. Here we proceed with a healthy degree of caution, pausing before ascribing resemblance between two species as mimicry: it might be shared ancestry, or the result of sharing a similar environment (convergent evolution). We also aim to avoid story-telling on the basis of perceived resemblance of the putative model and mimic to humans—who are unlikely to have similar visual and cognitive systems to agents of selection that lead to mimicry. Masquerade has attracted a great deal of interest in the past decade, and we now review a number of these recent studies, as well as their implications. By contrast, Batesian mimicry has long been understood theoretically, and its properties have been supported by lab and field studies. These studies not only demonstrate its occurrence, but also the fine-scale predictions of the theory (such as the nature of the underlying frequency dependence). Nevertheless, since the first edition there has been some exciting progress on a range of topics including uncovering the genetics of some polymorphic Batesian mimics, laboratory experiments on the ways in which natural predators classify profitable and unprofitable prey which differ in several traits, and experimental and theoretical work on how and why imperfect mimicry is not improved upon by natural selection. Fascinatingly, evidence thus far suggests that different mimetic species use very different underlying genetic architectures. The role this has in explaining differences in mimicry systems is an interesting avenue of discussion and future research.

In Chapter 10, we consider startle defences, providing a new definition and discussing such fascinating adaptations like eye mimicry. We also present a discussion on whether distress calls in prey species might be startle signals. Chapter 11 then looks into deflective signals which influence the position of the initial contact of a predator with the prey's body in a way that benefits the prey. Here the interesting questions are how this mechanism interacts with other defences and how predators allow themselves to be manipulated in this way. We next present a new chapter, Chapter 12 on dazzle camouflage, a putative anti-predator defence that was only briefly noted in the first edition, but that has now been subject to intensive experimental research such that we felt it deserved its own chapter. Readers may be familiar with dazzle from photographs of warships painted with high contrast geometric shapes. The idea is that bold patterns on moving prey may cause predators to misjudge the speed and trajectory of prey, hampering pursuit and capture. Experimental support for this is very mixed, and we are also not yet clear whether dazzle effects occur in natural systems. This chapter aims to review what we currently do and do not know about 'camouflage in motion'. Finally, in another chapter new in this edition, Chapter 13 discusses the intriguing defence of death-feigning, or tonic immobility, in which prey adopt a relatively immobile state that can often be visually reminiscent of a dead individual. Like deflection, the interesting questions are how this mechanism interacts with other defences and how predators allow themselves to be manipulated in this way.

Before moving on to these issues, we want to finish this introduction by returning to the concept of predation as a sequence of events. We feel that the sequential aspect of a predator–prey interaction is fundamental to a holistic view of linkages between all the defences that we are about to consider individually in the rest of the book.

The sequence of a predator–prey encounter and investment across multiple defences

As discussed in the previous section, this book will review the varied forms of anti-predator sensory

defences; examining, for each, aspects of natural history, of mechanism, of ecology and evolution. Before we do this in earnest, however, we first pause (briefly) to consider that prey defences are often multiple; camouflaged prey may also be good at rapid movements that lead to escape from nearby predators, chemically defended, warningly coloured, and mimetic prey may also use camouflage or employ tough integuments to protect themselves from predation. Hence here we briefly pose two general questions that should in our view permeate our thinking about prey defences: how many defences should a prey deploy, and what kind(s) should it use?

Perhaps the simplest explanation for the use of multiple defences is that they act at the same time, and tend to maximize survival of a prey from an encounter with a predator. Camouflage and nocturnal activity may both contribute to reduced apparency of prey to diurnal predators, hence combining them may decrease risk of detection, perhaps synergistically. Prey that deploy chemical defences often use multiple toxins that act simultaneously. Examples of simultaneously acting defences are not then hard to find, but they may not be the whole story.

A more complex reason for multiple defences could be that prey defences have a predictable sequential organization, designed around the structure of attacks. As discussed earlier in the chapter, John Endler (1991) argued that an attack by a predator on its (animal) prey is typically composed of a sequence of six stages:

1) encounter (spatial proximity)
2) detection
3) identification
4) approach
5) subjugation
6) consumption

At each stage in this sequence the prey organism can put up one *or more* lines of defence with the aim of preventing, interrupting and stopping the attack. Interested readers are recommended to read Malcolm Edmunds' (1974) and Tim Caro's (2005) texts which both develop the idea that defences evolve to reflect the typical sequences of events during an encounter between predators and prey.

An animal prey may, for example, hide to prevent encounter (stage 1), and detection (2), use masquerade and cryptic coloration to prevent detection (2) and identification (3), perhaps form aggressive defensive groups to prevent approach (4). They may alternatively have a startle display or use vigilance and rapid escape behaviours to prevent approach (4). They may violently retaliate, perhaps using stings, spines, or bites, and/or deploying irritating or toxic chemicals (secondary defences), to prevent subjugation (5) and consumption (6). At each stage in the sequence that Endler identified, one or more defences could be deployed by a prey animal, and they will often operate sequentially, as some defences are typically used only if earlier-acting defences have failed to stop the predation event. We can now ask whether there is one or more general framework(s) within which we can begin to evaluate how many and what kinds of defences prey need.

There are interesting parallels between the organization of biological and human military defences. Both concern protection of valuable yet vulnerable targets, seeking optimal deployment of costly defensive 'assets'. A relevant military tactic is 'layered defence' in which sets of defensive resources, such as border security, naval warships, and intercontinental ballistic missiles, are deployed in sequence; when a first line of defence fails against an incoming threat a second line of defence activates to minimize further risk, and after that perhaps a third or fourth defence. In the military theory literature, layered defence has been described as follows (Wilkening, 2000). Suppose that there are m defence levels and L_i is the probability that a warhead passes through the ith layer of an army's defence. The probability that the target is *not* destroyed, K_w, is

$$K_w = 1 - L_1 L_2 ... L_m \qquad (0.1)$$

The probability that the target survives an attack then increases with the number of defensive levels and the efficacy of each level. Suppose the defence at each layer is only moderately successful, $L_i = 0.5$. In the case of only one defence layer, the probability that the target is *not* reached and destroyed is $K_w = 1 - 0.5 = 0.5$; for two layers $K_w = 1 - 0.5^2 = 0.75$; when there are four sequentially acting defensive layers, however, K_W is now much greater $1 - 0.5^4 = 1 - 0.0625 = 0.9375$. Hence, layering of several moderately effective defences leads to a high probability of survival from

an attack. The maths here can, of course, be applied directly to sequentially acting organismal defences (including plants, microbes as well as animals). Defensive layers can operate at any of the stages of predation that Endler envisaged. There may, furthermore, be several defensive levels within a stage of predation. Thus a prey animal could use several defences in sequence to prevent subjugation—for example sharp spines, tough exterior, and secretion of irritating chemicals. We argue here that a sequential organization of defences provides a useful general framework within which to think about the organization of protective phenotypes.

Thus we can now ask: when should a prey disperse defensive investment across defensive levels or put all of its investment in a single level? Equation 0.1 illustrates the diminishing nature of incremental survival benefits as more and more defensive layers are added to a set of defences. The first layer in our example makes a much bigger contribution to survival (here 0.5) than the second (0.25), the third (0.125) and so on. If, as seems likely, there are costs associated with defences then we can see limits on the number of defences that it is optimal for a prey to invest in. Imagine, for example, that sequentially acting defences are relatively cheap to construct and maintain, but at their best each provides little protection, i.e. the survival benefit saturates at a low value. Here, multiple levels of defence could be optimal to provide high levels of protection in a cost-effective manner. In contrast, a very effective (and indeed a very cost-effective defence) may be sufficient on its own, and effectively block the evolution of any later-acting defences (this has been called strategy-blocking, e.g. Britton et al., 2007).

We suggest that this integrated approach to defences that act in sequence can be important in directing research questions. For example, much of the early part of this book focusses on mechanisms that generate camouflage and crypsis. We can now ask: how often is camouflage so effective that it blocks the need for later-acting (secondary) defences, such as repellent toxicity (and hence, how often are camouflaged prey edible)? Or, how often is camouflage cheap but relatively ineffective and hence followed by a sequence of further defences that act after camouflage has failed, such as rapid escape and repellent toxicity?

An integrated-sequential view of defences can be informative too in helping to determine what kinds of defence should be invested in, not merely the number of defences. An organism that protects itself by reliably deterring enemies before they make contact, for example by camouflage and crypsis or by defensive grouping or warning coloration, suffers no injuries to its tissues at an encounter. In contrast, an organism that relied only on chemical defence may survive encounters but would more likely pay costs that follow from having its tissues ruptured during the attack, or from loss of fluids and other resources. Hence, might we expect selection to favour investment in the *'before contact'* primary defences, over those that operate as secondary defences, during contact between prey and their predators? An additional point is that the 'lifestyle' of an organism is surely influential on the kinds of defences that are deployed. Sessile organisms, for example plants and some animals such as barnacles, cannot use movement as a defence, so hiding, evasion or rapid variation in group size is not possible. The intrinsic apparency of the organism in its situation may determine the overall investment and type of defences used (cf. plant apparency hypothesis, see Stamp, 2003). For example, many forms of cactus and adult barnacles are both exposed and highly vulnerable to predation, and both defend themselves with robust physical defences, shells or spines respectively that will prevent attack by a broad range of predators.

A final consideration with respect to sequential defences is that earlier-acting defences will, by definition, be deployed more frequently than later-acting defences (equation 0.1, and Endler, 1991). For a given effectiveness of defence, therefore, earlier-acting defences will reduce per capita mortality more than later-acting defences and therefore be under stronger natural selection. Can we then expect selection to work harder to make earlier-acting defences more efficient and effective than later-acting defences? Selection on predators for adaptations that overcome early-acting defences should also be stronger, making defences better in absolute terms, but relatively no more effective. Earlier-acting

defences may also tend to be less effective than later-acting defences.

The relative efficacy of multiple defences is perhaps more easily studied in plants than many animals, because risks from enemies can be more easily assessed over a long period in sessile organisms. Thus in a meta-analysis Carmona et al. (2011) found that variation in traits that act at the earlier stages of attacks (varied flowering times, growth and morphology, and physical defences) predicted susceptibility to herbivory more strongly than variation in forms of chemical defence, which offer a late stage of defence. This makes sense in our framework of sequential defences (equation 0.1); since effective chemical defences are then the last line of attack, variation in these traits will be less influential on survival than variation in earlier-acting defences. This interpretation would be consistent with John Endler's (1991) view that it is in the interests of victims to interrupt predation at early stages in order to save energetic and time costs, and to reduce risks of injury (and see Broom et al., 2010).

There are relatively few theoretical treatments of layered defence in the biology literature. Notably Frank (1993) considered the importance of 'sequential defence' models of hosts and their parasites. Here a parasite must successfully cross m barriers to gain any benefit from infection. Frank compares this to models of simultaneous defence in which there is only one defence layer, but with more than one defence in operation. Broom et al. (2010), in contrast, separate defence investment into the two general layers of primary and secondary defences. They use analytical and numerical solutions to determine the different scenarios that favour alternative outcomes (no defence, both primary and secondary investment, primary or secondary investment only). One intriguing prediction is that in some circumstances there may be very little fitness difference between alternative strategies, so that diversification in defensive organization within clades may be strongly influenced by processes of drift. Species that hide and use camouflage may be as persistent as those that use warning coloration and toxicity to defend themselves (and see Stamp & Wilkens, 1993). Recently, Bateman et al. (2014) looked at a two-prey, one-predator system where prey can differ in their investment in primary and/or secondary defences. They sought to identify the condition where one prey type could invade a system where the other initially dominated numerically, but found that this was critically dependent on the fine detail of how costs and benefits of different types of defensive investment were described.

So far we have considered the simple scenario in which prey defend themselves against a single type of enemy. Nature is unlikely ever to be that simple, however, and defence systems must protect prey against multiple enemies which may have different modes of attack, and require different kinds of defence (see the excellent review in Greeney et al., 2012). Animal prey may have to defend, for example, against predators of different sizes and levels of mobility and against enemies such as ants or parasitoids which have different modes of attack to large predators. Rojas et al. (2017), for example, recently showed targeting of alternative defensive secretions by wood tiger moth (*Arctia plantaginis*) toward different kinds of enemy (here ants and birds). There are field-based measures of animal prey survival based on alternative forms of protection (see for example Lichter-Marck et al., 2015; Carroll & Sherratt, 2013); however, the plant literature provides quantitative insight of relevance. As part of their meta-analysis, Carmona et al. (2011), determined whether alternative defences offered protection to different kinds of herbivores. Their results show support for the idea that multiple types of defence evolved to protect against different types of herbivore. Chewing insects, for example, were deterred from damaging plants (expressed as per cent leaf area lost, biomass reduction etc.) by variation in gross morphology (size, number of branches etc.) and by variation in life history variables (flowering time, growth rates), whereas herbivores that pierce and suck plant fluids were not deterred by these traits (see Table 1 of Carmona et al.). Specialist herbivores were particularly deterred from damaging plants by physical defences and life history variation, whereas generalists were not deterred by these variables.

We suggest then that an important empirical question is the extent to which different defences in natural systems evolved independently, each to protect against different types of enemy, or because they are deployed in sequence against enemies. What seems to lack systematic investigation is whether stages in the sequence of an attack (early or late) tend to be more generalized and function across sets of enemy types or more specific and function only against specific enemies. Endler (1991), and Broom et al. (2010) have argued that primary defences tend to be applied generally to sets of different predators, whereas secondary defences may be more specialized. Cryptic coloration may, for example, protect individual prey against visual detection by a range of predators, but secondary defences such as toxins could operate more specifically. We think that this is an interesting area for further research.

In summary, while this book is largely structured around individual types of defence: different forms of mimicry, disruptive camouflage etc., we do not wish to suggest that defences are best viewed in isolation. Hence at the outset we emphasize that it is common for prey to deploy more than one form of defence (see Blanchard & Moreau, 2017 for a highly integrative approach to the biology of defences). Three reasons for multiple defences have been identified: simultaneous action to maximize survival at a stage of encounter, organization across sequentially operating defences, and multiple types of enemy. There are no doubt others, and we encourage readers to keep in mind the potential for integration of defences as they read the rest of the text.

CHAPTER 1

Background matching

1.1 Introduction, definition, mechanism, and chapter overview

In this chapter we describe the phenomenon known as 'background matching' in which an animal's coloration resembles the background against which it is observed. The adaptive function of such matching in colour patterns is to reduce the likelihood that a prey will be detected by its predators, or a predator detected by its prey, though we focus on the former. Here we will review evidence that background-matching coloration works in this manner. We subsequently consider how natural selection causes the evolution of background matching, focussing on the nature of frequency dependent selection and polymorphism, including the role of so-called 'search images' in predators. Finally, we consider ecological considerations that apply generally in the study of background matching, notably how variation in colour patterns of backgrounds limits the effectiveness of this defence mechanism. We argue however that visual background matching need not be perfect in order for detection to be hindered. One of several reasons for imperfect background matching is that organisms can be viewed against a range of different backgrounds and so adopt appearance traits that offer some degree of matching against several of these. We start, however, with a brief discussion of the terminology that relates the mechanism of background matching to the wider categories of crypsis and camouflage.

In the recent edited volume '*Animal camouflage: mechanisms and function*' Stevens & Merilaita (2011) propose that the term crypsis is used for strategies that hinder detection by predators, whereas camouflage refers 'to all forms of concealment including strategies to prevent detection ... as well as those preventing recognition', which include for example, disruptive coloration (Chapter 2) and masquerade (Chapter 9).

Visual crypsis thus incorporates multiple strategies, including countershading, transparency and silvering (which we discuss fully in Chapters 3 and 4 respectively) and must also include aspects of behaviour that contribute to reduced likelihood of detection.

Most obviously, though, the term crypsis invokes the phenomenon known as *background matching*, which describes situations where 'the appearance generally matches the colour, lightness and pattern of one (specialized) or several (compromise) background types' (Stevens & Merilaita, 2011). The authors also offer an elaboration of this definition which both clarifies aspects of the definition and describes the underlying mechanism, stating: 'Comparison of local features over the retina is used in subsequent visual processing to distinguish an object from the background (called figure-ground segregation). Thus if the appearance of an animal does not match its background closely enough, a viewer will potentially detect a marked deviation in the local features between the animal surface and its adjacent surroundings. This facilitates the detection of the animal as something that is not a part of the background. Background matching is therefore an adaptation that decreases the deviation in features between the appearance of the animal and its background to counteract the figure-ground segregation.'

These definitions seem clear and effective to us; however, we make the following points.

First, many authors use the terms *crypsis* and *camouflage* interchangeably. However, we agree with the terminological framework of Stevens & Merilaita

Avoiding Attack: The Evolutionary Ecology of Crypsis, Aposematism, and Mimicry. Second Edition. Graeme D. Ruxton, William L. Allen, Thomas N. Sherratt, & Michael P. Speed, Oxford University Press (2018). © Graeme D. Ruxton, William L. Allen, Thomas N. Sherratt, & Michael P. Speed 2018. DOI: 10.1093/oso/9780199688678.001.0001

(2011) and keep some distinction between the terms in this book.

We feel there is utility in the definitions adopted in Stevens & Merilaita (2009a) where crypsis refers to traits that hinder detection, whereas camouflage is a broader term that refers to traits that hinder detection and/or correct recognition of a prey individual. By this argument camouflage would encompass all the mechanisms considered to function in crypsis, disruptive coloration plus masquerade. We note however that this definition could also encompass mimicry of conspicuous, warningly coloured prey, which is not usually considered a form of camouflage (Chapter 9). There may be merit in this use of the term camouflage since it is in a sense a form of concealment of identity. Other readers may, however, prefer to limit the use of the term camouflage to colour patterns that do not draw a predator's attention to the organism per se.

A second point is that the concepts of crypsis and background matching can readily be extended to modalities other than vision (as considered in depth by Ruxton, 2009), and indeed could be applied multi-modally. Conceptually, if the environment in the absence of the organism stimulates the senses of an observer in characteristic ways, then background matching in another organism involves aspects of the organism stimulating the senses of the observer in ways that resemble the background and thus hinder detection of the focal organism.

A third point we make here is that although the above description of background matching is couched in terms of a cryptic animal, there is good evidence that crypsis can apply to plants as well as animals (see Niu et al., 2017 for evidence of crypsis related to local background matching in different morphs of a plant), just parts of organisms, groups of organisms, or indeed to objects made by animals (e.g. see Bailey et al., 2015 for evidence that some avian nests are adapted for improved background matching).

Finally we agree with Merilaita et al. (2017) that camouflage, including background matching, consists of a suite of adaptations to the perceptual and cognitive systems of receivers that aims to reduce the signal-to-noise ratio of stimuli utilized by searchers to detect, localize, and identify targets. We also agree with them that a focus on mechanistic functions is likely to be more fruitful that a focus on describing appearance types, a perspective we aim to follow in all our chapters on camouflage.

In the next section (1.2) we consider the empirical evidence for background matching in natural systems and discuss the distribution of this adaptation. We then consider evolutionary and co-evolutionary aspects of background matching (1.3 & 1.4), before turning to how background matching is affected by the ecology of predator–prey interactions (1.5). Finally we consider unresolved questions and opportunities for further research (1.6).

1.2 Empirical evidence of background matching

Organisms that seem to humans to offer uncanny levels of background matching are staples of natural history TV programmes. However, to us, the most persuasive evidence for background-matching adaptations comes from species that behaviourally select their microhabitat and orientation so as to enhance similarity to features of the background, and species that change aspects of their appearance in ways that enhance background matching. For example, many freshwater fish are able to adjust their pigmentation to match the background (Kelley et al., 2017; Kelley & Merilaita, 2015). Similar observations have been made for cuttlefish (e.g. Buresch et al., 2011; Hanlon et al., 2009). Other bottom-dwelling fish actively choose substrates on which they achieve good background matching (Tyrie et al., 2015). Lovell et al. (2013) demonstrated that birds that lay their eggs in simple scrapes can preferentially select substrates that maximize background matching of their eggs. Moths landing on bark often orientate according to patterning of the background in a way that enhances background matching and hinders detection (Kang et al., 2012).

There is also experimental evidence that organisms that appear to humans to be highly background matching are also challenging for non-human observers to find (e.g. Merilaita & Dimitrova, 2014). The classic example of selection by predators driven by the degree to which prey match the background remains the peppered moth *Biston betularia*, a polychromatic species with light and dark morphs. As coal burning increased throughout the industrial revolution dark melanic morphs increased in frequency relative to the light morph, supposedly because they suffered lower predation at daytime resting sites on dark surfaces. Since 1970, with the introduction of pollution controls, there has been a rapid reversal with pale morphs becoming more frequent, thought to have been caused by visual predators selecting against melanics at rest on today's less sooty surfaces. After decades of uncertainty about the underlying mechanism(s) driving morph fluctuations, Michael Majerus conducted a properly controlled experiment to test the predator selection hypothesis, releasing 4864 moths of different morphs then recording resightings of them over a six-year period. This experiment, reported by Cook et al. (2012a) after the death of Majerus, confirmed that differential bird predation against melanic peppered moths occurred at a rate sufficient in magnitude and direction to explain the recent rapid decline of melanism in post-industrial Britain. These data provide the most direct evidence yet implicating camouflage through background matching and bird predation as the explanation for the rise and fall of melanism in moths.

From the argument above we can predict that background matching should be most prevalent in species whose natural range offers predictability and uniformity of background, but this is a difficult hypothesis to test. In large part the problem here is that characterizing backgrounds and measuring

Not all organisms are camouflaged. Part of the explanation for this is the costs of adopting a camouflage strategy. First, since background matching is enhanced by adoption of certain microhabitats and behaviours (e.g. benefits through stillness, or adopting certain orientations relative to background elements) then there may be 'loss of opportunity' costs to background matching. For organisms that seek out microhabitats that enhance background matching there may be additional locomotion costs in searching for and reaching these microhabitats. There are also physiological costs associated with production of the pigmentation required for background matching. These costs may be recurring for organisms that change appearance in order to track changing backgrounds, or in organisms that shed and replace their outer covering.

degree of background matching of different targets within them is difficult. Fully quantifying the appearance of even a very simple background (e.g. the sky or a snowfield) is challenging. We can describe the average colour and average brightness relatively easily, for example—but quantifying spatial and temporal variation in appearance is much more challenging (Allen & Higham, 2013). A complete quantification of the level of background matching of a single target for a single observer in a single snapshot would require measuring all of the different visual parameters on which objects can be separated from the background (e.g. colour, luminance, edges, and conjunctions of these such as shape and texture), for both the target and the background, as they would be perceived by the observer, and with reference to how parameter changes affect detection probability. This is hugely challenging, and yet it still comes short as it ignores the differences between the sensory systems of the multiple observers organisms aim to hide from, and the temporal variation in the relationship between foreground and background attributes (e.g. the sun passing behind a cloud, movement in the target or background, or target movement to a different background). The usual approach taken is to measure just one or two attributes known to be relevant (e.g. colour difference) and use this as a proxy for the overall level of background matching. This has the consequence of restricting the range of predictions that can be tested, so it is exciting to see approaches being developed that make collection of richer measurements more straightforward (e.g. Troscianko et al., 2017), though there are still considerable methodological challenges to overcome (Maia & White, 2018).

1.3 The evolution of background matching

1.3.1 Polymorphism of background-matching forms

We now consider how background matching can be reconciled with the frequent observation of polymorphic coloration in cryptic species. Ford (1940) defined genetic polymorphism as the occurrence in the same locality of two or more discontinuous forms of a species in such proportions that the rarest of them cannot be maintained only by recurrent mutation from the other forms. This section will consider a number of different mechanisms via which selection for background matching in a population might lead to polymorphism. The main mechanisms produce frequency-dependent *per capita* predation risk on prey morphs, and careful exposition of this requires a detour into the complex maze of terminology used to describe frequency-dependent selection.

1.3.2 Definitions related to frequency-dependent predation

There is a strong body of experimental evidence for frequency-dependent predation: increased per capita predation risk for the common morph increases relative to less common morphs (e.g. Allen, 1989a; Reid, 1987). Whilst there is general agreement that frequency-dependent predation often occurs, identification of the mechanism behind it in any particular case can be challenging because possible mechanisms may act together and because different mechanisms can produce very similar behaviours. Here we focus on the potential mechanisms related to detection of prey, though there are additional post-detection mechanisms that can also lead to frequency-dependent effects (Endler, 1988; Greenwood, 1984; Sherratt & Harvey, 1993).

Let us consider a situation where a predator includes two prey morphs of the same species in its diet (labelled prey types 1 and 2). Over a certain time period, the predator eats E_1 of type 1 and E_2 of type 2. The densities of the two prey types in the environment are N_1 and N_2. We follow Murdoch (1969) and Cock (1978) (amongst others) in defining the predator's preference between their two prey types (p_{12}) by:

$$\frac{E_1}{E_2} = p_{12}\frac{N_1}{N_2} \qquad (1.1)$$

Of course p_{12} is a relatively uninformative measure of preference for one morph over another: in practice it is often difficult to separate a genuine 'liking' for one prey type over another from circumstantial details such as the true availability of alternatives.

Nevertheless, if p_{12} is greater than one, then this indicates an overall 'preference' for prey type 1 over prey type 2; if it is less than one then the preference is reversed. In the boundary case, no preference is apparent. Most importantly, since p_{12} is equivalent to the ratio of the per capita mortality of prey type 1 to the per capita mortality of prey type 2, it can be used as a simple measure of the relative fitness of these two prey types in the context of predation.

Say in a series of experiments, the ratio of prey availabilities (N_1/N_2) is varied, the ratio of consumptions (E_1/E_2) is noted, and p_{12} calculated. If p_{12} is unaffected by the ratio of prey availabilities, then we have a frequency-independent preference. Otherwise, if p_{12} increases as N_1/N_2 increases, then the predator displays a frequency-dependent preference, such that per capita predation rate on prey type 1 compared to the per capita predation rate on prey type 2 increases as prey type 1 becomes more common in the overall population. If p_{12} (as a measure of relative fitness) rises from below one to above one as (N_1/N_2) rises, then when prey type 1 is rare it will be selected for over prey type 2, and vice versa when it is common (see Figure 1.1). We call this form of frequency-dependence (pro-) apostatic selection (selection for the rare form). Conversely, when the preference for a prey type decreases from above 1 to below 1 as it becomes more common then such an outcome is referred to as anti-apostatic selection (see Figure 1.1). Following Greenwood (1984), we reserve the term potential pro-apostatic selection for a subset of cases analogous to pro-apostatic selection where p_{12} increases with increasing N_1/N_2 but one form is consistently at a selective advantage over the other (and potential anti-apostatic selection in a similar way). Naturally, you could invent more names for other outcomes (e.g. when p_{12} falls over a range of values of N_1/N_2 and then rises—as predicted in a theory paper of Sherratt & MacDougall (1995)) but this is perhaps going too far.

Of course, it is possible to consider cases in which the two prey types are separate species. Once again, preference for one prey type may cross the value of one as a prey type gets more common. Here the analogous form of positive frequency-dependent predation has been called 'switching' (Murdoch, 1969).

Be warned when reading the literature that not everyone adopts the definitions given above. The term 'positive frequency-dependence' can be particularly confusing—it typically refers to cases where relative fitness of a prey type is positively related to relative frequency, but has sometimes been inappropriately interpreted in the converse: attack rates increasing on a prey type as it becomes more common (this is negative frequency-dependence). Most authors do not differentiate between 'potentially pro-apostatic selection' and 'pro-apostatic selection', and refer to both as apostatic selection (e.g. Endler, 1988). Bond (1983) has called particular forms of apostatic selection 'matching selection' and anti-apostatic selection 'oddity selection'. Pro-apostatic selection has also been called 'unifying selection' by Pielowski (1959). 'Reflexive selection' (Moment, 1962; Owen & Whiteley, 1986) is also a synonym for pro-apostatic selection (Allen, 1988a, b). Bewildered?

The abundance of terminology reflects the general importance of frequency-dependent predation, and the fact that it has been observed in many different contexts. Hopefully, Table 1.1 will help (modified from Allen, 1989a).

1.3.3 Positive selection for polymorphism

In order to examine the role of frequency-dependence in the evolution of polymorphic crypsis, Bond & Kamil (2002) designed an ingenious experiment using birds that were trained to search a computer screen and peck at images of moths in return for a food reward. The population of moth images from which individual images were randomly selected to appear on the screen was allowed to evolve over time via a genetic algorithm, with offspring being slightly mutated versions of their parents. Selection for crypsis was introduced by allowing greater

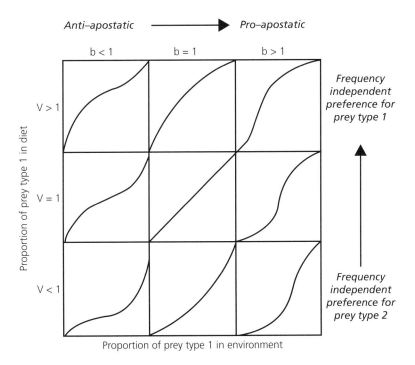

Figure 1.1 Graphs showing how the frequency-independent and frequency-dependent components of predator preference can combine to influence the relationship between the proportion of a prey type in the environment and its proportion in a predator's diet. All nine graphs have limits 0.0–1.0. Here we have considered just two prey types (1 and 2) and used the Elton–Greenwood model (Greenwood & Elton, 1979) to generate the graphs. The parameter b in the Elton–Greenwood model reflects the degree of frequency dependence in the preference ($b < 1$ anti-apostatic, $b > 1$ apostatic), while the parameter V reflects the underlying frequency-independent selection ($V > 1$ preference for type 1, $V < 1$ preference for type 2).

Table 1.1 A glossary of terms relating to frequency-dependent predation. Adapted from Allen (1989a).

Term	Examples of use	Meaning
Frequency-dependent selection (fds)	Fisher (1930), Ayala & Campbell (1974)	Selection that results in a positive or negative relationship between relative fitness and relative frequency
Positive fds	Levin (1988)	Fds where relative fitness of prey is positively related to relative frequency
Negative fds	Partridge (1988), Antovics & Kareiva (1988), O'Donald & Majerus (1988)	Fds where relative fitness of prey is negatively related to relative frequency
Apostate	Clarke (1962)	Rare morph maintained by apostatic selection
Apostatic selection	Clarke (1962)	Negative fds by predators in the absence of Batesian mimicry
Apostatic polymorphism	Clarke (1962)	Polymorphism maintained by apostatic selection
Pro-apostatic selection	Greenwood (1985)	= apostatic selection
Anti-apostatic selection	Greenwood (1985)	Positive fds by predators in the absence of Batesian mimicry
Potential apostatic selection	Greenwood (1984)	When the prey type taken is taken more often than expected by chance at all frequencies, and this selection increases with frequency
Switching	Murdoch (1969)	= apostatic selection (especially when prey are different species)
Matching selection	Bond (1983)	Apostatic selection (in matching backgrounds)
Oddity selection	Bond (1983)	= anti-apostatic selection
Reflexive selection	Moment (1962)	= apostatic selection (on massive polymorphisms)
Reflexive polymorphism	Owen & Whiteley (1989)	Massive polymorphism maintained by reflexive selection

representation of the least detected 'prey' in succeeding generations. Selection by the birds led to increased crypsis and greater phenotypic variance compared to control populations that were subjected to no selection. Bond & Kamil interpret their results as a suggestion that polymorphism is caused by the birds displaying pro-apostatic selection. That is, a given morph is less likely to be detected by the birds if the birds have little recent experience of that morph because it is rare in the population. This interpretation was strongly supported as both phenotypic variance and crypsis increased to higher levels in the experiments with avian predators than in experiments with computer-controlled selection that eliminated any frequency-dependent effects.

However, an earlier experiment using a similar set-up demonstrated that polymorphism was not an inevitable outcome of this paradigm (Bond & Kamil, 1998). If all phenotypes are too easy for the birds to detect then there will be no selection, since all phenotypes are always found when they are presented to the birds. Conversely if a morph is introduced that is impossible for the birds to

detect, then this will come to dominate the population, reducing diversity. These experiments are a convincing demonstration of predator-induced selection for prey polymorphism of appearance (albeit in an artificial system). Further, this polymorphism seems to be driven by pro-apostatic selection.

One of the most popular explanations for pro-apostatic selection is the formation of search images (Tinbergen, 1960). The term 'search image' has been used to describe an enormous host of activities related to foraging. This has led some (e.g. Dawkins, 1971) to suggest that the term is too imprecise to be useful. Whilst we have some sympathy with this viewpoint, we use the term here, but clearly define what we mean by a search image in the following section.

1.4 Co-evolutionary considerations

Simply speaking there are a range of counter-adaptations predators can adopt to combat background matching by prey. Firstly, and least interestingly, they could switch to focus on prey that have poorer background matching, or focus on foraging in places or under lighting conditions (e.g. specific times of day) when background matching is less effective. More interestingly, they could search the environment more carefully (likely at a cost of how quickly they can search the environment) and/or they could refine their search strategy to focus on the search for particular prey characteristics, forming a 'search image' for particular types of prey. We focus on these issues in this section.

1.4.1 Search image formation as a means of enhancing detection of cryptic prey

Following Dukas & Kamil (2001), we define a search image as 1.) a process of transitory attentional specialization that results in enhanced detection ability for particular cryptic prey types or characteristics (e.g. 'a triangular shape on a wing-like feature'), and that 2.) follows from repeated visual detection of an item over a relatively short timescale. This is predicted to result in increased capture success for prey that match the search image, but decreased capture success for prey which do not match the search image. While this definition remains somewhat imprecise about an exact mechanism, a key criticism of Dawkins (1971), it does allow us to be explicit about the types of processes underlying our conception of a search image. Equally importantly, we are explicit about the types of processes that do not lead to search images under our definition. These include behavioural changes such as modification of search paths, and strategic decision-making about which detected prey items to accept or reject.

A search image is formed after repeated detections of a particular prey type (Pietrewicz & Kamil, 1979), and requires further detections of that prey type in order to be maintained (Plaisted & Mackintosh, 1995; and we refer readers to the reviews of Ishii & Shimada, 2010 and Skelhorn & Rowe, 2016). Though a feature of individual cognition, search image formation may be influenced by observation of peers in social groups (White & Gowan, 2014). Search images should lead to relative protection for prey types when they are rare—promoting heterogeneity of prey types. This occurs because common prey types are more likely to be encountered so (providing prey types are equally cryptic in the absence

of search images) are more likely to induce and maintain a search image for that prey type. This will lead to prey types being subject to disproportionately high predation when common.

It is important to remember that formation of search images need not necessarily lead to stable polymorphism. Formation of a search image for a common prey type may reduce the per capita likelihood that a rare prey type is detected, but this reduction need not necessarily reduce this likelihood below that of the common morph, and so may simply slow rather than reverse a decline in the relative frequency of the rare morph. However, search image formation, and more generally pro-apostatic selection by predators on cryptic prey can provide an evolutionary pressure for the development and maintenance of polymorphism in prey species.

1.4.2 Control of search rate to enhance detection of cryptic prey

Frequency-dependent selection on cryptic prey can also arise from predators' control of their rate of searching the environment. There is likely to be a trade-off between how rapidly an area is searched for prey and the efficiency of detection (fraction of prey in the scanned area that is detected: for empirical support for this conjecture see Gendron & Staddon (1983)). Further, cryptic prey may require a slower search rate to obtain a specified detection efficiency, compared to a less cryptic prey type. Gendron & Staddon (1983) argued that it would be optimal for predators to reduce their rate of search when cryptic prey types are common in the environment. However, when cryptic prey types are uncommon, compared to less cryptic types, the optimal strategy is to increase search rate. Although this will lead to a greater fraction of cryptic prey being missed, this will be more than compensated for by the larger number of less cryptic prey discovered. Hence, optimal control of search rate leads to a reduction in the per capita risk of detection to individuals of the more cryptic morph when this morph is rare. In this way, predator strategies involving control of the rate at which they search their environment can often explain empirical results that had originally been attributed to a search image effect (Allen, 1989b; Guilford & Dawkins, 1989a,b).

1.4.3 Comparing search image and search rate mechanisms

Optimal control of search rate leads to a reduction in the *per capita* risk of detection to individuals of the more cryptic morph when this morph is rare. However, per capita detection risk is always lower for the more cryptic morph than for the less cryptic morph (in the absence of search image effects). Thus (in the absence of any other differences between the morphs), frequency-dependent predation acts to increase the rate at which the relative frequency of the more cryptic morph increases. However, its frequency would still increase in the absence of this effect. When the less cryptic morph is rare, it does not receive the same protection. In such circumstances the predator should slow up its search rate so as to improve its detection of the more cryptic prey. This will not decrease its ability to detect the less cryptic morph, indeed it may increase it. Thus, in contrast to search image formation, optimal search speed does not lead to protection for the less cryptic morph (which will always be discovered disproportionately often), and so polymorphism may not be maintained. This difference allows distinction between the search rate control and search image mechanisms, even though both provide the ability to produce frequency-dependent selection. Under the search image mechanism, the rank order of per capita detection risk of different morphs can change; this is not true of the search rate mechanism. Hence, other things being equal, in a situation where search images do not occur, then the more cryptic of the two morphs will always have a fitness advantage and will steadily increase its frequency in the population. Should a random perturbation to the frequencies of morphs in the population or change in selection pressures mean that the more cryptic morph is at a low level, then control of search rate will act to strengthen the competitive advantage of this morph, and so may reduce the likelihood of stochastic extinction. In this very specialized set of circumstances, optimal control of search rate acting in isolation from other mechanisms (such as search image formation) may play a role in the maintenance of polymorphism.

An interesting comparison between the two mechanisms considered above involves comparing a monomorphic population to a polymorphic population where all the morphs are equally cryptic. The search-image mechanism suggests that frequency-dependent selection will occur in the polymorphic case, whereas the search rate hypothesis does not. The search rate model predicts entirely the same search rate in the two cases, and predicts that individuals of all prey morphs will be equally vulnerable to detection. Whereas the search image hypothesis suggests that polymorphism will reduce the predator's performance, search images will either take longer to arise or will be less successfully maintained because the encounter rate of individual morphs decreases compared to the monomorphic case. Hence, in this special case where the morphs have equal crypsis, polymorphism is advantageous under the search image mechanism, but not the search rate one (Guilford & Dawkins, 1987; Knill & Allen, 1995). Knill & Allen (1995) found that human 'predators' were less effective at detecting prey in the polymorphic case. Similarly, Glanville & Allen (1997) found that human subjects were slower to detect computer-generated prey displayed on a screen in trials where prey were polymorphic compared to monomorphic trials. The prey types were assumed to be equally cryptic, although this was not demonstrated explicitly. However, these results tend to support the suggestion that in these studies frequency-dependent selection occurred through search image formation rather than search rate control.

There is no reason why search image mechanisms and optimal control of search rate cannot occur simultaneously. Such a situation was explored theoretically by Dukas & Ellner (1993). Their model does not explicitly use the phrase 'search image'; however, such a phenomenon is implicit in the assumptions of the model. They assumed that prey detection requires information processing, and that the predator has a finite capacity for processing, which it must apportion to detecting different prey types. Increasing the processing ability devoted to a given prey type increases the chance that an encountered individual of that type will be detected. However, because of the finite processing ability available, increasing ability to detect one prey type can only be bought at the cost of reduced ability to detect other prey types. The other key assumption of the model is that the more cryptic a prey type is, the more information processing capacity is required to achieve a specified detection level. These assumptions are motivated by consideration of neurobiological experiments mostly on humans (see references in Bernays & Wcislo, 1994; Dall & Cuthill, 1997; Dukas & Ellner, 1993; Dukas & Kamil, 2001). This model predicts that when prey are very cryptic, then the predator should devote all its processing capacity to detection of one type. However, as the conspicuousness of prey increases, so the diet of the predator should broaden, as it becomes advantageous for it to spread its information processing capacity across a wider range of prey types. When the predator divides its 'attention' between several prey types, this division should not necessarily be even. If prey are cryptic, then the predator should give more attention to the least cryptic of the prey types; whereas if the prey are generally less cryptic then the predator should attend most to the least detectable type.

In summary, much heated debate has surrounded the concept of search images over the last 30 years. This debate has arisen because many other mechanisms (most notably control of search rate) can also produce very similar behaviour to search image formation. This had led some to overstate evidence of search image formation, without logically excluding plausible alternative explanations. However, more recent studies do seem to have demonstrated the existence of this mechanism, and it does appear that search image formation (perhaps working in concert with other mechanisms, such as control of search rate) could potentially select and maintain polymorphism in cryptic populations. This has been demonstrated in the laboratory, but not yet in the field.

1.5 Ecological considerations

The key ecological issue for background matching is the overwhelming majority of organisms will be viewed against a range of different backgrounds. Even species that have very narrow microhabitat use (e.g. living on the bark of a particular tree species) will be viewed by observers from a diversity of distances and angles, and under a range of different light conditions. There are two options here: we consider organisms whose appearance is more-or-less

fixed in ontogeny before turning to those organisms that have the ability to change appearance. Finally, we consider the effects of ecology on combining crypsis with other selection pressures.

1.5.1 Optimizing of background matching for a single appearance in visually variable backgrounds

A significant drawback to background matching as a strategy for crypsis is that almost all organisms will be viewed against a variety of backgrounds with varied appearances. Even if the habitat is physically homogeneous, then temporal change in light conditions will change the nature of the background that the organism must attempt to match. Indeed, an organism can be viewed against two backgrounds simultaneously when viewed by two organisms with different visual sensory systems. This raises the question, should the organism specialize by maximizing its matching to one particular background, or should it seek a compromise that provides reasonable crypsis against more than one background but which is not maximally effective against any one of them. This was addressed using a simple model by Merilaita et al. (1999). We outline a very slight generalization of their analysis below.

We assume two backgrounds (a and b). Let the probability of being viewed by a predator against background a be V_a. The corresponding probability that the potential prey is not detected whilst in the predator's view is C_a. The prey is always viewed against background a or b, so

$$V_a + V_b = 1. \tag{1.2}$$

Hence, the overall probability of being detected by a predator is

$$D = V_a(1 - C_a) + V_b(1 - C_b). \tag{1.3}$$

The probability of escaping detection is

$$E = 1 - D = 1 - V_a(1 - C_a) - V_b(1 - C_b). \tag{1.4}$$

In some cases, there is likely to be a trade-off between crypsis against the two backgrounds, and improved crypsis against one background can only be bought with reduced crypsis against the other. Mathematically, C_b is a declining function of C_a.

$$C_b = f(C_a)$$

$$\frac{df(C_a)}{dC_a} < 0. \tag{1.5}$$

It is trivial to show that E has a turning point (i.e. either a maximum or a minimum) if there is a value of C_a that satisfies

$$\frac{df(C_a)}{dC_a} = \frac{-V_a}{V_b}. \tag{1.6}$$

Further, this point is a maximum if, at that point,

$$\frac{d^2 f(C_a)}{dC_a^2} < 0. \tag{1.7}$$

Figure 1.2 describes this situation graphically. We assume that the prey individual is free to adopt any value of C_a from 0 (which maximizes crypsis against b but provides no crypsis against a), to a value A (which maximizes crypsis against a but provides no crypsis against b). If the shape of the trade-off curve is 'convex' like the solid line in the figure, and if there is a point where that line has gradient $-V_a/V_b$, then the optimal strategy for minimizing predator detection is a compromise value that provides some crypsis in both environments. Increasing V_a (or decreasing V_b) moves this compromise towards improving crypsis against background a, as we would expect. However, if no such point can be

found, or the trade-off line is concave, like the broken line in the figure, then the optimal strategy is to specialize and maximize crypsis against one of the backgrounds. In the figure, the optimal strategy is to maximize crypsis against a providing

$$AV_a > BV_b, \tag{1.8}$$

where B is the value of C_b corresponding to $C_a = 0$, i.e. the maximal probability of not being detected whilst in the predator's view against background b. Otherwise crypsis against b should be maximized. That is, crypsis should be maximized in the environment where the product of the probability of being viewed by a predator and the probability of that view not leading to detection is maximized.

Hence, both background specialization and compromise can be predicted, a key determining factor being the shape of the trade-off between crypsis in one environment and another. We now look to give a more ecological description of this shape. Imagine a prey organism that specializes in crypsis against background b, and so adopts strategy $C_a = 0$. If it can increase its crypsis against a by losing a relatively smaller amount of its crypsis against b, then we have the type of convex shape of trade-off curve that can lead to evolution of an intermediate level of crypsis that provides some protection in both environments. However, if a little crypsis against a can only be bought with a relatively large decrease in crypsis against b, then we have a concave shape and background specialization is favoured. Merilaita et al. (1999) give simple examples of abstract background combinations that might lead to these two different types of situation; these are reproduced in Figure 1.3.

Of course, the situation can become much more complex than the simple example considered here. The costs associated with being detected need not be the same for detection against both backgrounds (since, for example, the predator may be more effective at capturing prey against one background). The trade-off curve can have both convex and concave segments, or can incorporate straight line segments and/or discontinuities, and many more backgrounds than two can be considered. In such circumstances, the adaptive landscape will be much more complex and there will be considerable potential for evolutionarily stable polymorphisms. However, the essential

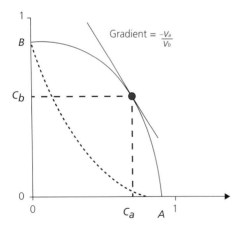

Figure 1.2 The probability of not being detected when viewed against background b (C_b) as a function of the equivalent probability against background a (C_a). If the function is convex (like the solid line) and has a point where its gradient is $-V_a/V_b$, then a compromise level of crypsis is optimal, otherwise the organism should maximize crypsis against one of the backgrounds.

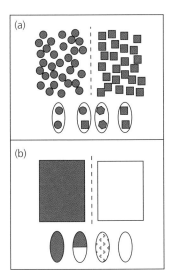

Figure 1.3 Two hypothetical examples of heterogeneous habitats and animals relying on crypsis through background matching. In (a) the habitat consists of two different microhabitats, one with circular and the other with square elements. The two outermost of the four animals have adapted to the microhabitats with respective patterns only. The two animals in the middle, one with a circle and a square to the left and one with two hexagons to the right, represent compromised adaptations for crypsis in both microhabitats. Successfully compromised colorations give the trade-off between crypsis in these microhabitats a convex form. In (b) one microhabitat is black and the other is white. Again the outermost animals represent adaptations to one microhabitat only. However, this time the compromised colorations in the middle are apparently very poor, making the trade-off between crypsis in the two habitats concave.

points of our analysis above will remain unchanged, specifically:

1) The optimal cryptic strategy for an organism may be one that does not maximize its crypsis against any one of the backgrounds against which it is viewed, but rather provides some measure of crypsis against a suite of the backgrounds that it is viewed against.

2) In some cases, specialization against one background will be favoured, in which case this will be the background against which the organism can maximize the rate of occurrence of viewings of it by a predator that do not lead to detection.

The case where specialization is predicted warrants further scrutiny. Let us return to our simple case of the two environments a and b. Let us imagine that AV_a is only slightly bigger than BV_b. The opti-

mal strategy is to maximize crypsis against background a. However, if there is a small change in one of the parameters, such that AV_a becomes slightly smaller than BV_b then this has dramatic consequences and now the optimal strategy is to maximize crypsis against b. This has important ramifications:

- Two types of organisms could have very similar ecologies but adopt very different appearances because the difference in their ecologies means that they lie on different sides of the knife-edge described above and so adopt colorations that maximize their crypsis against different backgrounds.
- If the environment inhabited by a population changes (perhaps only slightly), such that it moves to the other side of this knife-edge, then individuals in that population that previously had the optimal choice of coloration can now find themselves with a coloration that is far from the optimal. Evolution towards the new optimum may be particularly challenging as colorations only slightly different from the 'old' optimal may still be selectively worse than the current situation. An example of this is shown in Figure 1.4. Here we have a concave trade-off function shown in Figure 1.4a. This means that the turning point in D is a maximum, and so the optimal strategy is one of the extremes. Imagine, first, that we are in a situation represented by the solid line in Figure 1.4b, $C_a = A$ is the optimal strategy. However, imagine now that V_a is changed such that we move to the broken line: now $C_a = 0$ is the best strategy. However a small mutation away from $C_a = A$ produces an increase in detection rate D. Only a macromutation producing a C_a value less than β in the diagram would be selected over the strategy $C_a = A$.
- Polymorphism within a population may be maintained by even slight fluctuations in the proportions of different backgrounds that individuals experience, if those fluctuations continually move the system across the knife-edge.

In a follow up paper, Merilaita et al. (2001) reported the results of experiments with captive birds searching for artificial prey. Birds were faced with one of two background types (differing in the size of patterning). There were three types of prey: one which matched the small background pattern, one which

(a)

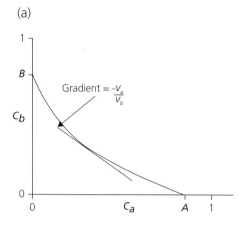

(b)

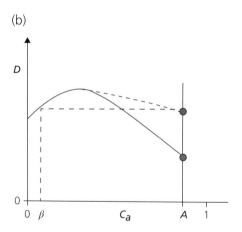

Figure 1.4 (a) The trade-off between C_a and C_b is concave so specialization is predicted. (b) When V_a is such that detectability (D) is given by the solid line then $C_a = A$ is the optimal choice, but if V_a changes slightly so we move to the broken line, then $C_a = 0$ is optimal.

matched the larger patterned background, and an intermediate-sized pattern. On the small-patterned background, small-patterned prey were most cryptic (measured by mean time between prey discoveries), followed by the intermediate pattern, which in turn was more cryptic than the large pattern. On the large-patterned background, the small-patterned prey was least cryptic, but there was no difference between intermediate and large-patterned prey. This means that in a situation where both backgrounds were encountered with equal frequency, the intermediately patterned prey would be the best protected of the three, and generally that circumstances exist where some intermediate type

between the two extremes of background matching would be optimal. These results can be seen as supportive of the theory presented in this section; however, further empirical study of this theory would be very welcome.

Houston et al. (2007) extended the modelling approach of Merilaita et al. (1999) by allowing predators control of which background they chose to search against. However, the key result of Merilaita remained—under some combinations of parameter values background specialism was selected in the prey, under other parameter value combinations compromise solutions flourished.

An interesting situation arises when backgrounds themselves can evolve in response to predator–prey interactions occurring against them, as is the case of flowers on which predators lurk to prey on the flower's pollinators. This situation was modelled by Abbott (2010), with results predicting that flowers should match the appearance of predators when predator densities are high to prevent predators overexploiting the plant's pollinators, and when pollinators are less discriminating as they are willing to land on a flower even if it might contain a predator. Floral polymorphisms could evolve that both concealed and revealed ambushing predators.

The arguments in this section have many parallels with the arguments of Edmunds (2000) on why some mimicry systems display poor mimicry such that humans can easily tell model and mimic apart. He suggests that under some circumstances evolution will drive a mimic species to specialize on one particular model species, whereas in other circumstances generalized mimicry of a number of different models (necessitating imperfect mimicry of each one) will evolve (see Barnard (1984) for a similar idea). Further discussion of the evolutionary pressures on the 'perfection' of mimicry can be found in Chapter 7.

Empirical demonstration of a natural organism having compromise appearance is challenging. We have already discussed why quantification of the degree of background matching to one particular background in challenging. In this case, we must add that exploration of compromise appearance requires identification of the backgrounds involved and the frequencies at which the organism is viewed by the observer against these different backgrounds.

However, we have valuable empirical studies utilizing more artificial situations. Merilaita & Dimitrova (2014) trained wild-caught birds to search for artificial prey against two different backgrounds in an aviary and found that a compromise design offered as much protection as prey designed to provide a good match against one of the backgrounds (but sometimes found against the other background). Sherratt et al. (2007) explored a situation where humans were tasked with detecting prey presented on examples of two different backgrounds. They found that in general habitat specialists performed better than compromise prey, but the extent of this benefit varied considerably depending on the nature of the two backgrounds and the difference between them, and there were some situations where a compromise appearance could gain very similar protection to a specialist. In a very similar study, Toh & Todd (2017) found situations where compromise prey had lower net detectability than background specialists.

While most studies have considered very simple situations in which prey can be found on one of two backgrounds, Michalis et al. (2017) considered a more realistic situation where there is continuous variation in background appearance. They argue that, logically, in such a situation the optimal choice for prey to maximize protection through background matching should be to match the background that they are viewed against most frequently. They support this prediction with evidence using artificial prey pinned to trees and predated by wild-living birds, and with experiments using human searching for targets on a computer screen. They note that Bond & Kamil (2006) demonstrated that when birds were trained to search for cryptic prey against backgrounds that showed fine-grained variation in background (i.e. such that the background changed on a spatial scale similar to that of individual targets), birds selected for a prey population composed of a shifting portfolio of different morphs through apostatic selection. Initially this seems counter to the arguments of Michalis et al.; however, they suggest that the difference results from the importance of predator learning in Bond & Kamil's experiments whereas theirs were designed to minimize learning by having prey at low densities and in different locations. In the absence of learning, therefore, match-

ing the most-common background is an effective strategy for prey, but once predators respond to this by forming a search image for that prey the conditions exist for prey with alternative appearances to become advantaged via apostatic selection. The experiments of Bond and Kamil have another interesting aspect; when the background was changed to a more coarse-grained structure where it was homogeneous on the spatial scale of targets, then predators switch from a strategy of extensively searching the whole environment to one where they focussed primarily on looking for prey on the parts of the environment with the currently most-profitable background. This behaviour reduced the effects of apostatic selection and led to a population that mainly just shifted in the relative frequency of morphs that offered a good match to one of the background types.

It is also worth remembering the issue that some backgrounds are fundamentally more challenging to detect targets against than others, regardless of background matching issues, as demonstrated elegantly by Merilaita & Lind (2005) in experiments with wild-caught birds searching an aviary for artificial prey against control backgrounds. Further, prey can preferentially select such habitats over those that would confer a benefit through background matching (Kjernsmo & Merilaita, 2012).

1.5.2 Changing appearance to enhance background matching

Some organisms (e.g. cuttlefish) are known to change appearance in ways that seem to improve background matching on timescales of seconds. Other species can take minutes, hours, or days like some fish (Akkaynak et al., 2017; Smithers et al., 2017) and frogs (Kang et al., 2016). Others, such as crustaceans and insect larvae, change ontologically in a way influenced by environment but controlled by the timescale of their moults; still other species (such as many Arctic vertebrates) show seasonal variation (Stevens, 2016). With the exception of cuttlefish, this form of crypsis has been very understudied, as emphasized by a recent review by Duarte et al. (2017). From our perspective, there is much similarity between issues raised by such within-individual change and within-population change discussed elsewhere in the chapter. However, additional questions worthy of more consideration relate to costs. First, exploring the costs of maintaining the ability to sense background (or some correlate—such as available food), and changes in appearance in response to this. Second, the costs of a time lag between a change in the background and a change in responsive appearance; or the costs of decreased reliability of correlates of background (such as the link between time of year and snow cover for vertebrates that turn white in winter). Lastly, once these costs are better understood, we can ask how unpredictable change in background has to be for such within-individual change in appearance to bring a net advantage. As the review of Duarte et al. emphasizes, we are only just beginning to scratch the surface of these questions.

Eacock et al. (2017) describe how larvae of the peppered moth (*Biston betularia*) respond to match the appearance of the twigs that they rest on when not feeding. They observed that larvae that were subject to exposure to a diversity of twig backgrounds evolved a good match towards the appearance of one twig type rather than a compromise. The flexibility in appearance is suggested by the authors as being linked to wind-based dispersal and ability to feed on a range of tree species, which means that the background an individual will experience cannot be predicted over ontological timescales. However, the lack of compromise might be linked, we suggest, to reasonably strong fidelity to a given tree and thus individuals uncommonly experiencing a range of backgrounds (twig colours) on shorter timescales.

1.5.3 Combining background matching with other functions

Since effective background matching depends on the visual abilities of the viewer, this can mean that an organism that does not appear cryptic to us is cryptic to its predators. One obvious example arises from inter-specific differences in colour vision. If an organism's main predators are colour-blind, then matching intensity of light is important to crypsis but spectral matching is not. Similarly, detail in the background that falls below the acuity threshold of predators need not be matched. This can present an opportunity for signalling to conspecifics without

impairing crypsis, if the cryptic species has higher acuity than its predators or if signalling to conspecifics occurs under higher ambient light levels than predation events. For example, conspecific signalling can involve small multicoloured spots that do not match the pattern of the background but which blend over the coarser spatial frequency of predator acuity to provide a good match to the background (for fuller discussion see Endler, 1991; Marshall, 2000). It has also been demonstrated that background matching can be combined with aposematic signalling by using striped patterns that blend to a cryptic colour at a distance, but are effective warning patterns when viewed close up (Barnett & Cuthill, 2014). Similarly, it has been suggested that certain avian alarm calls decay rapidly over distance, so nearby conspecifics can be warned about an overflying bird of prey without alerting the predator to the caller's position (Klump et al., 1986; Klump & Shalter, 1984).

Some deep-water organisms may use red light for intraspecific signalling without giving their position away to their predators (a 'private communication channel'), since red light rapidly attenuates in water and (because of this) many deep-water organisms are not receptive to red light. Cummings et al. (2003) have suggested that UV may have similar advantages to aquatic organisms as a short-range intraspecific signal, because of high signal degradation over longer distances due to UV's high propensity to scattering by water molecules. In such situations, we must explain why the predator has not evolved its visual system to detect intraspecific signals between prey individuals. This is relatively easy for generalist predators, since a diversity of intraspecific signals combined with physical trade-offs in the visual system would prevent evolution of a system that could effectively detect the different signals of the predator's suite of prey types. Prey individuals on the other hand have no need to detect the signals of other species that share the same predator, and so do not have the same constraint on their visual system. Hence, for these predators being a generalist presents sensory constraints that may prevent enhanced detection of 'private channels of communication'. In contrast, explaining why a specialist predator has not developed the ability to detect the intraspecific signals of its prey is more challenging, but not impossible.

One reason that specialist predators do not enhance their capacity to detect 'private channels of communication' is that prey generally have shorter generation times than their predators, producing different rates of evolutionary change. Furthermore, intraspecific signalling may be essential for reproduction, where a prey individual that does not signal does not reproduce. In contrast, predator individuals that can detect intraspecific signals would do better than those that could not, but those that cannot would still be able to contribute to the next generation (if the prey can also be detected by other means). Both this 'life-dinner' argument and the difference in generation times might allow the prey to evolve signals faster than their specialist predator can evolve to detect them. It is also worth remembering that specialist predators are uncommon, and generalization over a number of prey types is the norm for predators (Brodie & Brodie, 1999a,b).

While the ways organisms have managed to combine crypsis and conspicuousness are fascinating and illuminating, the more usual situation is for conflict. Many studies have shown that more exuberant intraspecific signalling leads to higher predation rates (e.g. Roberts et al., 2006). A neat example of the trade-offs organisms often make to balance camouflage and signalling was recently demonstrated by Kelley et al. (2016) who showed that male western rainbowfish (*Melanotaenia australis*) who darken their colour to improve their degree of background matching receive more aggressive interactions from rival males because skin darkness is also used to signal dominance. Subdominant males must therefore pay a cost in aggression from rivals to achieve better background matching. As subordinates were observed to change colour less than dominant males when placed on a dark background, this cost is apparently more severe than the cost of increased predation risk.

1.6 Unresolved issues and future challenges

An outstanding basic question is how the visual appearance of the background selects for different anti-predator strategies. Dimitrova & Merilaita (2009, 2011) show experimentally in a lab environment that background matching is more effective against more visually complex backgrounds, a result confirmed in field experiments (Xiao & Cuthill, 2016). This implies that simple backgrounds should select for alternative strategies such as warning coloration. Whether this predicted relationship between background complexity and defence strategy is seen in across-species patterns remains to be investigated.

CHAPTER 2

Disruptive camouflage

2.1 Introduction and overview

Background matching coloration has several limitations in protecting prey from detection and recognition. For example the alignment of patterning between the organism and background may be offset, shadows may give away its form, and the outline of an animal may therefore remain quite visible. This creates opportunities for alternative strategies for improving crypsis, one of which is known as disruptive camouflage. Disruptive camouflage involves using coloration to hinder detection or recognition of an object's outline, or other conspicuous features of its body. A classic example of disruption is coloration patterns that extend across the folded legs of a frog (Cott, 1940; Figure 2.1). By forming a continuous pattern between the upper and lower legs and then across the foot, disruptive theory proposes that the frog is harder to find or recognize. This antipredatory benefit arises because the high-contrast patterns make the 'true' interior and exterior edges that visual predators use to find and recognize prey less apparent.

The most recent formal definition of disruptive camouflage is 'a set of markings that creates the appearance of false edges and boundaries and hinders the detection or recognition of an object's, or part of an object's true outline and shape' (Stevens & Merilaita, 2009b, p.1). This definition highlights the dependence of disruption on the perceptual and cognitive processing of receivers. Whereas countershading can be defined on the basis of phenotype alone, and background-matching camouflage by the relationship between the target and background, disruptive camouflage *must* be explained with respect to how it works to prevent predator detection or recognition. Detection involves finding locations in a visual scene that contain features corresponding to the object class of interest (where is it?), and recognition involves determining if a set of visual features corresponds to an object in a particular class (what is it?).

Like other anti-predator coloration strategies discussed in this book, the concept of disruptive camouflage originated in the work of Poulton (1890) and Thayer (1909), and went on to be substantially developed by Cott (1940). Writing about the caterpillar of the privet hawk moth *Sphinx ligustri*, Poulton noted that its 'stripes increase protection by breaking up the large green surface of the caterpillar into smaller areas.' (Poulton, 1890, p.42). Thayer (1909) observed that objects with a very high level of background matching are often still easily detectable by their outline. For example, shapes cut from a patterned wallpaper can be picked up off a sheet identical to the one that they were cut from fairly easily because they have a conspicuous cut edge and pattern elements will likely misalign. There was early appreciation of the critical importance of animals' outlines to detection and recognition, and therefore the advantage that might be gained by using coloration to interfere with the perceptual processes involved in finding object outlines (Cott, 1940). While the task of finding and picking out an object with a clear outline seems quite straightforward to us as actors, in terms of visual computation it is enormously complex (Marr, 1982), and it is no great surprise that the colour and pattern of the object potentially makes the task more difficult in a number of distinct ways. A major topic in this chapter is therefore explaining the multiple mechanisms by which colour markings can create 'false' information and obfuscate 'true' information to improve camouflage.

Avoiding Attack: The Evolutionary Ecology of Crypsis, Aposematism, and Mimicry. Second Edition. Graeme D. Ruxton, William L. Allen, Thomas N. Sherratt, & Michael P. Speed, Oxford University Press (2018). © Graeme D. Ruxton, William L. Allen, Thomas N. Sherratt, & Michael P. Speed 2018. DOI: 10.1093/oso/9780199688678.001.0001

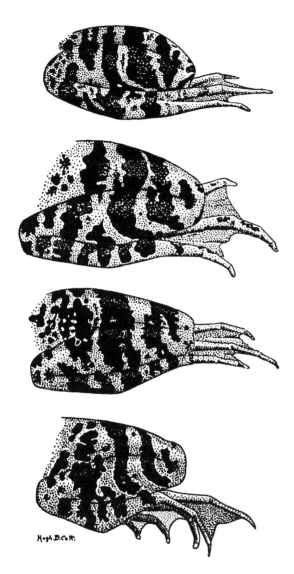

Figure 2.1 Coincident disruptive camouflage on the common frog *Rana temporaria* . From Cott (1940).

The traditional approach (*sensu* Cott) has been to organize disruptive mechanisms into 'principles of disruptive camouflage' (Cott, 1940). Each principle describes how an aspect of disruptive camouflage design might work to make detection or recognition of the object more difficult. Cott proposed his principles with little by way of formal investigation and testing, and the predictions from each principle remained largely untested for several decades. Hence, as recently as 2005 'disruptive camouflage'

was accurately described as a 'little studied principle of concealment' (Merilaita & Lind, 2005, p.665), with no definitive experimental verification that disruption could work to improve camouflage. This changed rapidly, however, as a flurry of studies tested Cott's ideas experimentally, providing support for the main hypothesis that disruptive principles can improve camouflage, independent of background matching (Cuthill et al., 2005; Merilaita & Lind, 2005; Schaefer & Stobbe, 2006). This line of research continues, and new studies that test and refine Cott's proposals enter the literature annually. After consideration of these recent theoretical and empirical developments, Stevens & Merilaita (2009b) reduced the nine distinct principles of disruptive coloration that can be identified in Cott's writing down to five: differential blending, maximum disruptive contrast, disruptive marginal patterning, surface disruption and coincident disruptive coloration (see summary of principles in Table 2.1, illustrated in Figure 2.2).

There is much value in the 'principles' approach described in Table 2.1. Here, however, while continuing to use the terminology, we structure the discussion of disruptive camouflage mechanisms using an emerging contemporary approach which replaces the traditional focus on the design of disruptive patterns (Figure 2.2) with a focus on receiver visual processing (Cuthill & Troscianko, 2011; Fraser et al., 2007; Kelley & Kelley, 2013; Merilaita et al., 2017; Osorio & Cuthill, 2013; Stevens & Merilaita, 2009b; Troscianko et al., 2009; Webster, 2015). Several studies have begun to focus in detail on the perceptual processes that disruptive camouflage targets (Egan et al., 2016; Troscianko et al., 2013; Webster et al., 2013), and this approach represents the likely direction for future work on disruptive mechanisms. Hence, we feel that a receiver-centred approach—rather than a prey-phenotype-centred approach—is more suited to providing an explanation of disruptive camouflage, because it is more appropriately integrated within the sensory ecology of predator–prey interactions.

Investigators are also beginning to examine the taxonomic, ecological, and behavioural correlates of disruptive camouflage strategies, and work on the relationship between disruption and other forms of protective coloration, as well as develop advanced approaches to quantifying disruption in images of real animals. These topics are all discussed in detail later in this chapter.

2.2 Examples of disruptive camouflage

Identifying examples of disruptive camouflage is challenging because confirming disruption requires establishing the perceptual response of receivers, consequently disruption cannot be defined on the basis of appearance alone (see section 2.4). However, many likely examples of disruptive camouflage have been proposed, some with good circumstantial or experimental evidence.

Examples of disruptive camouflage suggested by Cott (1940) include woodcock, giraffe, jaguar, pipits, and plovers, all animals that feature high-contrast markings across parts of their bodies that do not seem to be used in signalling. Other likely examples include the contrasting black, white and grey panels slashing across giant anteaters' (*Myrmecophaga tridactyla*) bodies (Caro & Melville, 2012), and the dorsal stripe that bisects the two halves of many distantly related frog species such as the African bullfrog *Pyxicephalus adspersus*, the Western toad *Anaxyrus boreas*, and Mascarene grass frog *Ptychadena mascareniensis*. Moths are the model organism when it comes to testing disruptive mechanisms, with many of the experimental studies discussed in the next section featuring triangular paper 'moths' as artificial prey of wild birds. This is not by coincidence, as many real moths are good cases of apparent disruption, with the garden carpet *Xanthorhoe fluctuata* and oak beauty *Biston strataria* moths featured as examples by Cott (1940).

An interesting example of putative disruptive camouflage is bioluminescence creating disruptive patterns on the undersides of plainfin midshipmen *Porichthys notatus* (Harper & Case, 1999) and splendid ponyfish *Leiognathus splendens* (McFall-Ngai & Morin, 1991), as well as many other bioluminescent fish. Here bioluminescence is produced such that bioluminescing elements match the intensity of downwelling light while other ventral regions that do not bioluminesce contrast with the background when viewed from below, disrupting the true outline of the fish.

Disruptive camouflage has been most intensively investigated in colour-changing animals such as the bullethead parrotfish *Chlorurus sordidus* (Crook, 1997), chameleons (Stuart-Fox & Moussalli, 2009), and especially cuttlefish (Hanlon et al., 2009; Mathger et al., 2007). These animals are of particular interest because they allow experimental investigation of the ecological features that select for different camouflage strategies, such as habitats with different appearances, or predators with different visual systems. This research is discussed in detail in section 2.7.

The presence of disruptive coloration in plants is relatively unstudied. There is good evidence that seed coloration has a role in crypsis, and some seeds

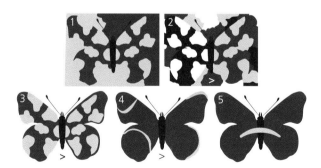

Figure 2.2 Sub-principles of disruptive camouflage identified in Thayer (1909) and/or Cott (1940) and retained in Stevens & Merilaita's (2009) reorganization. In panels 2–4 the '>' indicates which half of the moth should have greater camouflage according to disruptive principles. 1.) Differential blending applies as all tones of the moth's wings match those present in the background. 2.) Maximum disruptive contrast is achieved on the left wing as internal tones are highly contrasting, whilst still commonly found in the background. 3.) Disruptive marginal patterns on the left wing break up the true edge, while the identical but internally shifted patterns on the right wing may make the edge more conspicuous. 4.) Internal markings on the left wing break up the recognizable shape of the moth, whereas identically shaped markings on the right wing margin do not. 5.) The coincident marking across the two wings and body creates false edges that hinder recognition. See Table 2.1 for further detail.

Table 2.1 Sub-principles of disruptive camouflage identified in Thayer (1909) and/or Cott (1940) and reasons for rejection or retention in Stevens & Merilaita's (2009b) reorganization.

Principle	Description	Reason for retention or rejection	Recent investigations	References
Retained as a principle of disruption:				
1 Differential blending	Disruption most effective when some or all tones match the background.	Conceals the real object boundary.	Highest survival when all object tones match the background.	Thayer (1909); Cott (1940)
2 Maximum disruptive contrast	Disruption most effective when pattern contrast is high.	Conceals the real object boundary.	Studies generally find survival highest when patterns have high contrast, but only within the range of background matching colours.	Cuthill et al. (2005); Stevens & Cuthill (2006); Stevens et al. (2006); Kelman et al. (2007); Stobbe & Schaefer (2008)
3 Disruptive marginal patterns	Disruption improved when markings intersect the outline of the body.	Conceals the real object boundary.	Strong support for markings that intersect the outline increasing camouflage.	Cuthill et al. (2005); Schaefer & Stobbe (2006); Stevens & Cuthill (2006); Stevens et al. (2006); Fraser et al. (2007); Todd et al. (2015)
4 Surface disruption	Creating false internal edges and shapes that do not correspond to the true outline.	Creates false edges and boundaries within an object.	Some evidence that internal markings can increase survival.	Stevens et al. (2009); Seymoure & Aiello (2015)
5 Coincident disruptive coloration	Disruptive marking that span across otherwise conspicuous body parts.	Creates false edges and boundaries within an object.	Two-tone coloration crossing conspicuous features can improve survival.	Cuthill & Székely (2009)
Rejected as a principle of disruption:				
6 Background picturing	Disruptive patterns should be background matching.	Synonymous with background matching.	Optimal camouflage does not necessarily maximize degree of background matching.	Fraser et al. (2007); Kang et al. (2015); Phillips et al. 2017; Wilkinson & Sherratt (2008);
7 Regularity avoidance	Disruption most effective when pattern elements are irregular.	Does not involve concealment of shape or outline.	Repeating and symmetrical patterns more likely to be interpreted as a single object.	Merilaita & Lind (2005)
8 Irregular marginal form	Morphological complexity conceals the shape of an animal.	Uses morphology rather than colour patterns.	No evidence for an interaction between shape and markings on survivorship.	Webster et al. (2015)
9 Distractive/ deflective markings	Markings that attract predators' attention towards parts of the body in a way that facilitates escape.	A distinct anti-predator strategy from disruption (see Chapter 11).	Colour markings have been shown to attract predator attention and deflect attacks to facilitate escape.	Section 2.5.3 and Chapter 11.

and seed pods have two-tone coloration which might provide disruption (Aviezer & Lev-Yadun, 2015; Lev-Yadun & Ne'eman, 2013). Variegated leaves, which often feature markings distributed around the edges, might also be good candidates for disruptive camouflage against herbivores (Givnish, 1990; Lev-Yadun, 2015). Similarly, the variegated patterns of some unripe fruits might camouflage against frugivores until seeds are mature (Lev-Yadun, 2013). However, to date all these hypotheses remain untested.

The possible presence of disruptive camouflage has been comprehensively reviewed in mammals (Caro, 2013) and insects (Théry & Gomez, 2010), with the suggestion that disruption is a taxonomically widespread anti-predator strategy. However, compared to background matching and countershading camouflage, it does not seem to be as ubiquitous. Some taxonomic groups such as pinnipeds, for example, do not seem to contain any species with potentially disruptive coloration (Caro et al., 2012).

2.3 The multiple mechanisms of camouflage by disruption

In an influential review 'Camouflage and visual perception' Troscianko et al. (2009) introduced the perspective developed here and by others (Cuthill & Troscianko, 2011; Kelley & Kelley, 2013; Merilaita et al., 2017; Osorio & Cuthill, 2013; Webster, 2015) of disruptive camouflage as a group of strategies that work by exploiting features of the visual system to make the encoding of an object's characteristic outline and shape more difficult.

The major task of complex visual systems is to process the responses of millions of photoreceptor inputs in a way that results in a coherent perceptual representation of the environment that supports behaviour, and does so with energetic efficiency in terms of how computational resources (neurons) are deployed and utilized (Vincent et al., 2005). Intensively studied complex visual systems, mainly mammalian, achieve this by 1.) constructing a hierarchy of features, beginning at retinal cells where phototransduction converts light of a specific frequency bandwidth from a specific point in the visual field to an electrical signal, through to 2.) early visual edge detection processing that summarizes this input by encoding the location, polarity, and orientation of small edges, then on to 3.) mid-level perceptual grouping processes that join edges spatially and temporally (detecting motion) and segment edges that belong to a single object from the background (Goldstein & Brockmole, 2016). Finally, 4.) high-level object recognition processes identify and classify objects in the scene and their properties, including whether they are potential prey. These broad stages in the hierarchy of visual processing, from primitive feature and edge detection, to feature grouping and segmentation, and then on to object identification and classification (Palmer, 1999), effectively summarize the information in previous stages using both top-down and bottom-up processes, until it is in a form useful to subsequent processing and behaviour. The 'short-cuts' visual systems take to compress this information are what enable coloration to disrupt the encoding of object outline and shape (Merilaita et al., 2017), and each stage is likely to have specific mechanisms via which disruptive camouflage influences processing.

Here then, we review the small but growing body of work investigating the different perceptual mechanisms via which disruptive camouflage strategies have been proposed to exploit processing of object shape and identity at each stages of visual processing, and make links with traditional principles of disruption. As almost all work to date has been done on visual disruption, we leave discussion of disruption in other sensory modalities to the concluding section of this chapter. Work on visual system organization across species points to commonalities (Gibson et al., 2007); though there are many important differences too (Qadri & Cook, 2015). Future work should aim to expand knowledge of interspecific differences in how perception of object outline, shape, and other characteristic features can be disrupted, but for now we assume that the disruptive mechanisms identified apply to a broad taxonomic range of receivers.

2.3.1 Disruption of edge detection processes

One reason for the primacy of edges in visual perception is that they define spatial change in the environment, commonly where an object meets the background it is viewed against. Defining where edges occur in a scene is a necessary step towards defining object outlines; this is one of the most characteristic aspects of objects. However not all edges in a scene are parts of object outlines. Markings and textures created by pigmentary and structural colours are, along with shadows, intensity edges that are not coincident with physical object outlines. How visual systems tasked to locate and recognize objects by finding their outlines 'ignore' these types of edge is a classic problem in computational visual neuroscience (Marr & Hildreth, 1980) and a key cause of failure for artificial visual systems (Vondrick et al., 2013) and technologies that rely on these, such as driverless cars. The disruptive hypothesis predicts that an animal's coloration can decrease the salience of true edges and generate and increase the salience of false edges, making separation of 'false' and 'true' outline edges at subsequent stages of processing more difficult. Merilaita et al. (2017) frame this as a signal-to-noise manipulation, where disruptive camouflage increases the salience of false edge noise and decreases the salience of true edge signal.

In non-specific suppression, the responses of simple edge detectors are suppressed when strong edges are detected by other nearby simple cells (Brouwer & Heeger, 2011). Troscianko et al. (2009) therefore suggest that non-specific suppression should be important in disruptive camouflage, since strong coloration edges should make nearby true outline edges less visible. This idea is clearly relevant to Cott's (1940) principle of maximum disruptive contrast (#2, Table 2.1), which states that disruptive patterns should be more effective when they are high contrast (i.e. edges are strong).

The principle of maximum disruptive contrast has been tested empirically several times, most notably by Cuthill et al. (2005) in a study that introduced an influential experimental paradigm where triangular paper 'moths' printed with different colour patterns were pinned to trees with mealworms attached. Targets were left exposed to wild avian predators and the survival rate of mealworms in each treatment was recorded. This provides an effective measure of camouflage effectiveness that allows for both tight control of experimental stimuli (e.g. standardizing the degree of background matching across treatments in terms of avian vision), whilst measuring the responses of real predators in a relatively natural foraging task. Cuthill et al. found that targets with high-contrast disruptive patterns survived better than those with low-contrast disruptive patterns, supporting the prediction of maximum disruptive contrast and the idea that strong false edges can make nearby true edges less conspicuous. Following up on this result, Schaefer and Stobbe (2006) investigated whether maximizing contrast could be effective when it led to a decrease in background matching and found that there was room for inclusion of non-background matching colours when they increased contrast. Rapidly changing, cuttlefish dynamic-disruptive camouflage sometimes exceeds the level of contrast in backgrounds (Hanlon et al., 2009). However several studies have since demonstrated that the effective degree of contrast is strongly limited by the level of contrast in the background, as targets using much lighter tones than are present in the background suffer increased predation (Fraser et al., 2007; Stevens et al., 2006; Stobbe & Schaefer, 2008), presumably because this also increases the contrast between the true outline and the background. The emerging picture is that higher contrast maximizes disruptive 'noise', but this is balanced by opposing selection for background matching.

Also following from the phenomenon of non-specific suppression, edge detectors in early stages of vision (such as the mammalian primary visual cortex) tuned to respond to the true outline should be inhibited by the responses of other neurons responding to edges that run perpendicular to their receptive fields (cross-orientation suppression; Brouwer & Heeger, 2011; Priebe & Ferster, 2006). Camouflage markings that intersect an animal's outline should therefore weaken the responses of the edge detectors that are tuned to edges of the animal's true outline. This process relates to the disruptive marginal patterning principle (#3, Table 2.1.), which states that disruptive camouflage is improved when markings intersect an object's outline. This prediction was also tested by Cuthill et al. (2005); in an 'edge' treatment, markings randomly sampled from the oak bark background were placed on a triangular target's outline, in a second 'inside 1' treatment the exact same markings were shifted inward so they no longer intersected the outline, and in a third 'inside 2' treatment randomly selected markings were placed so they did not intersect the outline (Figure 2.3). Survival was best when markings intersected the outline, supporting the marginal patterning hypothesis. While the treatments were designed to be equivalent in terms of background matching (all markings were a random sample from the background) it remained possible that the conspicuous straight edge created by the 'inside 1' manipulation enhanced the true outline, or the increased density of inside markings increased their conspicuousness. This issue was resolved by Fraser et al. (2007) who repeated Cuthill et al.'s experiment but added a third 'inside' treatment that had the same low pattern density as the edge treatment. While the low-density inside treatment did survive better than the high-density one, indicating that the inside treatments in Cuthill et al. did have worse background matching than the edge treatment, the edge treatment still survived best, indicating a specific effect of intersecting the edges. In an experiment on human predators searching for camouflaged prey on a computer monitor that replicated the disruptive

| Edge | Inside 1 | Inside 2 | Black | Brown |

Figure 2.3 Stimuli used to test the principle of disruptive marginal patterning in Cuthill et al. (2005). In the edge treatment markings are placed on the outline. These targets have higher survival against wild avian predators than the inside treatments and plain black or brown treatments.

marginal patterning effect, Fraser et al. demonstrated generalizability of disruption across species and contexts. The finding of increased survival for prey with markings that intersect edges has now been replicated several times under different experimental designs for human and avian predators (Cuthill et al., 2006b; Merilaita & Lind, 2005; Schaefer & Stobbe, 2006; Stevens et al., 2006, 2009; Webster et al., 2013, 2015).

The hypothesis that these results are specifically a consequence of disruption of edge detection processing in early stages of vision is supported by studies that present disruptively patterned stimuli to computational models of early vision. Stevens & Cuthill (2006, following Osorio & Srinivasan, 1991) demonstrated that various edge detection algorithms based on properties of the vertebrate visual system detected false internal edges but not true outline edges in Cuthill et al.'s (2005) disruptive stimuli more often than in the background matching stimuli. The consequence of this was that a line detector based on the Hough transform located the sides of the triangular disruptive targets less often.

2.3.2 Disruption of perceptual organization: grouping and segmentation

Perceptual grouping of simple features (i.e. edges and collections of edges) involves determining which parts and regions of a visual scene belong together in a mid-level unit (Brooks, 2015; Peterson et al., 2013). Segmentation involves determining which of these units belong to which objects and deciding that a feature is not part of the background (figure-ground organization). Gestalt laws of perceptual organization, first developed by perceptual psychologists in the 1920s, codify how simple image features are processed, leading to grouping or separation and the perception of organized patterns and objects (Figure 2.4; Wagemans et al., 2012).

The Gestalt law of grouping by colour similarity clearly relates to Cott's (1940) differential blending principle (#1, Table 2.1): If the colours of an animal contrast with each other, but match the background, then in addition to reducing the conspicuousness of the outline, the different colour patches are more likely to be grouped with different elements of the background than with each other.

Differential blending and disruptive marginal patterning also combine to create artificial gaps in an object's outline when true edges blend with different parts of the background. This can be considered as targeting the Gestalt law of good continuation, which specifies how edges link to form a continuous line. Detecting object outlines in natural scenes is made difficult by non-camouflage factors such as occlusions, shadows, and complex backgrounds. Disruptive camouflage mechanisms can be thought of as adding to this list of problems that result in incomplete object edges and the detection of many edges that do not correspond to object outlines. To resolve this problem and 'fill-in' missing edges, contour completion mechanisms have evolved that aim to connect the correct edges together. Animals use heuristics based on the assumption that missing edges are likely to span gaps with smooth lines and curves (mammals: Kanizsa et al., 1993; birds: Regolin & Vallortigara, 1995; fish: Sovrano & Bisazza, 2008). While it remains to be specifically investigated, it is likely that disruptive patterns specifically target the use of these heuristics by offering parsimonious solutions to image ambiguity that lead to incorrect interpretations of a scene.

The principle of surface disruption, which proposes that camouflage can be improved with contrasting markings spanning body surfaces as well as intersecting edges (Cott, 1940; #5, Table 2.1), also likely targets mid-level perceptual grouping processes. There is good evidence that surface disruption can be effective. In an artificial moth field

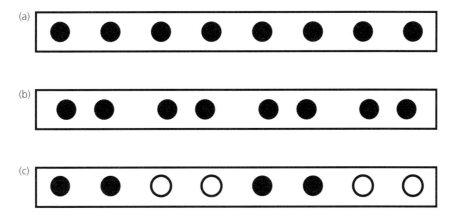

Figure 2.4 (a) shows how eight evenly spaced dots are grouped into a single unit. In (b) the proximity between dots is altered resulting in grouping into four groups of two (law of proximity). (c) shows grouping by the law of similarity, where "like parts band together", in this case by tone/colour (Wertheimer, 1923).

predation experiment, Stevens et al. (2009) found that treatments with high-contrast internal markings but low-contrast edge markings survived better than other combinations. As high- and low-contrast markings were equally common in the background, and image analysis following the computational edge detection approach used in Stevens & Cuthill (2006) showed that there were no differences in the number of edges identified between treatments, this indicates the survival advantage is not being granted by low-level edge detection mechanisms. Instead the authors suggest that high-contrast internal markings worked because the most salient (internal) edges did not correspond to the true outline. Similarly, Seymoure and Aiello (2015), following up on Silberglied et al. (1980), found that naturalistic models of *Anartia fatima* butterflies, which display an internal wing stripe, survived better than models that had been manipulated to have a continuous or broken stripe at the wing edge. While this result may have been a consequence of a post-detection preference for the novel manipulated prey, as birds generally show a neophobic response to novelty, a decrease in camouflage effectiveness is the more likely explanation.

A class of internal disruptive patterns given special consideration by Cott (1940) are what he termed coincident disruptive patterns (# 5, Table 2.1). Here patterns overlap different elements of an internal feature such as the sclera and iris of an eye, or different portions of a limb, and are proposed to interfere with determining the shape of the internal feature (Cuthill & Székely, 2009). This interference distinguishes coincident disruption from simple background matching of potentially conspicuous features using coloration (e.g. a black eye surrounded by black body). To test the coincident disruption hypothesis, Cuthill and Szekely performed an artificial moth field experiment and a replicate human computer search experiment, using prey with disruptive wings and a central 'body' feature with a non-background matching shape. When the body was two-tone and the surrounding wing coloration was coincident (i.e. both the two body tones matched the surrounding wing colour), survival was higher than when the body was monotone and matched the surrounding wing coloration, and when the body was two-tone but the wing only partially matched. The specific perceptual mechanisms behind this and the surface disruption results have not been investigated. However it seems likely that surface and coincident disruption work by interfering with perceptual grouping processes, such that groupings are based on coloration rather than features such as the true outline, the eyes, or the three segments of a frog's folded limbs.

Gestalt laws of perceptual grouping are often in competition with each other. For example differential blending may produce an incorrect assignation of a bicolour prey to two or more separate groups, where each tone of the prey is grouped with a different background tone due to the law of similarity.

However, if the disruptive patterns are symmetric then grouping by symmetry will offer a competing explanation that groups the pattern elements together. Indeed, symmetric patterning has been shown to reduce the effectiveness of disruptive camouflage (Cuthill et al., 2006a, b). Grouping laws may also be disrupted synergistically, an idea investigated by Espinosa and Cuthill (2014) who tested the hypothesis that colour patches on animals will be most camouflaging when they are high contrast and match different colours in the background (differential blending), but especially when they match the colours in different background regions. Two-tone stimuli were positioned in a scene such that they spanned a region consisting of either one or two background objects, whilst controlling for local contrast. Results supported the prediction that the disruptive benefit is greater when the colours in the immediate background are perceived as belonging to different objects. While this and other similar experiments can appear quite abstract, plausible real-world examples include reef fish that differentially blend with two distinct coral types that are perceived by predators as being different objects (Marshall, 2000) and isopods that blend with brown algae and their epiphytes (Merilaita, 1998).

Disruption of shape and object processing in mid-level vision may also be achieved through a variety of visual illusions. Specific examples that have been demonstrated in the context of disruptive camouflage include edge enhancement, where tones of colour patches increase in contrast where they meet, a commonly observed phenotype on diverse species such as copperhead snakes *Agkistrodon contortrix* and many hawk moths in Sphingidae. Egan et al. (2016) showed that this creates the impression of pictorial depth between different pattern elements for human observers, and that this improves camouflage beyond just simply matching the level of pictorial depth in backgrounds. The likely explanation is that regions perceived as being in different depth planes are less likely to be grouped together (Rock & Brosgole, 1964).

Whether disruptive processes at the perceptual grouping stage are more or less effective than processes at earlier or later stages remains an open question. In one of the very few studies that attempts to identify which perceptual processes underlie

examples of disruption, Webster et al. (2015) measured human predator responses to artificial moth prey presented on a computer screen. Prey varied in whether their markings intersected edges or not and the shape of their outline (straight vs. jagged vs. wavy). The authors predicted that if disruptive mechanisms were operating at the shape completion stages of processing, then the survival advantage of disruptive markings should be dependent on object shape, with simple straight outlines gaining more from disruptive markings than jagged or wavy ones. Instead they observed that disruptive markings were effective irrespective of object shape, indicating that the advantage of disruption, in this experiment at least, lay in disruption of local edge information rather than shape information.

2.3.3 Disruption of object detection and recognition

From a computational perspective, object detection and recognition are closely related processes (Sermanet et al., 2013; Tu et al., 2005). As stated previously, detection involves finding locations in a visual scene that contain features corresponding to the object class of interest, and recognition involves determining if a set of visual features corresponds to an object in a particular class. Both detection and recognition can be involved in searching for camouflaged objects, during which time bottom-up information generated by the visual scene and processed by earlier stages interacts with top-down information (Gilbert & Li, 2013) specifying what features to search for, with this interaction guiding where attention is focussed (Wagemans et al., 2012).

Visual search for camouflaged targets guided by top-down information may be made particularly difficult for predators if disruptive coloration helps to prevent so-called 'search images' forming and being utilized (Guilford & Dawkins, 1987; Ishii & Shimada, 2010; Tinbergen, 1960a, Ch. 1.). A visual search image might be for a very specific combination of colour and shape that is highly characteristic of a favoured prey species, or it might be a more general representation that, for example, guides attention towards anything with a 'furry' texture. In the latter case, disruptive coloration that interferes with processing features that define a broad class of prey might be an effective strategy. For example, if most frogs among all those that a predator encounters over its life do not possess coincident leg stripes, then the image feature created by frogs' folded legs is a useful diagnostic feature of class membership and likely to form part of the search image for frog prey. Disruptive patterns, in this example coincident leg stripes, may therefore work specifically by interfering with detection of the key features predators are using in their search images; using search images to become better at finding frogs without leg stripes has the consequence that predators become worse at finding frogs with leg stripes. There is potential here for frequency-dependent effects where invariant disruptive patterns themselves become a valuable search cue if they are sufficiently common. Most experimental investigations use stimuli that are partly random (e.g. Cuthill et al., 2005), but variability of disruptive patterns may be low in natural populations. Testing for frequency-dependent effects in disruptive patterning would be a valuable avenue for future research.

As taxonomists will attest, colour, size, and texture are rarely identifying features of a particular species; however, shape often is. Indeed, despite variation in shape due to viewpoint and translation, visual systems in general (not just those of taxonomists!) extensively utilize shape information to detect and recognize objects (Hummel, 2013), especially animals (at least in humans; Elder & Velisavljević, 2009; Lloyd-Jones & Luckhurst, 2002). Thus disruptive camouflage that manipulates how shape is perceived should particularly affect target recognition, though to date there is little evidence of this mechanism (Webster et al., 2015). Interesting computer vision work demonstrates how images containing animals can be reliably classified on the basis of edge curvature alone (Perrinet & Bednar, 2015), suggesting that disruption of low-level but characteristic aspects of stimuli, such as edge curvature, may be a good strategy to prevent classification of high-level object categories.

Disruptive camouflage may also target perceptual learning and the initial formation of search images, particularly those for finer classes of stimuli such as a particular target species. Testing this hypothesis, Troscianko et al. (2013) set up experiments using humans as model predators to test learning rates for background-matching, disruptive, and distractive camouflage types. In general across the experiments, results showed that high-contrast disruptive treatments survived best initially, but that this advantage was eroded across trials. These results imply that the value of different camouflage types is dependent on interactions between detection, recognition, and learning processes, and that disruptive markings, at least those used in the study, while hard to detect initially, are learnt quickly and so become easier to detect over time. This is likely because while disruption is promoted by high contrast, high-contrast stimuli are also learnt more easily. While this effect may be a consequence of encountering many stimuli in quick succession, if results translate to natural systems the implications are intriguing. Understanding why increased rates of predator learning do not wipe out camouflage advantages of disruption is a necessary avenue for future research.

2.4 Identifying and quantifying disruptive camouflage

As previously mentioned, determining whether a particular animal or artificial stimulus has disruptive effects on receiver perceptual processing is not a trivial task. A common approach is to base quantification of disruption on features of the camouflage pattern. For example, Stevens and Merilaita (2009b) argue that disruptive patterns should have a different distribution of contrast to the background to satisfy the principle of maximum disruptive contrast, and suggest that disruptive markings should be dependent on body shape and size while background matching markings should not (Merilaita & Lind, 2005). Other predictions include the idea that animals pursuing disruption should have a higher proportion of markings touching the outline of the body, over and above the distribution that would be predicted by matching the background (Merilaita, 1998). Establishing whether these metrics are a valid proxy for the presence of disruptive coloration, let alone the degree of disruption, at minimum requires establishing a correlation between behavioural performance on search tasks and stimulus properties—something that the psychophysical studies reviewed in the previous section have begun to do. Moreover,

Non-disruptive   Edge detection disruption   Feature detection disruption

Figure 2.5 Disruptive mechanisms may be identifiable on the basis of appearance. From Webster (2015).

in theory there should be a correlation between properties of the disruptive camouflage pattern, and the perceptual process(es) that it targets. For example, an untested hypothesis is that the position of markings for edge disruption should intersect long edges to break them up, whereas disruption of feature detection processes should involve edges intersecting areas that provide the most information about object shape and identity, such as the corners of a triangular object (Figure 2.5; Webster, 2015).

Testing these hypotheses experimentally may establish associations between disruption and appearance that in the long term have the potential to identify specific disruptive mechanisms deployed by an animal on the basis of appearance alone. However, testing this and other similar predictions is difficult, as it is unclear what measurable behavioural difference will result from disruption of different perceptual mechanisms. As eye movements reflect internal cognitive processes (Findlay & Gilchrist, 2003), one promising approach is to use eye tracking to investigate the stage of visual processing affected by a particular disruptive pattern. Webster et al. (2013) tracked the eye movements of humans searching for targets with and without edge-intersecting patterns, quantifying the period of time that subjects viewed the target area overall, how frequently targets were passed over, and how long they were fixated immediately prior to a detection response. All three measures were correlated positively with the number of edge-intersecting patches on targets, which the authors argued reflected object recognition processing rather than disruption of lower-level processes.

An alternative solution to quantifying disruption that gets around the problem of establishing a suitable assay involves presenting targets designed to disrupt different processes to both animal observers and banks of computer vision algorithms that have been designed to either incorporate or exclude models of different visual processes. If the performance of an algorithm that incorporates particular perceptual processes correlates with observer performance on search for camouflaged targets better than an algorithm that excludes it, then this might be suggestive that the process is targeted by the camouflaged pattern. To date this approach has been applied to demonstrate disruption of edge detection in humans when viewing artificial prey, using metrics such as 'GabRat' which employs banks of Gabor filters (Troscianko et al., 2017). A Gabor filter determines the spatial frequency content in a particular direction at a particular image region. GabRat uses the responses of many filters tuned to different frequency bandwidths and orientations to calculate the ratio of coherent to incoherent edges around an object's outline. In principle, if correlations between different visual models, different observers, different visual environments, and different target types are sufficiently clear, the mapping could be applied to identify and quantify disruptive mechanisms operating in real animals on the basis of appearance alone.

An interesting approach to quantifying examples of disruptive camouflage in nature is to observe how animals select particular resting locations in the environment that offer the best camouflage (Jaenike & Holt, 1991; Kjernsmo & Merilaita, 2012), and measure the degree to which they improve background matching or disruption. Kang et al. (2015) investigated two species of moth, *Hypomecis roboraria* and *Jankowskia fuscaria*, that alter their resting positions on a very fine scale to improve camouflage (Kang et al., 2012). Captive moths were released and allowed to settle on tree trunks; they were then photographed in their initial landing position and then again after one hour, by which point some individuals had made fine adjustments to their positions. Comparing differences in background matching and disruptive camouflage measures derived from image analysis of pre and post repositioning images showed that in both species, as well as improving their degree of background matching, the amount the

edge was broken up also increased as a result of increased differential blending and phase matching between the moth and bark texture. This suggests that moth behaviour optimizes disruptive as well as background matching camouflage, with the implication that moth camouflage employs disruption. Similarly Webster et al. (2009) show that the alignment behaviour of two moth species can be explained by improved differential blending, as human predation on moths is highest when the moth is at an atypical orientation relative to the tree.

A similar strategy was taken in a study that demonstrates Japanese quail *Coturnix japonica* select egg-laying sites that improve the camouflage of the particular egg maculation phenotype they lay (Lovell et al., 2013). Interestingly, in this species there seems to be intraspecific variation in camouflage strategy adopted. Females that lay lightly maculated (spotted or stained) eggs select sites that match their eggs' background colour, suggesting a primarily background matching camouflage strategy. Females with heavily maculated eggs instead select sites that match their maculation colour, which image analysis suggests improves outline disruption as well as background matching.

2.5 The relationship between disruption and other forms of protective coloration

2.5.1 Background matching (Chapter 1)

Several empirical results have demonstrated that an insufficient degree of background matching makes disruptive camouflage ineffective (Fraser et al., 2007; Kang et al., 2015), likely because key features such as outlines that disruption might be expected to act on are so conspicuous. Wilkinson and Sherratt (2008) conclude that the relationship between background matching and disruption is asymmetric; background matching can work without disruption, but disruption cannot work without sufficient background matching. This seems to hold true in most situations, though observations of cuttlefish displaying high-contrast disruptive patterns when hunting prey in the open may suggest that using coloration to disguise body shape can still be effective when background matching cannot be achieved (Hanlon et al.,

2009). In a signal-to-noise ratio framework (Merilaita et al., 2017) background matching reduces the information in the signal, while disruptive camouflage both reduces the true edge signal and increases the noise by creating false edges.

2.5.2 Aposematism (Chapter 9)

A longstanding idea is that disruptive patterns can have a dual function in aposematism. The two strategies seem incompatible, since disruption is thought to reduce conspicuousness while warning colours are generally most effective when they are highly conspicuous. However the visual effect of factors such as changing viewing distance, utilization of backgrounds with different appearances, or differential targeting of receivers with high- and low-acuity visual systems, creates the potential for multiple strategies for a single colour pattern that operate in different circumstances (Tullberg et al., 2005). This idea is supported empirically in a study of the wood tiger moth *Parasemia plantaginis* (Honma et al., 2015), which has effective aposematic coloration when resting in foliage. When approached this species sometimes feigns death and drops to the ground, at which point it becomes hard to spot on a background of leaf litter. Manipulating the moth pattern into 'edge' and 'inside' treatments (Figure 2.3), Honma et al. found 'edge' moths survived better on some leaf litter backgrounds, with the inference that survival was improved by disruption of edge processing. Examining distance-dependent effects in aposematic firebugs *Pyrrhocoris apterus*, Bohlin et al. (2012) found that making bugs disruptive by moving patterns to the margins did not increase detection times except on certain backgrounds at a long distance, suggesting that long-range disruptive effects are possible in warningly coloured species. European vipers have been suggested as deploying both aposematism and disruption in their zig-zag dorsal pattern; however, experimental studies suggest that the function is purely in warning coloration (Valkonen et al., 2011). The emerging picture is that warning signals and camouflage can be balanced and that a dual function is possible (Barnett & Cuthill, 2014; Barnett et al., 2017), but whether a dual strategy is frequently implemented in nature is as yet unclear.

2.5.3 Distractive, divertive, and deflective markings (Chapter 11)

Distractive markings are conspicuous patches of colour that draw a predator's attention away from distinctive features of the prey to facilitate camouflage, and divertive or deflective markings manipulate where a predator directs its attack toward a region that facilitates escape or reduces mortality (Dimitrova et al., 2009; Kjernsmo et al., 2016; Kjernsmo & Merilaita, 2013; Merilaita et al., 2013, 2017; Stevens et al., 2012). Cott (1940) presented examples of distraction under a disruptive camouflage heading; however, because these markings aim to increase visibility and attract attention, it is now generally treated as a separate strategy.

2.5.4 Dazzle (Chapter 12)

Dazzle coloration affects a receiver's judgement of speed or trajectory while in motion. Since both disruptive and dazzle coloration are proposed to be most effective when patterns are high contrast, one possibility is that some patterns have a dual role, being disruptive when the animal is stationary, and then dazzling when the animal is moving (Stevens, 2007; Stevens & Merilaita, 2009b). This suggestion is not fully supported by the available evidence: Stevens et al. (2011) found that striped patterns prevented capture of moving targets most effectively when low contrast, arguing that high contrast patterning facilitated object tracking. Furthermore striped targets were poor camouflage when motionless, and non-striped disruptive targets were not effective in motion, with the latter finding replicated in Hall et al. (2013). Multiple studies have found high contrast patterns to be effective at preventing capture of moving targets (e.g. Hall et al., 2016; Hogan et al., 2016b; Scott-Samuel et al., 2011; Chapter 12), but these have not demonstrated a disruptive effect while stationary. Thus demonstration that coloration can be both disruptive when stationary and dazzling in motion remains elusive.

2.6 The ecology of disruption

Identifying under what ecological circumstances animals pursue disruptive camouflage strategies is challenging, primarily because identifying the type of camouflage used and the degree of protection it gives for a large number of individuals, populations or species is difficult (section 2.4). With this caveat, numerous studies have attempted to make progress on this subject by using appearance-based measures of the degree to which an animal utilizes disruption.

Observation of colour changing animals in response to different stimuli is one of the best ways of understanding the ecology of disruption (Duarte et al., 2017). Most work has been done on common cuttlefish *Sepia officinalis*. Cuttlefish can rapidly adapt their camouflage over very short timescales (0.5–2 seconds) and use visual cues to determine what colour pattern to display (Chiao et al., 2015). Cuttlefish switch between background matching (uniform or mottled coloration) and apparently disruptive camouflage featuring blocks of high-contrast colour, largely depending on substrate appearance. As well as providing insight into cuttlefish vision, manipulating the substrate and observing the colour response also allows investigation of what environmental features select for apparent disruption.

Testing cuttlefish responses on substrates that vary in spatial scale, including checkerboards (Barbosa et al., 2007, 2008; Kelman et al., 2007; Mathger et al., 2006; Zylinski et al., 2009a), semi-naturalistic (Chiao et al., Hanlon, 2009) and natural backgrounds consisting of rocks of different sizes (Mathger et al., 2007), shows that apparently disruptive patterns are elicited when check sizes are between 40 and 120 per cent of the largest disruptive skin patch that the cuttlefish can switch on or off (Figure 2.6). This demonstrates that disruption is dependent on the scale of the background, and supports the idea that disruption is mainly effective when pattern elements match the background to a sufficient degree. Similarly disruption only emerges when backgrounds are of sufficiently high contrast (Barbosa et al., 2008; Zylinski et al., 2009b), specifically the contrast of short-edge fragments (Chiao et al., 2013; Zylinski et al., 2012), suggesting that disruption is most effective when backgrounds are full of fragmented incomplete contours. More surprising is that disruptive patterns are more likely in cuttlefish when backgrounds are darker (Chiao et al., 2007). Interestingly, this matches the result in Japanese quail where dis-

ruptive eggs tended to be laid on dark backgrounds while background-matching eggs were laid on light backgrounds (Lovell et al., 2013). The presence of visual depth in the background is also a key stimulus element that elicits disruptive patterns (Kelman et al., 2008). While noting that many of these relationships are also what would be predicted by adaptation for improved background matching, it is a reasonable assumption that the patterns which have all the hallmarks of design principles such as differential blending, edge enhancement, and maximum disruptive contrast (Hanlon et al., 2009) actually have disruptive effects when perceived by cuttlefish predators.

Accepting that cuttlefish do use disruptive camouflage, if the relationship between expression of disruption and environmental structure is not species-specific, it is reasonable to extrapolate the rapid responses of cuttlefish to other species whose colour adapts to the environment over evolutionary timescales. Comparative evidence in felids suggests that species with patterns that appear to use disruptive principles, such as ocelot, clouded leopards, and jaguar, are those that occupy closed habitats (Allen et al., 2010) where dappled light creates the high-contrast fragmented background that favours expression of disruptive patterns in cuttlefish, although this relationship is also predicted by selection for background matching. Similarly tentative conclusions can be drawn in chameleons, where species with ventral line markings are associated with arboreal habitats where this marking might break up the silhouette of animals when they are viewed from below (Resetarits & Raxworthy, 2016).

The ecological correlates of some potentially disruptive markings, such as the stripes seen on the flanks of several artiodactyl species, associated with stotting and leaping behaviour, suggesting that these markings instead have a signal amplification role (Allen et al., 2012; Caro & Stankowich, 2009). Similarly, in cetaceans, high-contrast markings are no more common in well-lit environments where they might have a more disruptive effect (Caro et al., 2011). Diagonal stripes on butterflyfish have evolved in tandem with increased dietary complexity, but not habitat diversity (Kelley et al., 2013), and in cichlids vertical stripes occur in species using rocky environments (Van Alphen, 1999). As selection for

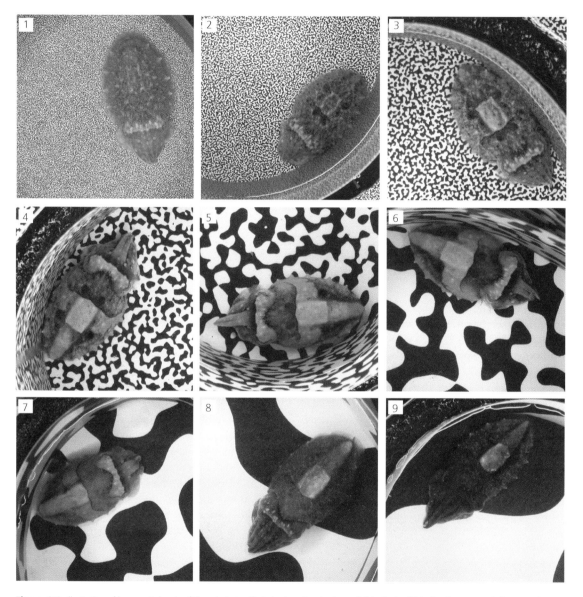

Figure 2.6 Illustration of how spatial scale of the substrate affects body patterning in cuttlefish *Sepia officinalis* . From a mottled pattern the disruptive pattern emerges and strengthens (S2 to S6) as the spatial scale of the substrate increases, before declining as it changes to a uniform pattern (S7 to S9).

disruption cannot straightforwardly explain these patterns, this suggests that at macroecological scales these markings, which seem to conform to disruptive principles, have not been selected solely or chiefly for their disruptive effects. Alternatively, disruption may not predict strong ecological associations of these types. More work needs to be done

to establish a set of testable predictions on the ecological circumstances under which the presence of disruptive camouflage is expected, for example Wilkinson and Sherratt's (2008) suggestion that disruption should emerge when animals use a greater variety of backgrounds, as disruption may still be able to provide protection when mismatching the background to some degree. However there is little evidence supporting this prediction and the few other similar hypotheses, so until more experimental work has been conducted, comparative predictions like this are on shaky foundations. More generally there are few predictions regarding whether disruption should evolve in, for example, species pursuing particular life history strategies, or occupying particular ecological niches.

Two recent studies on egg camouflage in several avian species have taken a rigorous image-processing-based approach to quantifying disruption in natural scenarios in the context of the local background (Stoddard et al., 2016; Troscianko et al., 2016). Neither study found support for various measures of disruption predicting survival. In nightjars, high-contrast eggs only survived better when they were on a high-contrast background, suggesting background matching being the primary mechanism. In snowy plovers *Charadrius nivosus* and least terns *Sternula antillarum* no camouflage metric predicted clutch survival, including those specifically designed to identify disruptive principles.

2.7 Unresolved issues and future challenges

While our understanding of disruptive camouflage has progressed rapidly in the last 15 years, a focus on mechanistic questions has meant progress has been uneven, and many open questions remain (including mechanistic questions). A major issue highlighted as a barrier to progress throughout this chapter is lack of knowledge about whether and to what degree organisms have evolved a disruptive camouflage strategy. Common cuttlefish are the best understood species, yet even here disruption in terms of the severity of effects on predator visual processing has not been quantified.

In studies which measure the survival of artificial or manipulated disruptive prey, the issue of controlling for equivalence in background matching is still not effectively resolved. Part of the problem is in quantifying the degree of background matching (Chapter 1), but many manipulations deployed are likely to affect both background matching as well as disruption. For example, 'inside' treatments (Cuthill et al., 2005; Figure 2.2), even if pattern density is

controlled (Fraser et al., 2007), by focussing on the dark pattern element, the potential change in background matching of the light pattern element is neglected, when it is likely that the light border created makes a shape that does not occur in the background, and is therefore less background matching.

Additionally, as many experimental studies use simple triangular 'moths' as stimuli, it is currently quite unclear how disruption of features involved in object recognition operate, or under what circumstances disruption of high-level processing is effective compared to disruption of edges. One interesting idea discussed in Webster (2015) is to use targets that have the same local and global properties as real animal outlines. Hopefully future work will pick this up to add to our understanding of high-level disruption. Other avenues for future work in this area include investigation of whether disruption targets different levels of visual processing when prey are targeted by different predators. Low-level edge disruption may be a good general-purpose strategy against a wide variety of predators, whereas disruption of higher-level processing may be deployed when targeted by specialist predators and those which develop search images for particular prey types.

One problem for this idea, and another major area for improved understanding of disruptive camouflage, is whether disruptive patterns are learnt by predators more or less easily than patterns that do not produce disruptive effects in natural settings (Skelhorn & Rowe, 2016). The experiments of Troscianko et al. (2013), which found an eroding advantage for high-contrast disruptive markings over time, suggest that disruptive patterns are learnt more rapidly than pure background matching patterns. This result predicts that disruption should be most effective when predators detect and recognize prey infrequently, or encounter many different prey types, making search image formation more difficult, whereas if prey are common or targeted by specialized predators, less easily learnt patterns may be more advantageous. This hypothesis remains to be tested in comparative or experimental work. Predator learning should also constrain the form of disruptive camouflage, with pattern elements selected for being indistinct, irregular, and hard to learn. Another intriguing area for future work is whether there are examples of selection for visual adaptations that work to counter the effects of disruptive camouflage, for example perceptual ability to better discriminate true and false edges. Related to this is how disruption might work on predators with visual systems very different to those of highly visually oriented humans and birds, the groups on which most work has been done so far.

Further theoretical progress on defining what is and what is not disruption is also required. The prevailing view of disruption as a family of adaptations that exploit features of how object outline and shape are processed to improve camouflage incorporates uses of defensive coloration not typically thought of as disruptive, for example a gradation of pigmentation towards the edges of an object to make the true outline blurry, and thus weaken the response of edge detectors (Troscianko et al., 2009).

A final subject for future work is investigation of non-visual disruptive camouflage (Ruxton, 2009). There are multiple examples of camouflaging effects in other modalities that could accurately be described as disruptive. For example the high-pitched 'seet' calls of many small passerine birds have been suggested to provide camouflage not by being undetectable by predators, but because they target predators' sound localization processing to make localizing the call difficult (Jones & Hill, 2001). There is, however, poor understanding of this issue and whether theoretical developments from studies of visual disruption can be usefully applied to other modalities. Given that different modalities share similar signal processing strategies (Stevens, 2013) it seems likely that examples of mechanical, magnetic, electrical, thermal, and chemical disruption will be established in the future.

CHAPTER 3

Countershading

3.1 What is countershading?

'Animals are painted by nature, darkest on those parts which tend to be most lighted by the sky's light, and vice versa.'
(Thayer, 1896, p.125)

Perhaps the most common coloration pattern observed across species is that exterior surfaces most exposed to light, typically dorsal surfaces, are more darkly coloured than those oriented away from light, typically ventral surfaces. This phenotype is commonly known as countershading and sometimes as dorsal pigmentary darkening (Kiltie, 1988).

There is a great deal of variation in appearance of countershading patterns, including subtle transitions of cryptic browns, greens, and greys across the surface of the body, sharp transitions between black and white, and dorso-ventral gradients created by altering the size, density, or colour of spots or other pattern elements (Figure 3.1).

Countershading is most widely discussed as an anti-predator camouflage defence, and it is this function the chapter focusses on. However several other functions may also select for countershading, including thermoregulation, abrasion resistance, and protection from ultraviolet light. Countershading may also occur simply as a result of lack of selection for colour production on ventral surfaces. These alternative explanations are discussed where relevant in order to evaluate the general applicability of anti-predator hypotheses.

The idea that countershading has a camouflage function was developed towards the end of the nineteenth century (Behrens, 2009). Two principal mechanisms were proposed: first that countershading works to offer improved background-matching camouflage when viewed from above against dark backgrounds such as the ocean depths, or below against generally light backgrounds such as the sky (Wallace, 1877). Likely examples include the coloration of many pelagic animals such as manta rays (Figure 3.1C). The second mechanism, generally termed self-shadow concealment (Kiltie, 1988), but also referred to as obliterative shading (Cott, 1940), was first put forward by Edward Poulton (1888) then more fully developed by Abbott Thayer (1896). Their idea was that countershading improves camouflage by reducing the visual effect of self-cast-shadows. Since natural illumination almost always comes from above (the sun, moon, and stars), shadows typically fall on ventral surfaces. Countershading coloration was proposed to 'cancel out' the appearance of these shadows that appeared to increase the animal's conspicuousness.

Thayer vigorously promoted the self-shadow concealment theory using strategies such as presenting audiences with animal models painted with and without countershading to persuasively demonstrate the camouflaging ability of countershading (Behrens, 2009). The perhaps apocryphal story is that the neatness of the idea and intuitive appeal of the background matching and self-shadow concealment effects meant countershading rapidly became a widely accepted text-book example of adaptive coloration (Cott, 1940; M. Edmunds, 2009; Gould, 2010; Sheppard, 1958), which had the consequence of deterring scientific research into countershading camouflage for several decades despite limited empirical support.

Few of the small number of early studies investigating countershading have had a lasting influence (De Ruiter, 1956; Turner, 1961; Young & Roper, 1976), and recognition of methodological and conceptual shortcomings led the reviews of Kiltie (1988) and

Avoiding Attack: The Evolutionary Ecology of Crypsis, Aposematism, and Mimicry. Second Edition. Graeme D. Ruxton, William L. Allen, Thomas N. Sherratt, & Michael P. Speed, Oxford University Press (2018). © Graeme D. Ruxton, William L. Allen, Thomas N. Sherratt, & Michael P. Speed 2018. DOI: 10.1093/oso/9780199688678.001.0001

Figure 3.1 Examples of countershading, counterillumination, and reverse countershading. (a) Countershading created by the size, distribution, and colour of pattern and background pattern elements of an Argentine horned frog *Ceratophrys ornata*. (b) A herd of chital *Axis axis*. The individual in the foreground is made conspicuous by the light ventral surface showing as it rears. Individuals in the background remain inconspicuous, likely partly due to self-shadow concealment. (c) Aquatic species such as this reef manta ray *Manta alfredi* likely countershade to match the background radiance, which is much higher looking up the water column than down. (d) This naked mole-rat *Heterocephalus glaber* is countershaded though it will spend the vast majority of its life in complete darkness. (e) A juvenile pufferfish in Tetraodontidae that countershades by altering the size of light spots. (f) European hamsters *Cricetus cricetus* possess reverse countershading. (g) Hatchet fish have ventral photophores to produce counterillumination that matches the radiance of downwelling light. (h) 'Living fossils' such as nautiluses possess countershading. Left. *Nautilus pompilius*; centre. *Allonautilus scrobiculatus*; right. *Nautilus macromphalus*. (i) Countershading is widely employed in military camouflage. Here the barrel of a Sherman Firefly gun has been countershaded (see Plate 1).

Ruxton et al. (2004a, c) to emphasize that most key questions remained unanswered, to the extent that the last edition of this book noted that there was 'no conclusive evidence that countershading *per se* provides any enhancement of crypsis' (Ruxton et al.,

2004a). Reassuringly, since then the basic question of whether countershading can reduce predation rates has been resolved experimentally. Edmunds and Dewhurst (1994) introduced a paradigm that exposes to avian predation artificial prey—pastry 'caterpillars' constructed with either countershaded, reverse-countershaded, dark or light coloration. The results of Edmunds and Dewhurst's experiment demonstrated a survival advantage for counter-shaded prey, but with a small sample size. When the experiment was repeated by Speed et al. (2004) with an additional manipulation of background colour to isolate camouflage effects on detection from post-detection preferences for different prey types, results were mixed. Several experiments showed no effect of countershading while in another counter-shading only reduced attack rates for some species of predator. This conflict and the suggestion that birds may have habituated to prey types and/or formed search images because of long presentation times in Speed et al.'s study, led to a third refine-ment by Rowland et al. (2007b), this time finding a strong survival advantage of countershaded prey, both when prey were displayed on colour-matched boards and on lawns. The conclusive result, how-ever, came from an ingenious set of experiments where the pastry caterpillars were pinned to either the upper or lower surfaces of branches (Rowland et al., 2008). Countershaded types survived best when placed on upper surfaces, but reverse-countershaded types survived best when placed 'upside-down' on the underside of twigs so that they became countershaded with their dark ventrum uppermost. This body of evidence enabled the last comprehensive reviews of countershading (Rowland, 2009; updated 2011) to be confident in the camou-flage potential of countershading.

Demonstration that countershading coloration can have an anti-predator function has allowed researchers to begin addressing more advanced questions, and it is these that this chapter focusses on. Recent experimental work has begun to investi-gate how countershading might work to reduce predation (Cuthill et al., 2016; Kelley et al., 2017). Comparative approaches are formally analysing the distribution and drivers of countershading phenotypes in taxa including ruminants, primates, and bats (Allen et al., 2012; Kamilar, 2009; Kamilar & Bradley, 2011; Santana et al., 2011), adding to numer-ous single-species observations of countershading appearance interacting with behaviour and envir-onmental variability (e.g. Kekäläinen et al., 2010). These investigations have been supported by computational approaches that model counter-shaded objects in virtual lighting environments to provide insight into the form of optimal counter-shading in different circumstances (Penacchio et al., 2015a, b). Our understanding of the developmental genetics of countershading coloration is also rapidly developing (Ceinos et al., 2015; Manceau et al., 2011). After decades of research neglect, the last ten years have been an exciting time for understanding this basic feature of so many observed colour patterns.

This chapter reviews several key examples of countershading and counterillumination before dis-cussing six distinct mechanisms via which counter-shading might decrease predation. Following this are sections on the evolution and genetics of counter-shading, with the chapter finishing with ecological considerations and suggestions for future research.

3.2 Examples and taxonomic distribution of countershading camouflage

The basic countershading phenotype of surfaces oriented away from illumination being lighter than surfaces oriented towards illumination is extremely common with examples frequent in almost all major lineages, including vertebrates, sponges, echino-derms, bryozoans, molluscs, brachiopods, arachnids, crustaceans, insects, worms, coelenterates, fungi, and plants. The presence of countershading in bacteria and cyanobacteria is as yet unknown, though recent technical breakthroughs now make this question possible to address (Yokoo et al., 2015). While their small size affords limited scope for differential inter-action with light across the cell body, pigments can be non-uniformly distributed and countershading is possible (Vermaas et al., 2008). Protists also have organized pigment distributions, though counter-shaded patterning has not been specifically identi-fied (Lobban et al., 2007).

In photosynthesizing organisms, including plants, countershading is common but likely to be primarily the result of distributing chlorophyll and carotenoid pigments for efficient photosynthesis and warming whilst minimizing photooxidative damage (Archetti et al., 2009). It has been suggested that leaf counter-shading works to increase predation on insect herbivores, as by presenting two different back-ground colours the effectiveness of herbivores' background matching camouflage may be undermined (Lev-Yadun et al., 2004); however, this hypothesis remains untested. Stink bug *Podisus maculiventris* eggs laid on the upper surface of leaves are darker than those laid on the underside, with leaf reflectance being used by females to select egg colour (Abram et al., 2015), suggesting that leaf colour on upper and lower surfaces can select for insect coloration. However, in this case the explanation seems to lie primarily in protection from ultraviolet radiation rather than predation (Torres-Campos et al., 2016). A direct camouflage effect for countershaded leaves to hide them from herbivores may also be possible given emerging evidence for the advantage of leaf camouflage (Niu et al., 2014).

In animal lineages the presence of countershading is more likely to be a camouflage adaptation. Losing leaves to herbivory is a relatively minor consequence in comparison to a predator that fails to capture its prey, or to prey that is consumed following detection. In terrestrial environments many animals are countershaded. For example in mammals, 106 out of 114 species of ruminants were observed to have some degree of countershading (Allen et al., 2012), as were high proportions of primates (95/113) (Kamilar & Bradley, 2011), sciuromorphs (164/276) (Ancillotto & Mori, 2017) and lagomorphs (61/71) (Stoner et al., 2003). Such systematic surveys have not been conducted for other classes of terrestrial animal, but a cursory review indicates that proportions are similar for other tetrapods while terrestrial invertebrates are less often countershaded, perhaps because ventral surfaces are generally not as exposed. The question of how often countershading in terrestrial environments represents an anti-predator adaptation is considered further in section 3.3, but the likelihood is that many examples will have been selected either fully or partly for non-camouflage functions. For example, human tanning, which results in countershading, has evolved multiple times in populations from middle latitudes which experience high variance in annual levels of ultraviolet radiation. Here the primary function is to prevent overproduction of vitamin D in summer while avoiding deficiency in winter (Jablonski & Chaplin, 2010). Likewise it has been suggested that penguin countershading has a thermoregulatory function, creating a temperature switch that warms the penguin when the dark back faces the sun and cools it when the pale belly is turned towards the sun (Chester, 2001), as well as providing background-matching camouflage while in water (Bretagnolle, 1993).

Countershading is observed in some unexpected situations. Naked mole-rats, for example, can be quite strongly countershaded despite spending the vast majority of their lives underground and so out of the influence of light coming from above. It is possible that melanic pigments protect mole-rats from abrasion as they pass each other in tunnels but Braude et al. (2001) did not find this explanation satisfactory because dominant animals scrape against tunnels more than the subordinate individuals they pass over, but are more lightly countershaded. Instead the available evidence suggests that camouflage during the high-risk period of overland dispersal at night is sufficiently important for countershading camouflage to be selected for.

Reverse-countershading, where surfaces facing the light are more lightly pigmented than those facing away from the light is relatively common too. Striped skunks *Mephitis mephitis* and honey badgers *Mellivora capensis* are well-known examples. In these cases coloration is aposematic (Stankowich et al., 2011) and light from above likely increases the conspicuousness of the signal. Adult male bobolink (*Dolichonyx oryzivorus*) birds likely employ reverse countershading to increase the conspicuousness of its mating signals (Mather & Robertson, 1992). Other examples of reverse countershading such as European hamster *Cricetus cricetus* (Figure 3.1F) and muskox *Ovibos moschatus* await explanation.

Countershading is perhaps even more common in aquatic environments, though no surveys of prevalence have been conducted. Interesting examples include parasitic marine isopods, which attach to fish with their heads facing forward, that countershade by darkening whichever side is uppermost depending on whether they attach to the left or right flank of their host fish (Körner, 1982). Countershading is also seen in catfish of the family Mochokidae (Hiroshi et al., 1989), a cichlid *Tyrannochromis macrostoma* (Stauffer et al., 1999) and sea slugs in Glaucidae (Miller, 1974). These organisms are unusual in that dorsal surfaces are lighter than ventral surfaces; however, because they utilize upside-down body positions as they feed at the air–water interface, they conform to the standard countershading rule of surfaces towards the illuminant being more darkly pigmented. A neat demonstration of the orientation dependence of countershading is the reflexive colour change response of cephalopods to body rotation so that countershading is maintained (Ferguson & Messenger, 1991).

The most vivid examples of differential dorsoventral coloration occur in bioluminescent species (Figure 3.1G). Bioluminescence is extremely common in mid-water species including protists, algae, crustaceans, fish, and squid, as well as some aerial insects (Haddock et al., 2010; Young & Roper, 1977). In addition to signalling functions, bioluminescence is employed to create counterillumination where light is produced only on ventral surfaces (Haddock et al., 2010). This is suggested to camouflage an organism's silhouette against predators viewing from below better than would be possible using pigmentary or structural coloration alone (Johnsen, 2002). Several lines of evidence support a camouflage function of counterillumination, including appropriate bioluminescent responses in squid to improve silhouette camouflage against ships' downwelling lights (Young & Roper, 1976) and variation of the intensity of bioluminescence with depth to match the strength of downwelling light in sergestiid shrimp (Latz, 1995), squid (Jones & Nishiguchi, 2004; Young & Mencher, 1980) and fish (Denton et al., 1972). The 50 or so species of bioluminescent shark emit a fixed luminescent output (Straube et al., 2015) and so instead move up and down the water column in diel migrations to follow circadian fluctuations in the intensity of downwelling light in a way that maintains isoluminance between the shark and its background when viewed from below (Claes et al., 2010, 2014). Counterillumination is currently thought to be absent in rivers and lakes, perhaps as these environments are much less optically clear than oceans (Haddock et al., 2010). Counterillumination is also absent from flowering plants and terrestrial vertebrates, lineages not known to be bioluminescent. Many fungi are bioluminescent. Species with bioluminescent basilodomes such as bitter oyster mushrooms *Panellus stipticus* produce bioluminescence only from their gills, creating counterillumination. However the function of fungal bioluminescence is likely to be in attracting dispersers or the predators of fungivorous organisms rather than in camouflage (Desjardin et al., 2008).

Countershading is frequently employed in military camouflage, beginning with efforts between Abbot H. Thayer and George de Forest Brush to countershade naval vessels in the 1898 Spanish–American War (Behrens, 2002, 2009). There have also been several military trials of counterilluminating technology on ships and aircraft, which were conducted in the Second World War, for example the US's Project Yehudi placed lights on the leading edge of planes that automatically adjusted brightness to that of the sky (Duntley, 1946). Trials showed promising results on visibility reduction, but practicalities and shifts in military priorities meant the concept never entered operational usage.

3.3 Countershading camouflage mechanisms

The two main mechanisms through which countershading might achieve camouflage, background matching, and self-shadow concealment, are really umbrella terms for what have been hypothesized to be several distinct (but not necessarily exclusive) camouflage mechanisms.

The most straightforward mechanism is background matching from above or below against backgrounds that are dark underneath and light overhead. This form of background matching needs no special explanation in addition to that provided in Chapter 1. Optical considerations suggest that background matching when viewed from below will be necessarily imperfect as a lightly coloured ventrum will rarely reflect as much light as there is radiating around the animal, simply because there is much less upwelling light to be reflected back off the ventrum. This is apparent from viewing countershaded marine animals or birds that are nevertheless silhouetted when viewed from below, and explains the prevalence of counterillumination in mesopelagic environments. However a marginal silhouette contrast reduction may still be effective: in predatory gulls, individuals with undersides painted black catch fish at lower rates (Götmark, 1987). Environmental factors that increase the diffusiveness of light also increase the potential for improved background matching. Producing counterillumination by bioluminescence to create additional downward radiating light, however, has the potential to solve this issue. Background matching downwelling light when viewed from below, and dark backgrounds when viewed from above, is almost certainly the camouflage mechanism through which counterillumination primarily works (Johnsen, 2014).

A rarely considered possible camouflage mechanism is that countershading may work to increase the level of background matching for animals viewed from the side in environments where the lower field has a lighter tone than the upper field such as a savannah's mix of light dry grass mixed with dark shrubbery. This is suggested by the apparent 'overcountershading' of some species in these habitats, where a dark–light dorso-ventral gradient remains even after illumination from above (Allen et al., 2012). The chroma of light reflected off the substrate will also contribute to the perceived colour of light undersides viewed from the side, and this may improve the colour match of ventral surfaces against backgrounds of different colours (Norris & Lowe, 1964). Another possibility suggested by examples of overcountershading, and other rapid changes in colour between dorsal to ventral surfaces, is that countershading may work via disruptive camouflage mechanisms (Allen et al., 2012; Harper & Case, 1999; Penacchio et al., 2015b; see Chapter 2).

The self-shadow concealment hypothesis proposes that countershading works by counteracting the effects of light from above which creates, on average, stronger shadows on ventral surfaces. The self-shadow concealment idea itself potentially involves multiple distinct camouflage mechanisms. The simplest is that it improves camouflage via background matching the three-dimensional (3D) form of animals when they are viewed from the side (Cott, 1940). In the absence of countershading, patterns resulting from shading created by directional illumination are likely to be dissimilar to the colours, tones, and pattern elements of the background. Self-cast shadows may also increase detectability by increasing contrast between the target outline and background. Countershading coloration potentially resolves these problems by presenting light tones on areas where shadows fall and dark tones on areas receiving more irradiance, resulting in the reflected radiance over the entire body potentially being a better match to the background, a mechanism termed here and elsewhere (Penacchio et al., 2015b) as 3D background matching.

A separate self-shadow concealment camouflage mechanism involves reduction of cues to 3D form itself. The 3D form of objects is a major source of information that enables separation of objects from the background and recognition as a particular object, such as potential prey. The form and identity of 3D objects can be acquired from several distinct cues. For example, binocular cues allow the depth of surfaces comprising an object to be estimated from the difference in relative position of the surface on each retina or the angular difference of the eyes' rotation when focussing on a surface. However, binocular cues only provide useful information over short distances and in species with forward-facing eyes (Blake & Wilson, 2011). A more generally important source of information on 3D shape comes from patterns of shading created by directional light interacting with object form to create self-cast shadows that are characteristic of object form (Figure 3.2a) (Kleffner & Ramachandran, 1992; Lovell et al., 2012; Rittenhouse, 1786), and it is this information that self-shadow concealment is proposed to obscure.

The utilization of information from self-cast shadows in 3D shape perception has been demonstrated in most species that have been studied to date (rhesus macaques: Arcizet et al., 2009; pigeons: Cook et al., 2012b; starlings: Qadri et al., 2014; humans and chimpanzees: Tomonaga, 1998; cuttlefish: Zylinski et al., 2016). Humans, for example, perceive 2D discs shaded on top as concave and shaded on bottom as convex, a demonstration of the perceptual assumption that light comes from above (Ramachandran, 1988; Sun & Perona, 1998). Search for convexity defined by shading is a successful computer vision strategy for detecting camouflaged objects without countershading (Tankus & Yeshurun, 2009), and it is likely predators also search for either simple or complex patterns of shading that are indicative of potential prey (Skelhorn & Rowe, 2016). Countershading may therefore work to reduce predation by minimizing or eliminating shading cues to 3D form. Here we follow Penacchio et al. (2015b) and resurrect the term 'obliterative shading' to refer specifically to this unique mechanism

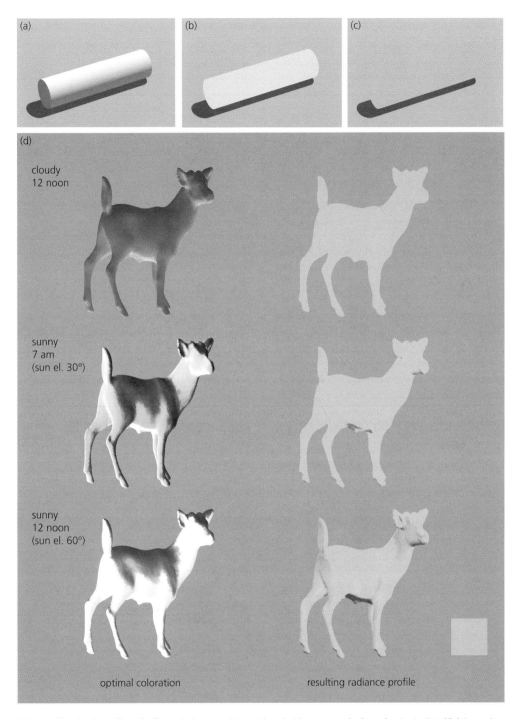

Figure 3.2 Modelling the visual effect of self-cast shadows on objects with and without countershading coloration in virtual lighting environments, where a.) illustrates self-cast shadows on a cylinder lit from above, b.) uniform reflectance when the cylinder's tonal gradient is the inverse of self-cast shadows (obliterative shading) and c.) when the cylinder's overall tone additionally achieves background matching as well as obliterative shading. b.) shows in the first column of images optimal coloration for 3D background matching and obliterative shading in different lighting conditions. The background colour is displayed in the lower left corner. Note how the strength of countershading required increases in sunny conditions and with increasing height of the sun. The second column of images illustrates the resulting radiance profile, highlighting that in sunny conditions a 'perfect' solution is impossible as pigmentation cannot counter the magnitude of irradiance gradients, since reflectance cannot be greater than 100% or less than 0%. Figures adapted with permission from Penacchio et al. (2015b).

of camouflage through reduction of shape-from-shading cues to 3D form (*obliterative shading of shape*). An object with perfect obliterative shading will appear optically flat to an observer using shape-from-shading cues to perceive the 3D form of objects (Figure 3.2). A conceptually distinct extension of the obliterative shading mechanism has been proposed by Penacchio et al. (2015b), who suggest that obliterative shading may also work by reducing cues to object depth relative to the background, thus hampering identification of prey as being in a different depth plane to the background (*obliterative shading of depth*).

Self-shadow concealment mechanisms may also work for objects when they are viewed from above or below as well as from the side. Rowland (2009) used modelling to demonstrate that illumination from above results in uniformly coloured cylindrical objects having non-uniform reflectance profiles when viewed from above as well as from the side, being darker at the object's edges. Edge contrast is particularly important for object detection (Troscianko et al., 2009) so countershading may be selected to reduce this cue as well, resulting in improved camouflage via either 2D and 3D background matching, disruption, and/or obliterative shading.

Experimental investigation of which potential countershading mechanisms are actually effective at promoting survival, in what situations, and their relative contributions to camouflage, is a research task that is only just beginning: Cuthill et al. (2016) recently reported a field predation experiment that used paper tube 'caterpillars' printed with optimal self-shadow concealment patterns for either sunny or cloudy conditions, as derived from the computational modelling of Penacchio et al. (2015b; Figure 3.3.). Results showed that countershaded targets optimized for sunny and cloudy conditions survived best in sunny and cloudy conditions respectively, demonstrating that the survival benefit of countershading in this instance is dependent specifically on the degree of self-shadow concealment given environmental light conditions. Penacchio et al. (in review) went on to replicate this result in laboratory conditions. Well over a century after the self-shadow concealment mechanism was proposed, this experiment finally verifies its operation in camouflaging prey against predators. What remains to be understood is whether increased survival through self-shadow concealment is achieved through 3D background matching or obliterative shading. The modelling work of Penacchio et al. (2015b) demonstrates that both these mechanisms

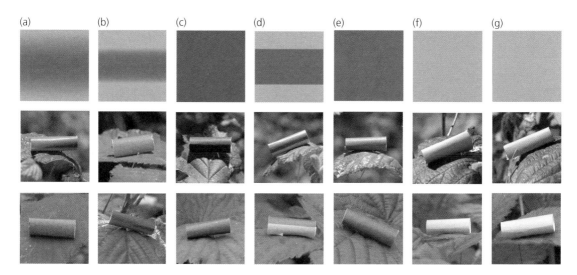

Figure 3.3 The experimental treatments used in Cuthill et al. (2016). The top row shows a plan view of each treatment. The second row shows examples of each treatment photographed in cloudy weather and the third row in sunny weather. Treatments are: (a) optimized for cloudy weather; (b) optimized for sunny weather; (c) dark green; (d) sharp dark–light transition; (e) mean colour of cloudy weather treatment; (f) mean colour of sunny weather treatment; (g) light green (see Plate 2).

are optimized by the same pattern and thus cannot be disassociated on the basis of colour phenotype alone.

In aquatic environments specific countershading mechanisms have been experimentally investigated for the first time by Kelley et al. (2017) in studies of the western rainbowfish *Melanotaenia australis*, a species that can make its pigmentation lighter or darker depending on conditions over the course of a few days. The authors measured colour change following manipulation of overhead light intensity and substrate colour in one experiment, and horizontal background colour in a second experiment. Background matching and self-shadow concealment hypotheses were teased apart by testing predictions generated by photographing a uniformly coloured model fish in the same conditions and observing what optimal countershading for each mechanism would be. On a light substrate the self-shadow concealment hypothesis predicts weak countershading, since light reflected off the substrate decreases the ratio of upwelling light to downwelling light, whereas a dark substrate, which reflects little light back up the water column, predicts strong countershading if self-shadow concealment is being optimized. Similarly when horizontal backgrounds are dark, sidewelling radiance will be low, increasing the strength of self-cast shadows and so predicting stronger countershading. Results showed that fish did not conform to these predictions—there was no effect of substrate or horizontal background colour on dorso-ventral gradients. Instead dorsal surfaces became darker on dark substrates and lighter on light substrates, replicating Kelley & Merilaita (2015). Ventral surfaces became lighter when overhead light intensity increased and lighter on light substrates, perhaps due to increased upwelling radiance. Overall the pattern is entirely consistent with 2D background matching mechanisms rather than self-shadow concealment mechanisms. Thus western rainbowfish, which inhabit rivers and lakes, seem to have evolved the ability to change countershading coloration in order to match the background from multiple viewing angles rather than hide self-cast shadows.

Together with the results of Cuthill et al. (2016), this provides experimental support for the hypothesis that background matching is the typical countershading mechanism in aquatic and aerial environments, whereas self-shadow concealment has a more important role in terrestrial environments (Johnsen, 2014; Kelley & Merilaita, 2015). Terrestrial animals inhabit a more 2D environment than pelagic animals and are thus assumed to be viewed from above or below less often and from the side more often than aquatic animals in general. Self-shadow concealment on land is challenged by enormous variability in the direction of light in terrestrial environments, such that for a static terrestrial animal diel variation in where self-cast shadows fall is enormous. This problem has led to some scepticism about whether countershading could be a camouflage adaptation in terrestrial environments, as it will only counter the appearance of shadows for a small fraction of each day (Kiltie, 1988). However the results of Cuthill et al. (2016) demonstrate that self-shadow concealment is resistant to some of this variability, as they found increased survival for countershaded prey left exposed to predation for long periods during which changing lighting conditions would have made self-shadow concealment suboptimal most of the time. Light in aquatic environments is much more homogeneous because of the scattering effects of water (though factors such as turbidity, the position and brightness of the sun and moon, and depth are still important: Johnsen, 2012), so the position and intensity of self-cast shadows should be more consistent, making self-shadow concealment mechanisms more effective. However initial observations and experiments are showing that improving background matching from above or below is the primary mechanism utilized by aquatic countershading and counterillumination (Claes et al., 2010; Kelley & Merilaita, 2015; Kelley et al., 2017).

In summary six mechanisms have been proposed to explain the anti-predator function of countershading: 1.) background matching against backgrounds that differ consistently depending on viewing angle (light backgrounds viewed from below, dark from above, and light–dark from the side); 2.) background matching following selection only on dorsal surfaces; 3.) disruptive camouflage that splits upper and lower sections; 4.) 3D background matching following self-shadow concealment; 5.) obliterative shape shading to reduce cues to 3D shape; and 6.) obliterative depth shading to reduce cues to overall object depth. With the exception of background matching following selection on only dorsal surfaces, but with the addition of non-camouflage mechanisms, these are non-mutually exclusive. This complexity partly accounts for the challenge of investigating the function and mechanism of countershading coloration.

3.4 Evolution

3.4.1 Evolutionary history of countershading

Given the prevalence of countershading on extant organisms and the fact that solar light has always come from above, it is possible that the countershading phenotype has been around and common for a very long time. For sedentary organisms, distributing costly pigments with respect to the sun is in principle straightforward from an evolutionary point of view. For motile organisms in shallow waters, being able to take advantage of countershading pigmentation requires orientation of body position with respect to the direction of light, which additionally necessitates an ability to sense the direction of light.

The use of countershading as a camouflage strategy could, at the earliest, coincide with the emergence of visually oriented predators in the Cambrian around 540 Mya (Parker, 2003). The Cambrian oceans contained numerous lineages that would plausibly have benefited from countershading camouflage mechanisms. Countershading that may function in camouflage is present on 'living fossils' with origins in the Cambrian such as nautiluses (Figure 3.1H), suggestive of an early origin for countershading camouflage, but by no means conclusive. To date the earliest examples of fossilized pigments with distributions that indicate countershading phenotypes are fish from the Permian and Carboniferous (Gabbott et al., 2016; Gottfried, 1989) around 300 Mya, and, on land, a taxonomically diverse array of dinosaurs from the Early Cretaceous starting at 100.5 Mya (Brown et al., 2017; Smithwick et al., 2017; Vinther et al., 2016). Vinther et al. (2016) demonstrated the presence of countershading in *Psittacosaurus* sp. by reconstructing coloration from the distribution of fossilized melanins in a specimen with very well preserved skin. Further, by reconstructing the 3D shape of *Psittacosaurus* from skeletal morphology,

the authors were able to infer the countershading phenotype, which featured a smooth and relatively shallow reflectance gradient, would have been more effective for self-shadow concealment in closed compared to open environments. With this conclusion corroborated by the presence of petrified tree trunks in the surrounding deposits, this represents a novel approach to understanding the ecology of a long-extinct species. It is equally interesting to speculate on why other examples of long-ago extinct reptiles do not appear to possess countershading (Lindgren et al., 2014).

As countershading has several potential non-camouflage functions, a plausible scenario for the evolution of countershading for camouflage is exaptation of countershading for ultraviolet protection, abrasion resistance or thermoregulation. Alternatively, predation may have been the primary driving force for the initial evolution of countershading. Establishing the sequence of evolutionary trajectories of countershading has received little attention and awaits future work.

3.4.2 Evolutionary history of counterillumination

Bioluminescence is produced following oxidation of a substrate luciferin triggered by a catalysing enzyme, either a luciferase or photoprotein, resulting in photon emission. Bioluminescence can generated intrinsically in photophores, or extrinsically through bacterially mediated symbioses.

In non-symbiotic bioluminescence, organisms possess a gene to produce luciferase or a photoprotein. Some species also produce luciferin directly while for others it is acquired through diet. The earliest origin of intrinsic non-symbiotic bioluminescence can be tracked to a split between two ostracod lineages 400 Mya (Haddock et al., 2010).

In organisms where bioluminescence is produced via symbiotic bacteria, the anatomy required to host the bacteria has evolved between 40 and 50 times among extant organisms (Haddock et al., 2010). In ray-finned fishes, which include several lineages containing examples of counterillumination, bioluminescence has evolved several times between the early Cretaceous (ca. 150 Mya) through to the Cenozoic (Davis et al., 2016).

3.5 Costs of countershading and counterillumination camouflage

It is expected that achieving effective self-shadow concealment camouflage entails significant opportunity costs, such as restricting an animal to particular lighting environments or activity periods, maintaining a fixed orientation with respect to the sun, or otherwise constraining potentially beneficial behaviours that might reduce the effectiveness of self-shadow concealment (Cuthill et al., 2016), particularly in terrestrial environments. Though costs have not been directly quantified, modelling work (Penacchio et al., 2015a, b) demonstrates the deleterious effect on conspicuousness that countershading has when an animal is viewed in suboptimal lighting conditions, potentially being more visible than uniform coloration. A recent experiment found that in simulated cloudy conditions, countershading optimized for sunny conditions survived no better or worse than uniform coloration, and in cloudy conditions sunny-optimized countershading only does slightly better (Penacchio et al., in review). Thus animals which rely on effective countershading to maintain acceptable level of predation risk would be strongly constrained by weather conditions. For bilaterally symmetric animals living in open environments under direct sun, not orienting to the sun will always have a negative effect on conspicuousness (Penacchio, Cuthill, et al., 2015a). These issues may be balanced if, as seems likely, countershading is a relatively cheap adaptation to produce and live with, such that even if it is of relatively limited effectiveness most of the time, it is still worthwhile to adopt. More work is required to investigate how selection optimizes trade-offs between camouflage effectiveness and other factors such as foraging, mating, and resting requirements, and the consequences this has for the form of countershading phenotypes observed (Allen et al., 2012).

From a purely energetic perspective, the costs of producing counterillumination are much higher than the costs of producing countershading. Young and Roper (1977) calculated that producing counterillumination accounts for 0.3–0.9 per cent of the squid *Histioteuthis heteropsis*'s resting metabolic rate at 400m in clear water. Because in the daytime downwelling light increases in radiance by a factor of 50 for every 100m decrease in depth, producing sufficient counterillumination in shallower waters becomes prohibitively costly, and indeed counterillumination in the daytime has not been observed in shallow waters. Claes and Mallefet (2010) discuss how the unique hormonal control of lantern sharks' (Etmopteridae) counterillumination evolved out of the mechanism to control physiological colour change in shallow-water sharks. While this innovation may have freed the family from being restricted to shallow-water niches and allowed a deep-water species radiation, the close association between different species' bioluminescent output and the depth at which they are caught also hints at the constraints this adaptation places on behaviour (Claes et al., 2014).

3.6 Developmental genetics of countershading

Most countershading phenotypes are, at least in principle, straightforward to produce from a developmental perspective. Mathematical models of pattern formation have proposed several processes by which dorso-ventral gradients in pigmentation could be produced (Gierer & Meinhardt, 1972; Murray, 2002), the simplest of which is diffusion of a morphogen from source cells in dorsal regions that degrades as it spreads towards ventral regions (Howard et al., 2011). Countershaded gradients with more complex shapes can be created by reaction–diffusion models, as can countershading patterns comprised of spots and stripes of varying widths and densities (Majumdar et al., 2014; Turing, 1952; Figure 3.1). Until recently the biological applicability of mathematical models was not well understood; however, the developmental genetics of countershading is beginning to be understood, with recent work concentrating on zebrafish *Danio rerio* (Ceinos et al., 2015) and deer mice *Peromyscus* (Manceau et al., 2011).

In mammals, pigment is produced by a single chromophore type, the melanocyte. Melanocytes originate from the neural crest during embryogenesis and pigment production within melanocytes is under the control of two loci, *Extension* and *Agouti*. The *Extension* locus encodes a receptor for α-melanocyte-stimulating hormone (α-MSH) at melanocortin 1 receptor (*MC1R*). Stimulation leads to eumelanin (black/brown pigment) synthesis. The *Agouti* gene *ASIP* instead codes for production of agouti signalling protein (*asip1*) in hair follicles which inhibits α-MSH induced eumelanin production and promotes pheomelanin (yellow/red pigment) synthesis (Lu et al., 1994). Countershading is produced by regulation of *Agouti* expression (Millar et al., 1995), with *Agouti* expressed at low levels in the ventrum (Vrieling et al., 1994). In *Peromyscus* mice, this delays melanocyte differentiation, creating bicolour banded hairs on the ventrum and a countershading pattern (Manceau et al., 2011). Higher contrast countershading can be created via upward shifts of *Agouti* expression to create completely depigmented hairs, and the boundary between light and dark hair can be shifted dorsally or ventrally by changes in the location of *Agouti* expression, genetic changes which have been linked to camouflage adaptations in wild populations (Hoekstra, 2011; Manceau et al., 2011; Steiner et al., 2007).

In zebrafish countershading is instead formed by differential densities of two chromophore types; dark melanocytes are concentrated in dorsal areas and reflective iridophores in ventral areas (Bagnara & Matsumoto, 2007). Despite this different cellular mechanism the same ASIP gene as in mammals is responsible for creating countershading via *MC1R* antagonism (Ceinos et al., 2015). However instead of regulating eumelanin/pheomelanin production, *asip1* expression in ventral regions controls the balance of melanocytes and iridophores along the dorso-ventral axis. *Asip1* is also expressed at higher levels in ventral regions of goldfish *Carassius auratus* (Cerdá-Reverter et al., 2005), turbot *Scophthalmus maximus* and sole *Solea senegalensis* (Guillot et al., 2012). The developmental mechanism is also understood to be similar in amphibians (Mills & Patterson, 2009) and birds (Gluckman & Mundy, 2017). As well as *Agouti* and *MC1R*, variation in countershading pigmentation has also been observed to result from changes in expression of *KITLG* (Miller et al., 2007; Steiner et al., 2007), a conserved protein coding gene involved in many cell migration functions across species, including chromophore migration and differentiation (Wehrle-Haller, 2003).

In summary, the genetic and molecular basis of countershading adaptations are beginning to be well understood for focal species and the emerging picture is that developmental mechanisms are highly conserved across taxonomic groups, despite fundamental differences in pigment production. Evidence from invertebrate lineages is now required, as is understanding of how developmental constraints might affect the range of countershading phenotypes that can be expressed.

3.7 The evolutionary ecology of countershading

Building on illustrative examples such as the dark ventra observed on upside-down swimming catfish, comparative studies have begun to formally analyse how countershading and counterillumination correlate with the ecology and behaviour of organisms. In doing so they provide insight into how and why countershading might be adaptive in different groups of organism.

Seabirds that feed at medium depths, such as many penguins and diving petrels, seem to be more likely to be countershaded than those that feed on the surface or lower depths (Bretagnolle, 1993; Cairns, 1986), suggestive of background matching from above or below since these species will be viewed from these angles more by both predators and prey. In cetaceans both smoothly and sharply countershaded species are more likely to be small, and feed on squid and fish, than species lacking countershading (Caro et al., 2011). Additionally, smoothly countershaded species are likely to be shallow diving and thus in environments with stronger directional light. These associations are consistent with a camouflage function of countershading, with small species being under greater predation risk, and squid and fish being visually oriented prey. Sharp countershading was also more common in species that form larger social groups, suggesting a communication role for high-contrast patterning. In a similar study of pinnipeds, however, countershading coloration did not associate with any ecological traits analysed, including prey type, body size, and diving depth (Caro et al., 2012).

In terrestrial environments the ecological correlates of dark dorsal and light ventral pigmentation also remained enigmatic in a study of lagomorphs (Stoner et al., 2003) and in 914 species of bat (Santana et al., 2011). However Santana et al.'s study found that species with light-coloured neck bands generally roosted upside down in open environments, suggesting anterior regions display countershading camouflage. In rainforest birds countershading was observed to be ubiquitous by Gomez & Théry (2007), with no clear environmental drivers of variation in countershading phenotype. As in pelagic habitats the principle camouflage mechanism in forest birds was suggested to be background matching against the dark forest floor and vegetation when viewed from above, and the lighter canopy or canopy gaps when viewed from below.

Kamilar (2009) and Kamilar & Bradley (2011) investigated variation in the strength of countershading across primate species, measuring dorsal and ventral coloration from photos of museum specimens then associating the degree of countershading with behavioural traits in a phylogenetic context. They found no support for the prediction that stronger countershading should be found on species active during the day when the strongest self-cast shadows are cast and UV levels are highest. However countershading was found to be weaker on larger primates, perhaps because they experience less predation and consequently selection for crypsis has been weaker relative to other functions (Kamilar, 2009; Kamilar & Bradley, 2011). The clearest predictor of countershading strength was the positional behaviour of species; primates that spend a significant portion of their time in a vertical position have little difference between dorsal and ventral coloration, whereas those that adopt horizontal positions generally have strong countershading (Kamilar & Bradley, 2011). The inference is that since self-cast shadows are generally only cast on ventral surfaces when in a horizontal position, and dorsal and ventral surfaces will be viewed from above or below more often when in a horizontal position, species that adopt horizontal positions will benefit more from countershading camouflage, whether through self-shadow concealment or background-matching mechanisms (Kiltie, 1989). In squirrels the presence of countershading has similarly been found to associate with utilization of arboreal habitats, though species differences in positional behaviour were not directly studied (Ancillotto & Mori, 2017).

One of the difficulties in drawing conclusions on how ecological circumstance affects the evolution of anti-predator countershading is that other potential countershading functions such as thermoregulation and UV protection often predict the same associations, for example that countershading should be stronger in open lighting environments, or stronger when horizontal postures are typically adopted (Cuthill et al., 2016). Allen et al. (2012) attempted to separate the predictions made by different countershading functions by taking more detailed measurements of countershading phenotypes from museum specimens of 114 species of ruminant, including the abruptness and position of change between dark and light tones, with alternative camouflage functions making unique predictions across multiple measures. The pattern of associations with ecological and behavioural traits observed were all consistent with the self-shadow concealment hypothesis; species in open environments had stronger countershading and more abrupt transitions from dark to light tones, and stronger countershading was found on species with smaller body sizes (Kamilar & Bradley, 2011) and also on species living closer to the equator (Penacchio et al., 2015b). The pattern of associations did not fit the predictions made by alternative functional hypotheses including thermoregulation (which predicts stronger countershading at high latitudes) and ultraviolet protection (which does not predict an effect of lighting environment on the abruptness of change between dark dorsal and light ventral tones). 2D background matching countershading mechanisms were also supported as countershading strength and overall tone correlated with lighting environment, but 2D background matching alone could not account for results since it does not predict the observed relationship between countershading strength and latitude. By analysing the strength of self-cast shadows on a uniform grey model deer photographed in open and closed environments, and at midday and in the evening, Allen et al. (2012) additionally showed that the degree of countershading across species was generally a good match to what would be required for effective self-shadow concealment. The virtual modelling of Penacchio et al. (2015a, b) extends this approach and allows greater control over factors such as the effects of sun altitude, body orientation, and the diffusiveness of light. Results agreed with the real-world observations of Allen et al., but cast some doubt over whether the form of countershading phenotypes can signify function, since predicted countershading for UV protection and, to a lesser extent, thermoregulation, was very similar to predicted countershading for camouflage.

In a similar attempt to identify countershading function and mechanism by testing the different predictions each makes for the ecological correlates of countershading phenotypes observed in nature, Kelley and Merilaita (2015) examined how intraspecific variation in western rainbowfish *Melanotaenia australis* countershading correlates with local lighting environments and predation pressure in the wild, and with substrate colour in the lab. Results showed no effect of downwelling or sidewelling irradiance levels, predation risk or substrate reflectance on the degree of countershading, suggesting against a self-shadow concealment mechanism. Instead there were important effects of both downwelling light and predation risk on dorsal and ventral colour separately, with fish from high irradiance environments having paler colours. Ruling out a role for ultraviolet protection by the lightening of dorsal surfaces in high irradiance environments and the association with predation risk, the authors concluded that only background matching camouflage from above or below was entirely consistent with patterns of variation. Overall this study lends further support to the idea that the principle countershading mechanism in aquatic environments is background matching. Analysing species with the ability to change colour over short timescales complements interspecies analyses examining colour change over evolutionary timescales. It will be interesting to apply the approach to additional species and environments. There is some suggestion that responses may not be consistent in different species, as in King George whiting *Sillaginodes punctatus* kept in laboratory environments under different light intensities the pattern of colour variation is opposite to that observed in wild western rainbowfish, with species in low light environments having lighter dorsal and ventral tones (Meakin & Qin, 2012).

The focus of many studies reported in this section has been on elucidating countershading function and mechanism from patterns of ecological and behavioural associations. More generally, it is clear from overall results that countershading phenotypes readily adapt to the visual environment and the behavioural ecology of individuals and species. However the selective forces on countershading appearance such as latitude or positional behaviour are found to explain only a small percentage of variation in countershading phenotypic diversity (Figure 3.1), leaving much about the evolutionary ecology of countershading to be discovered.

3.8 Countering countershading and counterillumination adaptations

For obliterative shading to be operational requires that an observer has ability to infer shape or depth from patterns of shading. This is expected to be a relatively basic visual ability, and indeed it has been observed in all species studied to date (section 3.3), though studies have so far focussed on highly visually oriented animals. Computer vision models designed to break other forms of camouflage by identifying patterns of convexity perform worse at identifying countershaded targets (Tankus & Yeshurun, 2009). Conversely, obliterative shading alone, whilst it removes shape and depth cues, does not necessarily affect contrast with the background unless it is combined with background matching, and edges may remain conspicuous without being disruptive. The only way to determine if obfuscation of depth cues is providing a camouflage advantage is through behavioural test (Penacchio et al., 2015b), but thus far this has happened for very few species. This is an efficient research activity as it also provides a route into understanding non-human object perception (Zylinski et al., 2016). If a consistent picture emerges then the presence of countershading camouflage can be used to infer the visual cognition of observers that cannot be tested directly, such as dinosaurs (Vinther et al., 2016).

Since the effectiveness of fixed countershading camouflage varies with factors such as time of day, weather, and microhabitat (Allen et al., 2012; Cuthill et al., 2016; Penacchio et al., 2015b), there is potential for interesting ecological and evolutionary predator–prey dynamics; predators may adjust their hunting time or place to instances where their favoured prey's countershading is least effective, and the prey may respond to this by adjusting behaviour or countershading (Bond & Kamil, 2006). However this possibility has not yet been studied.

Counterillumination can be broken by predators with acuity high enough that individual photophores can be detected (Johnsen, 2014; Johnsen et al., 2004). Counterilluminatiing species can respond by decreasing the size of photophores and increasing photophore density. This may explain why species in shallower water have finer photophores, since high acuity vision is only possible at high levels of illumination, and why many deep-sea fish have higher acuity for upward viewing than in other directions (Warrant & Locket, 2004). Deep-sea predators have a limitation on acuity placed by the requirement to spatially sum photoreceptor outputs to achieve sensitivity in very low light conditions (Warrant & Locket, 2004). Predators may also adapt to detect spectral mismatches between prey and downwelling light: counterilluminating species are generally monochromats, and thus are likely to be able to only sense the intensity of downwelling light, not its spectral distribution. Further, bioluminescence mechanisms constrain the spectrum of light that can be produced. The presence of yellow lenses in deep-sea predators may be an adaptation to enhance differences between the colour of downwelling light and that coming from counterillumination (Johnsen, 2014; Muntz, 1976).

3.9 Unresolved issues and future challenges

The increased level of interest in investigating countershading coloration since the last edition of this book will hopefully continue as many pressing questions remain unanswered.

First, and perhaps foremost, whether each of the six potential countershading camouflage mechanisms discussed can make an independent contribution to improving camouflage needs experimental verification. Unlike most experimental tests of anti-predator coloration which use 2D stimuli, tests of countershading coloration generally require 3D stimuli which presents an additional level of complexity for designing controlled experiments. Virtual modelling is generating predictions and specifying stimuli that can be used to tease apart predictions made by different mechanisms (Penacchio et al.,

2015a, b), but this work has only just begun with Cuthill et al. (2016). This study provides the strongest evidence to date of the anti-predator effectiveness of self-shadow concealment, but it remains to be investigated to what degree 3D background matching and obliterative shading mechanisms within the self-shadow concealment umbrella are operating to reduce predation. This will require behavioural tests of predators to determine what cues they are using to detect and identify prey, and the perceptual and cognitive processes employed (Merilaita et al., 2017), as on the basis of prey appearance alone these mechanisms cannot be distinguished (Penacchio et al., 2015b).

A further outstanding issue given recent experimental and theoretical results is how close to optimal countershading needs to be to provide effective protection (or avoid increasing conspicuousness). Due to variation in environmental conditions, whether there exists an optimal countershading phenotype for achieving camouflage through both 3D background matching and obliterative shading remains an open question. The answer is likely to be specific to the perceptual systems of the potential receivers and the ecological and morphological traits of the camouflaged target. Different receivers may weigh cues from imperfect background matching or imperfect removal of depth cues from shading differently, and within each mechanism different visual cues may or may not be used. For example, if predators primarily use object convexity as a cue to the presence of potential prey, then optimal countershading phenotype would differ to that expected if they primarily use optical flatness (Penacchio et al., 2015b, c).

Related to this is a need to improve understanding of how organisms might alter behaviour to improve countershading camouflage, and what trade-offs are made to do so. The area most likely to be fruitful in this regard is examination of orientation behaviour. Penacchio et al. (2015a) demonstrate that countershading camouflage is improved when bilaterally symmetric animals orient directly towards or away from a light source. Cuthill et al. (2016) suggest that for animals in the open, except for the exceptional instance when the sun is directly overhead, the consequences of not orienting to the sun are always deleterious for bilaterally symmetric animals, as only by orienting to the sun is the illumination on the animal symmetric. Indeed deviations of more than 15° in orientation to the light source result in a dramatic increase in human search performance for computer-generated countershaded prey (Penacchio et al., 2017). It has not yet been tested whether animals move to maintain equal illumination on both body halves, a behaviour that is likely to entail high costs. It is possible that animals orient to optimize countershading only on one half of their body at a time, such as the side most visible to predators. The orientation hypothesis also makes predictions that could be tested by comparative approaches: modelling shows that the advantage gained by orienting to the sun should be reduced for animals that are at risk of predation in more diffuse light, such as those that live in forests or under cloudy skies. Whether and how behaviour to maintain effective countershading operates in context of foraging, thermoregulatory, and social behaviour is a largely unexplored area (Claes & Mallefet, 2010), for which a fuller understanding of how countershading interacts with other types of coloration and the environment is required.

Another pressing need is better characterization of the backgrounds that countershaded animals are viewed against. There may be some backgrounds where non-uniform patterns of radiance from an animal created by countershading coloration and self-cast shadows are advantageous, for example when shadows create patterns which match the structure of a cluttered background full of shadows, or perhaps over-countershading improves background matching in environments where the background of the upper visual field is darker than the lower visual field. It is notable that little work has examined the appearance of countershaded animals in situ (Kiltie, 1989) Similarly, whether and how counterilluminating animals modulate counterillumination to deal with different angular, spectral, and intensity distributions of light is little understood, with only some suggestion that cephalopods use water temperature as a gauge to depth (Johnsen, 2014; Young & Mencher, 1980). Johnsen regards non-invasive in-situ observations of counterillumination in the oceans as the most pressing requirement. Fortunately imaging technology is coming online that allows in-situ high-speed recording of deep-sea bioluminescense (Phillips et al., 2016).

The message and conclusion of this chapter has changed from the previous edition more than any other chapter in this book. This is a reflection of the growth in the size of the evidence base and how little was well understood previously. It is also perhaps true that the subject of countershading and counterillumination for camouflage has greatest potential to develop in the future as it is still relatively understudied. This is not for want of interesting questions. Most light on earth always has and always will come from above; how life colours itself, adapts its vision, and modifies its behaviour in response to this basic feature is a wonderful puzzle to solve.

CHAPTER 4

Transparency

4.1 Definition and introduction

We begin this chapter by considering the ecological distribution of transparent organisms and, in particular, how they are found predominantly in one particular habitat type (open waters). A major aim of sections 4.2 and 4.3 will be to explain this trend. A perfectly transparent organism (or part of an organism) will not absorb or scatter incident light. Superficially, *transparent* might seem like a synonym for *visually undetectable*, but this is not quite true: we will discuss in section 4.3 how transparent organisms can still be visually detected by their predators or prey. Conversely, section 4.4 will demonstrate that there are circumstances where a little transparency can go a long way to reducing an organism's visibility. Some body parts cannot be made transparent, but section 4.5 argues that opaque body parts need not always significantly increase the detectability of a generally transparent organism. Section 4.6 will consider the distribution of transparency among natural organisms. The distributional observations we seek to explain are the greater prevalence of transparency among aquatic than terrestrial organisms, and the particular prevalence among midwater species. An alternative but related strategy to transparency, adopted by some midwater fish, and considered in section 4.4.2, is silvering of the body to provide crypsis by broadband reflection.

4.2 The distribution of transparency across habitats

Johnsen (2001) presented detailed evidence that the overwhelming majority of transparent organisms are in the pelagic zone (any water in a sea or lake that is neither close to the bottom nor near the shore). There are very few examples of substantially transparent species in terrestrial ecosystems, and even in aquatic environments they are uncommon at the surface of the water (the neustonic zone), on or near the underlying substrate (the benthic zone) or in very deep waters where the sun's light does not penetrate sufficiently to aid vision (the aphotic zone).

In section 4.3, we will argue that the different refractive indices of air and water suggest that transparency is a much more attractive option for an organism in water than in air, and that this is the major reason for the lack of transparent terrestrial organisms. The transparent wings of many flying insects are an interesting exception to this; for these there is little evidence that transparency has been selected to reduce visual detection by antagonists. It may be that transparency in this case can better be understood as a consequence of selection for very thin cross-sections and a lack of selection for investment in pigmentation.

In the very deep ocean (the aphotic zone), below 1000m, solar illumination is insufficient to aid vision. In these waters, animals are mostly red or black in coloration (Marshall, 1971; McFall-Ngai, 1990; Johnsen, 2005). The dominant source of illumination at these depths is bioluminescent light produced by organisms. Under these circumstances transparency seems to have been less often adopted than pigmentation designed to minimize the light reflected back by the organism. The reason for this is likely to be related, in part, to the directional nature of much of the bioluminescent light produced by organisms (as discussed in section 4.6). It may also be related to a generally lower rate of encounters between predators and prey at these depths, making continuous metabolic investment in transparency and/or the structural compromises

Avoiding Attack: The Evolutionary Ecology of Crypsis, Aposematism, and Mimicry. Second Edition. Graeme D. Ruxton, William L. Allen, Thomas N. Sherratt, & Michael P. Speed, Oxford University Press (2018). © Graeme D. Ruxton, William L. Allen, Thomas N. Sherratt, & Michael P. Speed 2018. DOI: 10.1093/oso/9780199688678.001.0001

required by this whole-body style of crypsis less attractive than crypsis based on pigmentation.

One reason for the rarity of transparency in benthic habitats might be that pigmentation is less costly when pigments are selected to match a relatively unchanging background, whereas in more sunlit waters background colour is subject to more change. Another reason may be that refraction of light passing through a transparent benthic organism causes a shadow to be cast on the benthos that might greatly reduce the effectiveness of transparency. Further, transparent animals tend to be more physically delicate and, thus, might be prone to damage by abrasion against the substrate in benthic habitats. Additionally, it may be that the backdrop of the substrate provides an opportunity for other types of crypsis, using pigments that avoid the structural compromises and/or continual maintenance requirements of transparency. Finally, it may be that the physical complexity of some substrates provides hiding places that offer an alternative to visual crypsis.

Transparency is also rare among those animals living on the surface film (the neuston). It may be that the complex light environment produced by the ever-changing surface of the water makes crypsis by pigmentation easier—although this hypothesis is currently untested (Hamner, 1995)—and such a strategy is physiologically cheaper and/or less constraining than crypsis by transparency. Additionally, or alternatively, for such animals the conflict between allowing UV radiation to pass through the body to achieve crypsis by transparency and minimizing cell damage by UV radiation may favour alternative methods of crypsis, based on pigmentation of the organism's surface (we discuss UV radiation more fully in section 4.3.3).

Hence it is understandable why the distribution of transparency in aquatic systems is predominantly where sunlight is present (but not in the surface layer), especially since alternatives such as hiding in physical structures are generally not available in open water. In section 4.4 we discuss the wide taxonomic spread of transparency across different organism types, and the indication that it has evolved and been lost on many occasions. This suggests that the key to understanding the ecological distribution of transparency is through the physical arguments discussed above and not through taxonomy. In the next section we will examine more closely exactly how transparency influences ease of visual detection and expand on some of the ideas introduced in this section.

4.3 How transparency influences ease of detection

4.3.1 Transparent objects still reflect and refract

A perfectly transparent object is one that does not absorb or diffusely scatter incident light. However, perfect transparency does not make an object perfectly invisible. Glass can be very close to perfectly transparent yet we can see glass window panes. To be invisible, an object must have no effect on any light that strikes it. Perfect transparency removes two physical mechanisms, absorption and diffuse scattering, which could affect any incident light. However, that still leaves several others. Whenever light passes from one medium into another, then that light is both reflected and refracted even if both media are transparent. It is this reflected and refracted light that allows us to see even the purest glass. Every material has a physical quantity called its 'refractive index'. When light moves between media of different refractive indices, some of the light is refracted and some reflected. The magnitude of both of these effects increases with the difference between the refractive indices of the two media.

We will deal with reflection first. At normal incidence (when the light beam is perpendicular to the interface between the media), the proportion of light that is reflected is given by the simple formula:

$$R = \left(\frac{n_1 - n_2}{n_1 + n_2} \right)^2 \qquad (4.1)$$

where n_1 and n_2 are the refractive indices of the two media that form the interface (Denny, 1993). These values vary with the wavelength of light, but representative values for different materials are given in Table 4.1.

Biological tissues have refractive indices between 1.34 and 1.55 (Johnsen, 2001). Using these values in eq. (4.1) suggests that the proportion of the light striking some biological tissue that is reflected is much greater when the animal is in air (2–5%) than

Table 4.1 Refractive indices for a selection of media. Modified from Denny (1993) and Johnsen (2001).

Medium	Refractive index
Air	1.00
Water	1.33
Cytoplasm	1.34
Densely packed protein	1.55
Glass	1.56

when it is in water (0.001–0.6%). These simple calculations go a long way towards explaining why transparency is much more commonly observed among aquatic organisms than terrestrial ones. A transparent organism in air reflects a fraction of incident light many times greater than that of the same organism in water. Hence, a transparent organism has a greater optical effect on the light that ultimately reaches the eye of its predators or prey when in air than in water. As Johnsen (2001) remarks, although ice is transparent we can see even fine detail in an ice sculpture without difficulty. This conclusion holds regardless of the angle between the light ray and the interface between the organism and the surrounding medium; indeed the further from the perpendicular the light strikes, the greater the fraction that is reflected (Johnsen, 2012, p. 146).

Refraction is manifested by the bending of the path of a light ray as it passes between media with different refractive indices. The greater the difference between these indices (and the less orthogonal the angle of incidence), the greater the distortion of the path. Hence, for biological tissues with refractive indices between 1.34 and 1.55, refraction is greater in air than in water. The practical consequence of refraction is that the image seen through a transparent object is not the same as the image that would be seen if the object were not there. This can be verified by holding a piece of glass at arm's length at a slight angle to the perpendicular. The view that you see through the glass does not perfectly match up with the view that you get from light that does not pass through the glass; the piece of the view that you get through the glass seems slightly shifted. This shift is due to refraction of light travelling through the glass. The magnitude of refraction would be less if we were viewing under water,

because the refractive index of water is closer to that of glass. This is easily observed whilst washing up drinking glasses—unless you wash glasses by switching on a machine!

Not only is the extent of refraction physically less extreme for an organism in an aquatic environment than in a terrestrial one, the apparent effect of any refraction is likely to be less easy to detect in aquatic environments. As a generalization, aquatic environments are relatively featureless compared to terrestrial ones (especially in midwater) and so the effect of image distortion will be harder to detect in such environments. This may help explain why transparency is particularly common among midwater aquatic species. For a transparent organism on a solid substrate in a non-uniform light field, refraction will cause a shadow to be cast on the substrate. This can be seen again if you place an empty drinking glass on a table in strong sunlight. As discussed in section 4.2, this mechanism may make transparency less attractive for benthic organisms in relatively shallow sunlight waters. However, as water depth increases so the downward shining light becomes more uniform regardless of the position of the sun (as discussed in section 4.4.2), and this may reduce this potential drawback to transparency.

In summary, transparent individuals can still be detected because, although they do not absorb or scatter visible light, they still reflect and refract it. In order to minimize these potentially crypsis-breaking mechanisms, the refractive indices of transparent tissues must conform as closely as possible to that of the surrounding medium. This is much easier to achieve if the medium is water rather than air. Reflection and refraction are two physical mechanisms that can allow a perfectly transparent organism to be visually detected. The next sections consider several more.

4.3.2 How transparent organisms influence polarization of light

Light waves oscillate in a plane perpendicular to the direction of travel of a light beam. Where on this plane a specific light wave oscillates is described by the polarization of that light. This matters to us because, as any scattered light becomes polarized (Mobley, 1994), ambient light is expected to become partially polarized as it passes through water. In addition, the polarization of light can be changed when it passes through so-called birefringent materials, of which muscle and connective tissue fibres seem to be examples (Shashar et al., 1998). Several fish, crustaceans, and cephalopods have been shown to be sensitive to the polarization of light, and there is some evidence to suggest that this may be used to enhance detection of transparent prey. Novales Flamarique & Browman (2001) found that juvenile rainbow trout appeared better able to find transparent zooplankton prey in a laboratory aquarium under polarized light compared to unpolarized light. This suggests that light is changed in polarization as it passes through the zooplankton, in a way that the trout can detect and exploit. Shashar et al. (1998) obtained similar results for juvenile squid preying on live zooplankton in an aquarium. Shashar et al. (1998) also demonstrated that adult squid showed a marked preference for preying on transparent glass beads that were birefringent, compared to beads that looked identical—at least to humans—but that had no effect on polarization. Hence, change in polarization may be another physical process that can allow a transparent organism to be visually detected.

Johnsen et al. (2011) report on polarization images of transparent zooplankton taken under natural lighting conditions. They argue that the small fraction of light scattered by transparent organisms may be important for detection from the side. Such detection could be achieved by contrast between the scattered light and the background in either intensity and/or polarization. Specifically, near the surface the intensity of downward-directed light will be orders of magnitude stronger than horizontal radiance. Thus the small fraction of this downwelling light that is scattered in the horizontal direction can provide a substantial intensity contrast with the darker background. Also, the horizontal radiance will often have a characteristic polarization, whereas the scattered light will be unpolarized. Johnsen et al. (2011) argue that, when considered separately, it is likely that horizontally orientated viewers will be able to detect transparent organisms more readily through difference from the background in intensity rather than in polarization; both because the intrinsic difference from the background will be stronger, and because viewers are expected to be more sensitive to small differences in

intensity of light than small differences in polarization. However, Johnsen et al. also argue that viewers may be able to effectively combine these two differences to aid detection of transparent prey using simple algorithms. For example, if the viewer views the scene through a polarizer orientated to minimize the background horizontal radiance, then this could well boost their ability to detect the intensity difference between the target organism and the background. It would certainly be valuable to explore whether such structures can be found in the visual systems of any predatory marine organisms.

4.3.3 How transparent organisms interact with light outside our visual range

An organism can be transparent to visible light but still absorb and/or scatter significant amounts of ultraviolet (UV) radiation. The ability to detect UV radiation is common in fish (Losey et al., 1999) and other aquatic predators (Johnsen, 2001). Browman et al. (1994) found that two species of zooplanktivorous fish appeared to find prey less easily in a laboratory aquarium when the UV part of the ambient light spectrum was removed. However, neither Rocco et al. (2002) nor Leech & Johnsen (2006) were able to replicate these results. Rocco et al. suggest two reasons for this difference between their study and that of Browman et al. First, Rocco et al. used natural sunlight, as opposed to Browman's artificial illumination; second, Rocco et al. used direct counts of rate of prey consumption rather than indirect estimates of detection distances. Hence, at the present time, there is no unequivocal evidence that UV vision is important in predator–prey interactions involving visibly transparent prey, but it certainly seems physically possible.

UV radiation is capable of damaging living cells (Paul & Gwynn-Jones, 2003), and can cause increased mortality among aquatic larvae (Morgan & Christy, (1996). This suggests that there may be a trade-off for an organism between transparency to UV radiation for crypsis reasons and having a UV-absorbing outer surface to protect internal organs from damage (Morgan & Christy, 1996; Johnsen & Widder, 2001). UV radiation is most intense at the surface of the water column and is rapidly absorbed by water and dissolved organic matter (Peterson et al., 2002), which explains why UV damage has seldom been reported at depths greater than 25m (El-Sayed et al., 1996). In contrast, it has been estimated that sufficient UV light for vision can be found at depths of 100–200m (Losey et al., 1999). Hence, it has been predicted that if a trade-off between radiative protection and crypsis influences the distribution of transparency to UV radiation among aquatic species, then UV opacity should be more common among species inhabiting surface waters (Johnsen & Widder, 2001), or that the pigmentation of an individual may decrease—and transparency increase—as organisms migrate vertically into deeper waters (Morgan & Christy, 1996). Certainly, UV light seems an important driver of diel vertical migration (DVM) in zooplankton (Leach et al., 2015), and it has been found that marine zooplankton change their level of transparency with depth and time of the day (Vestheim & Kaartvedt, 2006); thus UV light likely impacts crypsis and predation trade-offs.

In order to test the hypothesized trade-off between crypsis and avoiding radiative damage, Johnsen & Widder (2001) measured the transparency of zooplankton collected from epipelagic (0–20m) and mesopelagic (150–790m) depths. They report that the two groups did not differ in their transparency to visible light, but those from the surface layers were less transparent to UV, in line with their expectation from the trade-off argument of the last paragraph. However, these authors argue that the greater UV absorbency of surface-dwelling zooplankton need not necessarily substantially increase their detectability to predators. This is based on two arguments. First, absorbency in the UV-A range (320–400nm) was generally low in their study compared to that in the UV-B (280–320nm). This is important because only UV-A has been implicated in vision (Losey et al.,1999) and UV-B seems much more damaging to living tissues than UV-A (Johnsen & Widder, 2001). Second, Johnsen & Widder also report that many species with low UV transparency also had low transparency to visible light. UV radiation attenuates more rapidly in water than blue visible light. Hence, if an individual has low transparency to both blue visible light and UV radiation, then (all other things being equal) the visible light will permit its detection by organisms at a greater distance, compared to UV radiation. Consider a predator moving closer and closer to an as-yet-undetected prey item. If the prey item has low transparency to both blue light and UV light, then it will be detected initially because of its effect on the visible light reaching the predator. After its effect on the visible light has caused the prey to be detected, the effect of UV light, when the two organisms have drawn even closer to each other might matter little. Conversely, if the prey is highly transparent to visible light, then a predator could move very close to it without detecting it on the basis of its effect on the visible spectrum. In this case, even small changes in UV transparency may substantially affect the ease with which the prey organism can be detected. These caveats suggest that the situation may be rather complex, and that in order to understand the selection pressures on UV transparency we need a quantitative understanding of the effects of depth and degree of transparency on both risk of UV damage and risk of predation. We also need an understanding of the fitness consequences of UV damage. Such understanding is not currently available and should be the focus of future research.

4.3.4 Considering how Snell's window affects detection of transparent organisms

Another method by which visual predators could detect perfectly transparent prey near the surface in an aquatic medium is through using Snell's window. Figure 4.1a illustrates the light reaching a fish's eye. Whether light is reflected or refracted at the interface between two media (such as the surface of a body of water) depends on the angle of incidence of that light with the interface. A consequence of this is that all light reaching the eye from angles from the perpendicular less than 49° is light from above the water's surface. Indeed, the fish's whole view of the world above the water's surface is compressed into this so-called Snell's window by the properties of refraction discussed earlier. All light entering the fish's eye at angles greater than 49° is light that was travelling upward through the water column but was reflected at the surface of the water. The intensity of light entering the water column from above is generally much greater than the light from below that has been reflected back from the surface. Hence, a fish sees the world above the surface as a bright image in Snell's window surrounded by a much dimmer (reflected) image of the depths below. This could be a problem for a zooplankter situated just outside Snell's window. Light that would pass through that space if the zooplankter was not there can be refracted by the zooplankter so that its course is changed and it reaches the fish's eye (zooplankter *i* in Figure 4.1b). If the zooplankter were not there, that light would not reach the fish's eye. This light will show up as a bright spot against the dark background of the image of the depths, and so the zooplankter will be very visible to the fish. This will be less of a problem for a zooplankter well away from Snell's window (zooplankter *ii*), which will not refract light sufficiently dramatically to enter the fish's eye. Nor is there a substantial problem for a zooplankter in Snell's window, as light refracted through it will be seen against the rest of the bright light in the window, and so will not provide a strong contrast.

This theory was supported by the observations of captive blueback herring by Janssen (1981). These fish are visually feeding planktivores, and characteristically swim upwards towards their prey. This trajectory will cause the ring of visibility just outside

Snell's window to sweep inwards as the fish swims towards the surface, allowing the fish a chance to detect prey across a wide area of the surface waters. The median angle of attack by the fish in Janssen's observations was 54°. This is only slightly greater than the 49° that defines Snell's window and is consistent with attacking individuals detected just outside Snell's window. Such calculations can only be considered approximate since the position of the edge of the window will be affected by any surface waves (S. Johnsen, pers. comm.). Zooplankton can do little about their relative positioning with respect to predators; their best defence to this vulnerability is to minimize refraction by keeping their refractive index as close to that of water as possible. The issue of surface waves mentioned above may mean that this mechanism of prey detection is much more of an important factor in the calmer waters of lakes than in the open ocean.

The refractive index of any material is different for different wavelengths of light; this too may have ecological implications. Johnsen & Widder (1998) studied the effect of the wavelength of incident light on the transparency of 29 species of gelatinous zooplankton. In the overwhelming majority of cases, they found that transparency increased linearly with wavelength, albeit only very gradually. This means that the spectrum of light passing through such a zooplankton will be different from that of the background against which it is seen, potentially allowing detection in the same way that a green-coloured glass bottle is easier for us to see than a clear one. However, light of different wavelengths attenuate differentially in water such that the spectrum of ambient light is much broader in the surface layers than at depth (where blue-green visible light predominates). Hence, Johnsen & Widder argue that this potential weakness to crypsis by transparency is more of a problem when hiding near the surface. Even in the surface waters, we currently lack a demonstration that the visual systems of appropriate aquatic organisms can detect the relatively modest changes in light spectrum predicted by this mechanism.

In section 4.3 we suggested that transparent organisms could still be detected through the light that they reflect and refract. The last few sections have considered several other mechanisms that *might* allow an organism that is perfectly or partially

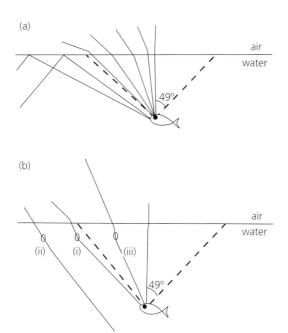

Figure 4.1 Snell's window.

transparent to visible light to be detected by either its predators or its prey: change in light polarization, differential absorption of different wavelengths (especially combined with UV vision), and Snell's window. We say 'might' because definitive experimental work is still lacking. Rocco et al. (2002) cast doubt on previous evidence that UV radiation helps predators capture visually transparent prey. Further studies using systems as similar as possible to natural field conditions are required to resolve this issue. Similarly, further experiments on the possible role of polarization in increasing the detectability of visually transparent prey under natural light conditions would be useful. Another related topic is the extent to which changes in hue caused by differential transparency of organisms to different wavelengths of visible light can be detected by potential predators. Janssen's fascinating idea and experiments on Snell's window are certainly worthy of further work exploring the ecological relevance of his results. We have now argued that transparency is not the perfect form of crypsis and that *transparent* is certainly not a synonym for *undetectable*. However, section 4.5 argues that despite these caveats, even less than perfect transparency can sometimes be an effective form of crypsis. Before that, in the next section, we consider transparency from an evolutionary perspective.

4.4 Evolutionary considerations

Johnsen (2001) provided a detailed overview of the phylogenetic distribution of transparency. Transparency has evolved on many separate occasions. Most transparent organisms belong to the ten phylogenetically disparate pelagic groups: cubozoans (box jellyfish), hydromedusae (predatory suborder of the Cnidaria), non-beriod ctenophores (comb jellies), hyperiid amphipods (a suborder of planktonic amphipod crustaceans), tomopterid polychaetes (planktonic suborder of polychaete worms), pterotracheid and carinariid heteropods (a branch of the marine gastropod molluscs), pseudothecosomatous pteropods (sea butterflies), cranchiid squid (glass squid), thaliaceans (class of tunicates), and chaetognaths (arrow worms). He also emphasizes that within classes, such as the arthropods and molluscs, there are many major groups containing both

transparent and non-transparent forms. Within vertebrates, a relatively small number of fish are transparent as adults but more are transparent as juveniles. Together, all this indicates that transparency has evolved on many occasions, and that ecology is a better predictor of transparency than phylogeny. Given that transparency (unlike so many other forms of crypsis) is a property of the whole body not just the integument, then its frequent evolution is especially noteworthy. Although achieving transparency may not be as challenging as it first appears, because the majority of organic molecules do not absorb visible light, the real challenge to transparency is not to avoid absorption, but to avoid scattering. Scattering is caused by discontinuities in refractive index. Thus, the challenge to making biological tissues transparent is mostly about making them homogeneous on the scale of wavelengths of visible light. As Johnsen (2001) points out, transparent tissues can have components of many different refractive indices, so long as the average index is constant over distances equal to half the wavelength of the incident light.

As mentioned above, unlike other forms of crypsis, transparency involves the entire body, not just the exterior (McFall-Ngai, 1990). We have argued that transparent organisms will gain an advantage by having the optical properties of their body (such as refractive index) as similar to those of the surrounding medium as possible. Many aquatic organisms achieve this by being gelatinous, where thick layers of extracellular watery material separate very thin layers of cellular material. Transparency of muscle and other dense tissues is more easily achieved if one dimension is very thin (McFall-Ngai, 1990; Herring, 2001). Clearly a gelatinous body plan involving greatly increasing water content leads to a huge increase in size. This increased size will have costs and benefits other than improved crypsis. For example, one added benefit of a gelatinous body plan will be decreased body density, and so increased buoyancy. For suspension feeders, a gelatinous body plan can bring increased capture rates without the increase in metabolism that would normally result from growing bigger (Acuña, 2001). Costs are likely to include increased drag and vulnerability to mechanical damage. These non-crypsis costs and benefits may have important implications for the evolution of crypsis by transparency through adoption of a gelatinous body-plan. We should also consider what constraints there might be on the evolution of transparency, and turn to this in the next section.

4.4.1 Constraints of transparency: some parts of an organism cannot be made transparent

Food is usually opaque, with the result that the stomach cannot generally be transparent after the organism has fed. Feeding on transparent prey does not seem to be a solution to this, as transparent organisms rapidly turn opaque after death (Herring, 2001). This change suggested to Herring, and to others (e.g. McFall-Ngai, 1990; Hamner, 1995), that transparency of cellular tissues can only be achieved with a continual investment in active maintenance. However, direct evidence of such a continuous metabolic cost to crypsis through transparency is currently lacking, and investigation of this would be very useful. It may be adaptive to have an opaque stomach if feeding on bioluminescent (or brightly coloured) prey, in order to reduce the visibility of stomach contents to potential predators or prey. Some prey continue to bioluminesce even when being digested (Johnsen, 2012, p. 159). Seapy & Young (1986) describe squid, heteropods, and hyperiid amphipods that are highly adapted to minimize the visibility costs of an opaque gut. They have a needle-shaped gut which always points vertically no matter the orientation of the animal, minimizing the silhouette of the gut when viewed from above or below. Further, the sides of the gut have a mirrored coating to reduce visibility when viewed from the side (see section 4.4.2).

Eyes must contain light-absorbing pigments and so cannot be wholly transparent. This can be a significant problem for deep-sea organisms that need large eyes to maximize photon capture at low light levels. However, adaptations can be found to minimize the conspicuousness of this. One tactic is to spread the light-absorbing surfaces over a wide area of the body, so that there is not a concentrated area of strong light absorption (Johnsen, 2012, p. 159). Other tactics include minimizing the volume of the opaque retina and using reflective structures (see section 4.4.2) around the eyes to reduce visibility of the opaque parts (Feller & Cronin, 2014). Recently

it has been demonstrated that a midwater squid uses photophores on the ventral surfaces of its eyes to emit bioluminescence to counterilluminate shadows cast by the eyes (Holt & Sweeney, 2016).

There has been considerable empirical investigation into whether the size of opaque parts increases predation risk for otherwise transparent organisms. As these studies take different approaches, we will try to provide a synthesis of their results. Giguère & Northcote (1987) found that the predation rate of juvenile coho salmon on transparent phantom midges (Diptera: Chaoboridae) increased with increasing stomach fullness of these insect larvae. Considering the relative sizes and mobilities of predator and prey, this is more likely to result from increased prey visibility following a meal, rather than decreased mobility. Tsuda et al. (1998) obtained similar results for herring larvae and salmon fry feeding on copepods. Similarly, Wright & O'Brien (1982) concluded that the detectability of mainly transparent midge larvae to predatory fish was based on the size of high-contrast structures within the larvae rather than the overall size of the prey individuals. Essentially identical conclusions were drawn by Hessen (1985) after experiments with roach feeding on zooplankton. Bohl (1982) offered predatory fish a combination of generally transparent daphnia with and without eggs in their dorsal pouch. Eggs are opaque and should act to increase the vulnerability of egg-bearing individuals. He found no preferences for egg-bearers under complete darkness and under bright light conditions, but a strong preference for egg-bearers under low light conditions. This seems to represent strong evidence that the egg-bearing individuals were more easily detected by predators than non-egg-bearers when visual detection was challenging, but not impossible. Brownell (1985) performed laboratory experiments to explore cannibalism in the (relatively transparent) larvae of cape anchovy. He concluded that the formation of pigmented eyes, at a certain development stage, triggered a substantial increase in larval detectability, and so vulnerability to cannibalism.

Zaret (1972) reported laboratory experiments where predation on a cladoceran appeared to be related to the amounts of pigmentation in their black compound eyes. These experiments used a species with two morphs that differ considerably in eye pigmentation; the morph with the higher pigmentation (i.e. the larger eye) was attacked more frequently than the morph with the smaller eye. Zaret took advantage of the fact that the stomach of these organisms lies near the eye to produce a 'super eye spot' by allowing individuals of the 'small eye' morph to pre-feed in water to which India ink had been added. This led to a reversal in relative predation rates in subsequent experiments, where the small-eyed morph with the 'super eye spot' created by black pigment in their stomach was preyed on more heavily than the larger-eyed morph without the India ink modification. Whilst alternative explanations based on behaviour or chemosensory changes are possible, the most likely explanation for these results is that the size of the opaque parts of the prey influences detectability by predators. A follow-up set of laboratory trials (Zaret & Kerfoot, 1975) demonstrated that variation in predation rate on a largely transparent cladoceran was better explained by variation in the size of its large compound eye than variation in the size of the whole body. However, a similar study by O'Brien et al. (1979) did not support these conclusions, finding body size to be a better predictor than eye size. Similarly, studies have found no effect of ink pre-feeding by daphnia on detectability by bluegills (Vinyard & O'Brien, 1975) or lake trout (Confer et al., 1978).

In conclusion, there is evidence that opaque parts to a generally transparent organism can have a substantial effect on predation risk, although this effect is not shown universally in empirical studies. However, we should not expect there always to be such an effect. If the mostly transparent organism can be readily detected by other means (and we have considered a number of methods by which a perfectly transparent organism can still be detected visually) then small opaque parts need not necessarily increase predation risk. Another important component to building an understanding of the selective pressures for the evolution of camouflage through transparency is to consider alternative forms of camouflage. One obvious alternative is background matching through pigmentation and this has been fully discussed in Chapter 1. Another alternative, which like transparency is particularly effective in aquatic systems, is silvering of the body

to provide camouflage by reflection: we now consider this in more depth.

4.4.2 Silvering as an alternative form of crypsis

In terrestrial ecosystems, there is generally strong directionality in the ambient lighting, since the sun takes up only a small part of the visible sky, and this part of the sky produces much more light than other equivalent-sized parts. This directionality can be seen clearly from the existence of shadows. As light passes through water, scattering by water molecules and suspended particles has the effect of reducing this strong directionality. Denton et al. (1972) suggest that 300m depth of clear water is required before direct sunlight is scattered sufficiently to become vertically symmetrical, reducing considerably (to perhaps around 50m) if the water is turbid or the sky cloudy. Thus, in the midwaters of the oceans and deep lakes, the light environment is potentially at least approximately symmetrical about the vertical axis. In these waters, there is also generally a complete absence of background features. We can plainly see the mirrors hanging on our walls of our homes, because they produce a reflection that does not match the background of the wall. However, in mesopelagic environments an animal with a body plan like a vertical mirror can make itself invisible to detectors viewing from side-on (see Figure 4.2). This is not an option for terrestrial organisms or organisms near the substrate, where background features mean that reflection is no longer an effective disguise.

One strategy to reduce the vulnerability of disguise by silvering to detection from above and below is to be laterally flattened, so as to restrict the vulnerable cross-sectional area. This is adopted by some silvered fish, such as hatchetfishes. This body form, though, restricts locomotive performance, and so muscular fish—such as herring, mackerel, and salmon—adopt a body form with an elliptical cross section, where the reflective scales act as tiny mirrors and each is aligned vertically rather than parallel to the curved body surface (see Figure 4.3). However, the problem of providing reduced detection from above and below still remains. Both Herring (2001) and Denton (1970) suggest that the tapered keels of many silvered fish act to reduce

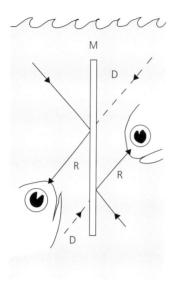

Figure 4.2 Diagram to show how a dark object can be camouflaged in the radially symmetric radiance distribution of the ocean by making it reflective. Vertical mirror (M) cannot distinguish between reflected rays (R) and direct rays (D), so the mirror (or silvered) fish is invisible (from Herring, 2001).

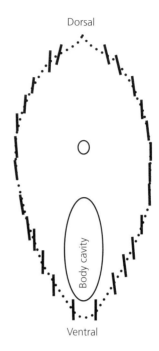

Figure 4.3 Reflectors aligned to the curve of the body surface would not be an effective camouflage for the flanks of a muscular fish of elliptical cross-section like a herring. Hence, the individual reflectors are aligned vertically, providing an effective vertical mirror surface (from Herring, 2001).

visibility from below, although the mechanism for this is not specified. Reflective scales could be angled so as to reflect light partially downward, and whilst this would reduce visibility from below, it would do so only at the expense of potentially increasing visibility from the side.

One feature of midwater environments is that the downwelling light is orders of magnitude more intense than the upwelling light (Johnsen, 2012). Thus, dorsally placed mirrors reflecting downwelling light would make an animal very visible against the dark background of the depths. Hence, mirrors are not the solution to reducing detection from directly above. Instead, silvered animals tend to have a dark pigmented upper surface, so as to provide a match to the background (see Johnsen (2012) for a full analysis of this). The reverse problem applies when being viewed from below, where reflected upwelling light provides a large contrast to the strong downwelling light. One solution is for the animal to produce its own downward-directed light to match that of the ambient downwelling illumination (for an overview see Widder, 1999; Haddock et al., 2010). Some species have been demonstrated to

alter the intensity and wavelength distribution of the light they produce appropriately as they change depth, so as to match the changing downwelling light (Case et al., 1977; Young & Roper, 1977; Latz & Case, 1982; Young & Arnold, 1982; Young, 1983). Organisms have also been demonstrated to produce light that matches the angular distribution of the ambient light (Denton et al., 1972; Latz & Case, 1982). Such counterillumination is only feasible at considerable depths when the intensity of downwelling light is sufficiently reduced that self-generated light can significantly affect the contrast (for further discussion see McFall-Ngai, 1990).

When light is reflected, its polarization changes. This may provide a mechanism by which the crypsis provided by silvering can be reduced. As noted in section 4.3.1, cephalopods are sensitive to the polarization of light, and Shashar et al. (2000) demonstrated in a laboratory experiment that cuttlefish targeted fish with normal polarization reflection over fish that, by means of filters, were made to not reflect the linearly polarized component of the light. However, the consensus of current research is that long-range detection of silvery fishes is unlikely to benefit from evaluation of polarization (Cronin et al., 2016; Johnsen et al., 2016; Feller et al., 2017). It appears that under natural conditions polarization of light is generally not strong underwater, and natural receptors are not strongly sensitive to small changes in polarization. It remains possible (but so far undemonstrated) that changes in polarization on reflection may aid in short-range processes of prey identification and precise localization after long-range detection has occurred.

Silvering seems largely confined to adult stages of fish. It may be that for juvenile stages and smaller plankton transparency is a more effective form of crypsis, but that this 'whole body' disguise becomes impossible for large fish. Large, adult fish need to move quickly, and must have large amounts of (relatively opaque) respiratory and connective tissues associated with their substantial muscle mass. The different depth distributions of larvae and adults may also have a role. Giske et al. (1990) suggest that juveniles tend to be distributed more in shallower waters than adult stages, and we have already discussed why silvering is more effective at greater depths where the light field is more homogeneous. Salmonid fish characteristically change from a dark and highly patterned body to a classical silvered appearance before migrating from stream and river to the ocean. This can be understood in terms of our argument above about the inefficiency of silvering in shallow water with visible background features, whereas their juvenile coloration would make salmonids highly visible when viewed from the side against the featureless ocean.

Silvering can also be a solution to enable crypsis of parts of an organism that cannot be made transparent in an otherwise transparent animal. Effectively, the eye or digestive organ can be thought of as a mini-organism that can be disguised by silvering in exactly the same way as described in this section for whole organisms.

Badcock (1970) and Denton et al. (1972) report that many silvery mesopelagic fish darken their silvery reflective surfaces at night. Recall that silvering is only an effective means of crypsis if the light field around the object is symmetrical. This will not be true at these depths at night, where the main light will not be downwelling star and moonlight, but rather the light generated by luminescent organisms. Under such circumstances, a dark, light-absorbing surface may be a more effective disguise than a silvery, reflective one. In fact, recent work (Johnsen et al., 2014) suggests that the light field to a depth of 100m is quite strongly asymmetrical for much of the daylight hours too, and thus the effectiveness of crypsis by silvering may be less than previously supposed.

Johnsen & Sosik (2003) discuss the effectiveness of crypsis by use of mirrored sides or cryptic coloration in near-surface pelagic habitats, before water has passed through sufficient depth for the light field to be uniform. In general terms, both methods have limited effectiveness in this region, which perhaps explains the particular prevalence of transparent organisms in the first few tens of metres and the vertical migration of other organisms so as to minimize exposure to the surface waters when they are well-lit. These authors took careful measurements of the light field at different depths in the open ocean, then used modelling of the visual capabilities of the Atlantic cod *Gadus morhua* as a model predator. For both cryptic coloration and mirrored sides, they found the optimal set of physiological traits for reducing detection under a given set of circumstances (e.g. time of day, depth in the water column, relative position of viewer), then explored how quickly crypsis generated by that particular optimum deteriorated with changes in the ecological conditions. Focussing on the ecological robustness of these two strategies for crypsis makes sense, because these broad strategies do not appear to offer strong flexibility with respect to adaptation to changing light conditions (except for the control of pigmentation achieved by many cephalopods). In comparison, transparency is trivially able to accommodate changing lighting conditions, and use of self-generated light (bioluminescence) is well known to be sensitive to prevailing illumination (Case et al., 1977; Young & Roper, 1977; Latz & Case, 1982, Young & Arnold, 1982; Young, 1983). Johnsen & Sosik (2003) reached a number of interesting conclusions. First, in stark contrast to benthic organisms that are viewed against the solid substrate, pelagic organisms using either mirrored sides or cryptic coloration will often not be able to find a robust optimum that offers effective crypsis across the range of ecological conditions experienced. Second, neither reflective nor pigmented strategies are clearly better than the other. In general, reflective strategies were more robust, but they could never be perfectly cryptic when viewed in the azimuth of the sun. Finally, both strategies showed strongest robustness when viewing from above was considered; this was because of the relatively unchanging nature of the background. So, perhaps these tactics might be particularly attractive to prey whose main predators are, for example, seabirds making feeding forays from the surface. These authors argue that viewing horizontally may present more of a problem for organisms in these upper waters than viewing from below. Essentially, wave-induced shadowing, internal reflection of light at the surface, and the possible presence of broken cloud all make the background against which prey in the surface waters are viewed from below very complex, and background complexity generally increases the difficulty of target detection.

As we have seen over this section, the distribution of transparency amongst living organisms seems easier to understand in terms of ecology than in terms of phylogeny. For this reason, we have been obliged to weave ecology into much of the material covered so far in this chapter; however, there are some interesting aspects to ecology that we have not yet mentioned, and we turn to these in the next section.

4.5 Ecological influences

An organism might well experience very different types of illumination, even on a daily timescale. This can be brought about either by the daily cycle of sunlight and/or (for many aquatic organisms) by vertical migration of the organism. Hamner (1995) contends that 'at sea, predation drives most, if not all, vertical migratory behaviour.' From the discussion earlier in this chapter, we can see that at different depths (or, more generally, under different lighting conditions) different mechanisms of crypsis will be favoured. Hence, it is no surprise to find that individuals can adapt their external appearance on

behavioural timescales so as to adopt the most effective mechanism for their current situation (Morgan & Christy, 1996). An alternative behavioural mechanism employed by some species is to adjust their depth, so as to remain within a range of preferred irradiances (Frank & Widder, 2002).

Illumination is not the only ecological variable that influences transparency. Bhandiwad & Johnsen (2011) subjected the transparent grass shrimp *Palaemontes pugio* to changes in ambient salinity and temperature and found that such changes could induce changes in transparency. This organism lives in shallow estuaries, and the authors suggest that the range of variation in the two parameters explored was within the range naturally experienced, due to variation in tidal cycles, evaporation, and runoff. However, the changes imposed were of a faster timescale than would naturally be experienced, and exploration of how the speed of imposed environmental changes affects changes in transparency would be very useful. This might be a useful model species for exploring the physiological mechanisms that allow tissues to be transparent but still functional within the organism. To be transparent, a material must have uniform refractive index on size scales greater than one-half a wavelength of the light applied. In this species muscle tissue can be transparent, but it appears that reduced transparency is caused by increased scattering. This results from a breakdown in the homogeneity of refractive index within the tissue, which itself is caused by the pooling of excess fluid in extracellular space between muscle fibres. Another underexplored issue is how variation in transparency translates to variation in ease of visual detection; we explore this issue next.

4.5.1 Imperfect transparency can be effective at low light levels

The visibility of an underwater object generally depends more upon its contrast than its size (Lythgoe, 1979). Contrast is the difference in intensity of light reaching an organism's eye that originates from the viewed object compared with the intensity of light reaching the eye from other sources. Contrast can be defined in various ways, but is generally expressed as a ratio. For our purposes we can think of it as the difference in light intensity coming from the organism and the background divided by the average of these intensities.

A predator's ability to detect prey (and vice versa) depends on the minimum contrast that it can detect: the contrast threshold. Specifically, prey whose contrast falls below the contrast threshold will not be visually detected. Contrast threshold decreases with increasing ambient light intensity. That is, an object can be seen at a greater distance as the ambient light intensity against which it is viewed increases. This means that aquatic prey will be more easily detected from below, against the bright (downwelling) light from the sky, than from above, against the much reduced scattered or reflected light directed upwards. However, downwelling light intensity decreases with depth, and so this should lead to an increase in contrast threshold (and thus a decrease in ease of detection) with depth.

Anthony (1981) estimated the contrast threshold for cod and suggested that this changed from around 0.02 at light intensities equivalent to those in the first 20m of clear water on a bright sunny day, up to 0.5 in light intensities more representative of depths of around 650m (Johnsen & Widder, 1998). Consideration of Figure 4.4, based on theoretical modelling of Johnsen & Widder (1998), suggests that this change should have a substantial ecological effect. At the top of the water column, when the contrast threshold is 0.02, transparency has to be greater than 85 per cent (i.e. 85 per cent of light passes through without absorption or scattering) to halve the maximum sighting distance of an object compared to a fully opaque counterpart. In contrast, at depth, when the threshold has increased to 0.5, transparency need be only greater than 30 per cent to more than halve the sighting distance. Indeed, at this contrast threshold, individuals with a transparency of greater than 50 per cent remain visually undetected no matter how close the predator or prey may approach. Johnsen & Widder (1998) measured the transparencies of 29 species of zooplankton as being between 50 per cent and 90 per cent. The effect of background illumination on ease of detection is strongly non-linear (see Johnsen, 2012) for a full discussion of this), because the detection distance varies with the difference between the natural logarithm of the inherent contrast of the object against the background and the natural logarithm

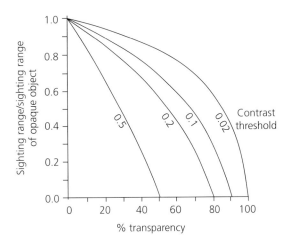

Figure 4.4 Sighting distance of a prey item versus its per cent transparency when viewed from below by predators with visual systems that have different minimum contrast thresholds for object detection. The sighting distance is divided by the sighting distance of an opaque object to control for water quality, prey shape, size, etc. The ratio gives an estimate of the advantage of transparency for crypsis in a given situation.

of the contrast threshold of the viewer. Effective camouflage involves minimizing this difference by lowering the inherent contrast (through transparency, silvering, pigmentation, or self-generated light) or by raising the contrast threshold by moving to darker waters. Thus, a 5-fold reduction in inherent contrast from 0.5 to 0.1 produces only a 1.7-fold reduction in sighting distance if the viewer has a contrast threshold of 0.01; whereas a 5-fold reduction from 0.1 to 0.02 produces a 3.3-fold reduction in viewing distance by the same viewer; and a reduction from 0.05 to 0.01 would render the animal virtually undetectable.

Utne-Palm (1999) demonstrated that of predacious gobies the maximum distance of prey detection was higher for opaque copepods than transparent ones. This is in line with our expectation that transparent organisms are actually more challenging to detect than opaque ones. Experiments to measure the contrast threshold and how it is affected by properties of the eye, the target, the medium, and the light field are technically challenging. However, the arguments in this section urgently require a stronger empirical foundation. So far, we have focussed on adaptations of organisms to evade visual detection, but of course we would also expect counter-adaptations by those seeking to detect them; we explore this issue a little in the next section.

4.6 Co-evolutionary considerations

Zylinski & Johnsen (2011) demonstrate dynamic change in the appearance of two cephalopods (*Japetella heathi* and *Onychoteuthis banksii*) that is consistent with changing cryptic strategies in response to the adaptations of predators at different depths in the ocean. In the upper waters of the ocean, many predators have 'tubular' eyes that are directed upward; these are effective in detecting the silhouettes of prey against the lighter background of the sky. Transparency is an effective defence against detection by such predators, and the two focal species of this study employ transparency of much of their body as their default.

However, in the deep ocean, sunlight is irrelevant and the main source of light is light generated biochemically by organisms themselves. In the waters below 500m, 70 per cent of fish and 65 per cent of decapod crustaceans are bioluminescent (Herring, 2001). Predators can use such bioluminescence as a 'searchlight' to illuminate potential prey. Such illumination is clearly strongly directional and transparency is of limited defence against this. As discussed previously, some light will be reflected back from a transparent organism, and the intensity of this reflected light will stand out from the dark background, allowing effective detection. Bioluminescent light is generally blue-green, and when individuals of the focal cephalopod species are illuminated with strongly directional light, they rapidly change appearance to being a blotchy dull brown-red. This can be interpreted as the best defence against bioluminescent searchlights; absorption of the light produced by the bioluminescent searchlight offers the prey the best chance of avoiding detection. However, this tactic makes the prey highly vulnerable to detection from below when in sunlit waters. Conversely, in the uniform light field of many sunlit waters, the small fraction of light that is reflected by a transparent organism may not greatly compromise crypsis. Thus, the control that cephalopods have over their appearance appears to offer them the ability to flexibly select the cryptic strategies required to thwart the predatory adaptions seen in different layers of the ocean.

Johnsen & Sosik (2003) discuss how searching behaviour by predators can most effectively undermine crypsis by use of mirrored sides or cryptic coloration in near-surface pelagic habitats, before water has passed through sufficient depth for the light field to be uniform. They argue that in such situations the effectiveness of both cryptic approaches will be dependent on the azimuthal angle at which the predator views the cryptic organism, and so circular swimming patterns while searching may be best for prey discovery in situations where prey are sparse and high value. Foraging when the sun is not high in the sky (Johnsen et al. (2014) will further increase the sensitivity of detection distance to azimuthal angle, and so hunting at these times may be particularly beneficial for the predators of prey that adopt these forms of crypsis. Finally, the effect might be further enhanced the higher in the water column the prey is, so predatory activity at depth that drives prey upward may be particularly effective in breaking these tactics. The authors provide references demonstrating the commonness of these tactics, but (since these tactics also have other benefits) their specific linkage to breaking crypsis remains unclear.

4.7 Unresolved issues and future challenges

In summary, visual detection of an organism requires that the light reaching a detector's eye has been modified by the organism. Perfect transparency means that light can pass through an organism without any modification through the processes of absorption and scattering. This does make transparent organisms challenging to detect. However, other physical processes (such as reflection and refraction) can also alter light as it passes between an organism and the surrounding medium (air or water). Hence, transparent organisms are challenging but not necessarily impossible to visually detect. The environmental conditions where transparency is likely to be particularly effective in reducing detectability occur in mid-depths of open water regions of lakes and seas. These regions also provide the conditions where a highly reflective skin can make an organism very challenging to detect. This chapter provides an overview of the current state of knowledge of crypsis by transparency or reflection. It seems that the fundamental physical processes are relatively well understood, but quantification of the importance of these processes to natural predator–prey systems is currently lacking. This may well reflect the bias of terrestrial-living scientists towards the processes they see in the world around them. However, this field does seem to have experienced renewed interest recently and we can expect further developments in the near future. We finish by highlighting two potential areas of investigation.

Johnsen (2012, pp. 146–9) points out that in some human-built optical devices the problem of reflection at the interface between two media is mitigated by a 'bumpy' surface structure to optical lenses; this means that as light approaches the lens, the bumps occupy a greater and greater fraction of the area, so that the 'average' refractive index (at the scale of wavelengths of light) increases smoothly rather than abruptly as light passes into the lens. This design feature reduces the level of reflection compared to abrupt transition at a smoother surface. He speculates that it would be worth exploring whether a similar approach is utilized by transparent marine organisms to reduce the amount of light that they reflect. For example, Binetti et al. (2009) suggest that such nano-sized protuberances act as an anti-reflection device on the transparent parts of the wings of the glasswing butterfly *Greta oto*.

Johnsen (2012, p. 157) also points out that our understanding of how many animals achieve transparency is very limited. In some cases (such as jellyfish and comb jellies) transparency is achieved by having gelatinous tissue with very high water content; in others part of the answer is simply being very thin. Light attenuation in tissue, whether due to absorption or scattering, is exponential; so we might expect selection for extreme body flattening. However, for many animals those two factors alone cannot explain transparency, and our understanding of how this is achieved is very limited. Such organisms generally become opaque soon after death, suggesting that transparency may require active (and presumably costly) metabolism.

CHAPTER 5

Secondary defences

5.1 Introduction and overview of the chapter

Much of this book covers the use of visual appearance to prevent attacks taking place. Defences that happen early in a potential predator–prey encounter affect the likelihood of the predator physically contacting the prey, and are generally called 'primary defences' (see the Introduction to this book). In this chapter we consider defences that are usually deployed during, or just before, contact between a prey and its predator: so-called 'secondary' defences. Secondary defences underpin defensive advertising described in other chapters, including aposematism, Müllerian mimicry, and deceptive forms of mimicry. Hence, we now explore the form of, and the evolutionary and ecological effects of variation in, secondary defences.

A key point to make at the start of this chapter is that secondary defences are found right across the tree of life from the simplest to the most complex organisms. Secondary defences therefore come in very many forms, including: 1.) chemical defences, that act variously as olfactory deterrents, irritating repellent secretions, and internally stored toxins that are transferred to predators during and after ingestion; 2.) mechanical defences, including changes in body shape to exceed the gape of predators, and the possession of piercing spines and/or tough integuments; and 3.) behavioural defences, such as forms of aggressive retaliation and perhaps social-defensive grouping. In the corresponding chapter to this from our original text (*Avoiding Attack*, Ruxton et al. 2004a) the lion's share of the material reviewed a large range of different defences and evaluated the evidence for costs associated with different defences. Here we change emphasis a

little. We will review selected examples that provide useful illustrations of the ecological and evolutionary characteristics associated with secondary defences. We will again discuss costs of secondary defences, but this time, place more emphasis on the consequences of such costs, especially as they relate to forms of social interaction. We show also that the acquisition of secondary defences may modify niche, life history, and habitat range of prey animals and review a well-known and significant study of predator–prey co-evolution of defensive toxins of prey and resistance to those toxins in predators. We do not limit our discussion to animals, but include a small selection of examples and ideas from the plant and microbe defence literature where we think a broader perspective is helpful. We begin the chapter by considering the evolutionary mechanisms that favour secondary defence evolution.

5.2 Evolution of secondary defence

5.2.1 Self and others: social evolution of defences

At first glance, the evolution of secondary defences by natural selection should be a simple matter in which the benefit of increased survival to individual prey outweighs the costs of acquiring, storing, and deploying the secondary defences. However the evolution of secondary defence is not necessarily so simple. Internally stored toxins may harm predators, but do so too slowly to protect the individual prey that bears them. For example, some toxins may only be liberated once the prey is killed and ingested, and hence provide no protection to the individual prey. Explaining how these apparently self-defeating defences can evolve brings us to one of

Avoiding Attack: The Evolutionary Ecology of Crypsis, Aposematism, and Mimicry. Second Edition. Graeme D. Ruxton, William L. Allen, Thomas N. Sherratt, & Michael P. Speed, Oxford University Press (2018). © Graeme D. Ruxton, William L. Allen, Thomas N. Sherratt, & Michael P. Speed 2018. DOI: 10.1093/oso/9780199688678.001.0001

the interesting aspects of secondary defences—that there may be a social component in their evolution.

In his classic text on natural selection and evolution R.A. Fisher (1958) described an early version of kin selection theory to explain secondary defences that do not protect individuals ('*nauseous flavours as a means of defence*', Fisher 1958). Fisher realized that through social aggregation the effects of attack on an individual prey would later be conferred on neighbouring prey, which gain protection from the subsequent disablement of the predator. If the aggregation were a set of siblings, hatched from a clutch of eggs from a singly-mated female then the 'selective potency' for each sibling would be '*only half as great as if the individual were itself protected. Against this has to be set that it applies to the whole of a possibly numerous brood*' (Fisher, 1958). Hence the alleles that confer '*nauseous flavours*' may not increase the likelihood of survival of the individual that is killed, but if the defence that they generate can protect a sufficient number of nearby siblings, the relevant alleles can be favoured by increases in indirect fitness.

We can easily conceive of different evolutionary routes to kin-selected influences on secondary defences (Ruxton & Sherratt, 2006). In one scenario, chemical defences that are inadequate to protect individuals (Fisher's 'nauseous flavours') may fail to evolve except in species that are socially aggregated. Here kin selection is the driver of defence evolution. In another scenario, chemical defence evolves first to protect solitary individuals, and this in turn causes a transition to social living and aggregation, in which benefits of defence are conferred both directly (protection of individuals) and indirectly (through enhanced survival of relatives). The nature of an individual's defence might then evolve so that it enhances the benefit to relatives, perhaps at the expense of some level of self-protection.

The influential phylogenetic work by Tullberg & Hunter on defence and aggregation (Sillén-Tullberg, 1988; Tullberg & Hunter, 1996), indicates that chemical defence in the macrolepidopterans typically came first, followed by social aggregation and warning coloration that advertises these defences. It is striking that more than 20 years after Tullberg and Hunter's key papers, there are almost no follow-up tests on lepidopterans or other arthropod groups; this is still a fruitful research area awaiting exploitation (see review in Ruxton & Sherratt, 2006). For insect species there is good evidence of associations between aggregation and chemical defence in general (Tullberg & Hunter, 1996), and a lot of experimental evidence that aggregation in prey groups facilitates avoidance by predators (Ruxton & Sherratt, 2006), but we note there is not always support for the prediction that survival increases with group size in chemically defended species (Lindstedt et al., 2006). Furthermore the rules may apply differently to other animal groups. Stankowich et al. (2014), for example, reported that presence of chemical defence in carnivores, manifest as noxious anal secretions, is associated with solitary not social living.

Fisher's theory for kin-selected defence is however supported in a fascinating example of clonal aphid defence. Aphid defences can operate through small tubes called cornicles, which point backward from the abdomen targeted at rear-approaching predators (example in Figure 5.1). When an aphid is attacked, the cornicles release defensive secretions that produce volatile alarm cues causing nearby individuals to cease feeding and perhaps to flee (Montgomery & Nault, 1977). Since individuals live in colonies of clones, this is reasonably interpreted as a costly kin-selected act. The secretion itself is waxy and rapidly hardens; so if it is smeared over an arthropod predator, the secretion will inhibit its movement, possibly clogging up mouthparts, thus

Figure 5.1 An oleander aphid (*Aphis nerii*) deploying defensive secretions from its cornicles, in response to attack from a ladybird larvae (see Plate 3).

preventing further attacks. The colonial grain aphid (*Sitobion avenae*) uses these cornicle secretions against enemies such as the parasitoid wasp (*Aphidius rhopalosiphi*) that lays its eggs inside the aphid. Wu et al. (2010) examined the fitness consequences of cornicle secretion for individuals and their clonal group-mates. In fact the secretion had no benefit for individuals, since aphids that used the defensive secretion were no more likely to survive the current attack, and were in fact less likely to survive a second attack. Smearing of the wasp with 'aphid wax', however, benefitted other group members because the wasp was diverted from attack behaviour by grooming behaviours to clean the wax off. Hence there is a benefit to group members from one individual defending itself, and no apparent direct benefit to the individual. A kin-selection explanation predicts greater investment in defence if the kin-group size increases, and indeed, after statistically controlling for group size, Wu et al. found higher rates of smearing the secretion as the number of clone mates in the group increased.

5.2.2 Costs

It is often assumed that defences are costly in the sense that in the absence of a predatory threat the presence of a defence reduces the prey's fitness. Potential causes of fitness costs are diverse. They may often follow directly from conflicts over resources within an organism, when production of a secondary defence requires re-allocation of energy and raw materials away from other key functions like growth or reproduction. There may also be self-damage (or 'design' costs), in which a defence interferes with other functions; for example, storage of defensive toxins may damage tissues, or defensive (trichome) hairs on plant leaves may reduce photosynthetic efficiency. Some defences may cause indirect ecological costs, if for example they were associated with decreased competitive ability or greater vulnerability to other kinds of enemy. Defences which are typically induced in relation to short-term threats may impose costs of plasticity, unlike so-called 'constitutive defences' that are constantly present. Finally some defences may incur costs of reduced opportunity, for example allocation of resource to defence at an early stage may

cause additional costs later, for example reducing growth rates (see Strauss et al., 2002).

In our earlier text we reviewed evidence for and against costs of secondary defences, often chemical defences. We would again refer the interested reader to M. Deane Bowers' (1992) excellent evaluation of costs of defence in arthropods; this remains a seminal review. Since 2004, we note greater depth to published reviews of costs of secondary defences, reflecting growth of empirical work. In a review of plant chemical defences, Cipollini et al. (2018) argue that the evidence they surveyed broadly favours the expectation that costs of defensive secondary metabolites can be consistently identified and quantified; and where they are not found a cause may be limitations in experimental conditions, rather than genuine absence of costs. Genetic mechanisms of growth–defence trade-offs in plant systems are now being described (Karasov et al., 2017), focussing on the genomic regulation of induced chemical defences. It seems likely that the coming years will provide very detailed information about costs of defensive secondary metabolites in plant systems.

For researchers interested in costs of animal secondary defences, Zvereva & Kozlov (2016) recently described a very informative meta-analysis, examining evidence for costs of chemical defences in insect herbivores. Readers wishing to review relevant experiments are recommended to see the list of source material used there. Robust meta-analysis involves filtering out papers that do not match carefully designed inclusion criteria. Given the remarkably widespread and diverse distribution of chemical defence in herbivorous insects, it is striking that after a very thorough review of the relevant experimental literature, Zvereva & Kozlov were able to include only 33 papers (including 22 species; but because of multiple investigations in each paper, up to 87 measures could be used to evaluate effect sizes). Studies were included if they measured performance variation caused in response to defensive secretion, or that measured correlations with variation in intake of defensive chemicals that were sequestered from the animal's diet. The studies retained covered lepidopterans (10 species), coleopterans (7 species), hymenopterans (3 species), and hemipterans (2 species). Zvereva & Kozlov found that sequestration of increasing amounts of plant allelochemicals was actually associated with increased herbivore performance. This indicates perhaps that specialist herbivores are adapted to metabolizing plant allelochemicals and used them as a nutrient, or alternatively that other nutritional components of the host plants correlate with increasing allelochemical concentration. Zvereva & Kozlov did find small costs of *de novo* synthesis (but sample size is only 3 here). Significant costs of external secretion of chemical defences (regurgitation, reflex bleeding, etc.), were clearly identified (excluding those that synthesize salicylaldehyde, for which glucose is a byproduct of biosynthesis). This is a notable result, as deployment of chemical defences in arthropods, whether biosynthesized or taken second-hand from components of diet, often involves secretion of a noxious substance that is lost to the prey when deployed. Higginson et al. (2011) for example found that defensive regurgitation in caterpillars (of the large white butterfly, *Pieris brassicae*) was costly in multiple ways, including reduced survival (probably through reduced immune function) as well as reduced pupal weight and adult female fecundity.

Zvereva & Kozlov (2016) present a robust and informative meta-analysis, but with a necessarily limited data set. There is in our view still considerable need for empirical evaluation of costs of chemical defences in insects. We would like to see this kind of analysis performed with a wider set of studies and species included before we can, for example, really understand whether allocation costs are usually negligible or nonexistent in sequestering animals, but more substantial in those that biosynthesize their defences. It would be very useful if researchers took note of and complied with the selection criteria used by Zvereva & Kozlov, because this analysis can of course develop as more studies are published. The studies included looked for costs which might be seen when toxin levels within prey vary; they do not count in the costs any constitutive phenotypes that are possessed by all members of the species, for example organs for storage and deployment of defensive toxins etc.

If we take Zvereva & Kozlov's results at face value we might conclude that 1.) *de novo* synthesis of and defensive external secretion of repellent chemicals are likely to be costly, and 2.) sequestration of plant defensive allelochemicals is associated with a lack of net costs and often increased performance in key components of fitness. This could be because herbivore dietary specialization is accompanied by strong selection to reduce costs, and perhaps bring metabolic benefits of plant allelochemicals. Certainly papers reporting positive or negligible effects of sequestration of chemical defence can easily be found (e.g. Cogni et al., 2012). In at least one species of sawfly (*Diprion pini*) that recycles diet-derived resins for its own defence, there appears to be no cost to survival or growth from increases in the resinous nature of the food (Lindstedt et al., 2011). *D. pini* larvae with a diet high in resin acid had a higher pupal mass and bigger defensive secretions than those with a lower resin component in their diets.

A good question, posed by Zvereva & Kozlov themselves, is the extent to which other kinds of cost are missed in experimental work. Hence it is common to look for correlations of defence with life-history parameters that indicate performance, such as body size, but not to look for ecological costs, such as decreased competitive ability. Gentry & Dyer (2002), presented a thorough investigation of vulnerability of caterpillars to enemies including parasitoids; they point out that the presence of chemical defences that protect against predators may be associated with increased attacks by parasitoid wasps. Zvereva & Kozlov's meta-analysis suggests that this kind of ecological cost may be general. Such indirect, ecological costs will not be seen in lab studies that focus on effects of chemical defence on the growth and maturation of insect prey.

We note also that some chemical defences are neither biosynthesized nor taken from the diet, but are instead supplied by microbial symbionts, and these are generally excluded from the kinds of study we refer to above. Symbiont-conferred chemical defences have been recorded in assorted organisms, including grasses that gain chemical defence from endophytic fungi that live between plant cells (Clay & Schardl, 2002), rove beetles that gain chemical defence from endosymbiont bacteria (Kellner, 2002), and numerous potential cases in aquatic invertebrates (Lopanik, 2014; and we recommend the review by Clay, 2014). The use of symbionts in this way may indicate reduced costs of secondary defences, since hosts do not need to invest in genetic or metabolic systems to create the toxins themselves.

But this must be set against any other fitness-influencing properties of these symbionts.

5.3 Consequences of variation in costs of secondary defence

Let's first of all assume that classes of secondary defence, such as dietary sequestration by insect herbivores, involve no net allocation costs; how might we expect an absence of costs to affect ecology and evolution of prey species? Cogni et al. (2012) consider this question as they discussed their findings of weak evidence for costs of pyrrolizidine alkaloid sequestration by a specialist arctiid moth, *Utetheisa ornatrix*. Here the only measurable cost reported was on larval development time. Cogni et al. (2012) point out that a lack of sequestration costs could affect co-evolutionary dynamics between prey and their predators, perhaps biasing co-evolution toward escalation, rather than fluctuating selection (see section 5.5). They also suggest that plant–herbivore relationships will be affected. If a specialist herbivore evolves cost-free resistance to the plant's chemical defences, and if that herbivore is one of the most commonly met, then the value of the plant defence declines, perhaps leading to relaxed selection and lower production levels if the plant's defences are costly (see Agrawal & Fishbein, 2008).

Now let's switch assumptions and ask: if there are allocation costs of secondary defences, what might be the evolutionary and ecological consequences of such costs? We can perhaps expect 'optimal deployment' of secondary defences (see Cipollini et al., 2018). This idea originated in the plant literature where it is known as 'optimal defence theory' (ODT), and predicts that deployment of costly defences are selectively targeted to the most valuable and vulnerable components of the plant. A meta-analysis of empirical tests reported some support for ODT across plant species (McCall & Fordyce, 2010). For example, young leaves which are more synthetically active than older leaves contain greater chemical herbivory deterrents. However, in this study there was no evidence that flowers are more strongly defended than leaves, yet it could be expected that reproductive organs are particularly worth defending from herbivory. Hence the case for optimal

defence theory in plants appears partly supported. There is a related idea in the animal defence literature, that chemically defended individuals store their defensive toxins in appendages that are most likely to be contacted first by a predator's mouthparts. Thus, monarch butterflies store their cardenolide toxins in their cuticles, not their internal body organs (Brower, 1988).

More generally, we can expect that investment in a costly defence is optimized, with varied costs traded off against the benefits of repellence. If the potency of a costly defence increased then (other things being equal) prey should use lower levels of that defence; or if marginal costs of production decreased then (again other things being equal) prey should use higher levels of the defence. In addition, we might expect that prey density is an important determinant of individual levels of investment. Aggregations of prey may, for example, dilute the risk that any individual will be 'picked off' by a predator. The average level of optimal individual toxicity could then be negatively related to local density, although this may need to be set against the potential for preferential targeting of larger groups by predators. Finally, an optimization perspective leads us to expect economical inducibility of costly defences. Many defences will not be 'always on', but rather switched on by stimulation indicative of a predatory threat.

5.3.1 Defence costs might lead to cheating

A second consequence of defence costs could be to incentivize 'defence cheating'. Some individuals may invest less in a costly defence than others. Assuming secondary defences protect individuals that deploy those defences (rather than just their kin, see section 5.2.1), such defence cheats would be more vulnerable if attacked, but would save on the costs of secondary defences, and so might have higher reproductive output. The extent to which defence cheats can invade and replace defended prey depends critically on the response of predators. If predators increase attack rates as cheats increase in frequency, then the mortality risk to cheats increases more quickly than it would for non-cheats because cheats are less likely to survive an attack. There is good evidence that the frequency of attacks can rise with the frequency of cheats, so this critical assumption is well supported, at least for bird–insect systems (Gamberale-Stille & Guilford, 2004; Skelhorn & Rowe, 2007a; Jones et al., 2013). If predators respond in this way, then theoretical models predict that a cheat that invades a population of defended prey will increase to a stable frequency at which the fitness of cheats (no cost of defence, high mortality risk from an attack) and non-cheats (cost of defence, but lower risk from attack) is equal. The equilibrium frequency of cheats is predicted to increase with the deterrent effect and the costs of the defence (Speed et al., 2006; Svennungsen & Holen, 2007; Sherratt et al., 2009).

The concept of defence cheats reveals the broadly social nature of secondary defence. An individual prey repelling a predator can confer a public benefit of protection, reducing risk to other potential victims if the prey kills, injures, disables or otherwise deters the enemy. This benefit may be conferred on close kin (as in Fisher's theory of kin-selected defences, see section 5.2.1) or it may be conferred on any individual prey in the same population, or even in other species. We can expect this 'public good of protection' to pertain whenever (1) predators are to some extent deterred from further attacks after an aversive experience with a prey individual, and (2) there are sufficient numbers of prey in the predators' locality to benefit from its aversion. If secondary defences protect individuals during attack and incur some non-trivial cost then we can expect cheating can be beneficial.

Theoretical models predicting stable cheating only require quite general assumptions leading to a prediction that the phenomenon of 'defence cheats' could be taxonomically widespread. It is certainly easy to find examples in which some members of a population have no investment (or very reduced investment) in chemical defences (see review in Speed et al., 2012). The 'classic' animal example is the monarch butterfly (*Danaus plexippus*) in which the toxicity status of each individual adult depends on the presence or absence of cardenolide secondary metabolites sequestered from the host plant of its larval stage (Brower et al., 1967b). Brower et al. coined the term 'automimicry' to describe this dimorphism in edibility, since the nontoxic

(automimic) individuals are assumed to gain protection from the presence of their toxic (automodel) conspecifics in a similar manner to that seen in Batesian mimicry, in which members of an edible mimic species copy the warning display of a distasteful model species. Dimorphism in the presence/absence of toxicity is not limited to the monarch butterfly; indeed the state of automimicry may be widespread across diverse taxa, life-history stages, and ecological niches. For example, automimicry has been noted in other arthropods (such as millipedes, Eisner et al., 1967; and beetles, Kellner & Dettner, 1996). Defensive responses are a likely source of automimicry in arthropods and other animals. Many insects can, for example, defend themselves by defensive secretion (by regurgitation, reflex bleeding etc.; Whitman et al., 1990; Eisner et al. 2005), and a non-trivial proportion of a prey population may be automimics in the sense that they do not produce any defensive secretion when threatened (Higginson et al., 2011), either because they choose not to respond, or because they do not have defensive fluids available. The 'classic' plant example is the dimorphism in cyanogenesis in white clover (*Trifolium repens*; Olsen et al., 2014) but other plant species show automimetic variation in cyanogenesis (e.g. Goodger et al., 2002). Well-studied microbial examples of automimicry can be found in the literature too (Jousset et al., 2009). What is lacking in most of these examples is a systematic demonstration of all the assumptions necessary for a public good explanation of cheating: that other individuals benefit, that predators increase feeding as non-defended prey increase in frequency, and that defence protects individuals and is costly. This omission is an important point because toxin level dimorphism could have other explanations, such as competition for dietary toxins (see review in Speed et al., 2012).

Understanding how and why defences evolve is not in principle particularly challenging; understanding the stunning diversity of such defences is. Further, we find this variation over ontogeny, between sexes, members of a population, as well as between species. There is variation in patterns of complexity of multiple defences too, and in how such defences interact with other life-history traits (such as aggregation). Thus there is a great deal of fundamental research still to do in this area; to make substantial progress we perhaps need especially to identify some tractable model systems, and develop a theoretical framework of defensive investment that will produce predictions that we can test in these model systems.

5.4 Ecology

5.4.1 Ecology-defence correlations

Given the vast array of alternative secondary defences seen in organisms, an important question is how the precise forms that defences take reflect the diverse nature of the threats that organisms encounter. Recent improvements in the resolution of phylogenies and advances in the sophistication of phylogenetic comparative methods are beginning to allow a systematic understanding of the relationships between secondary defence and ecology. An excellent example of this phylogenetic-ecological approach is the study of carnivore secondary defence by Stankowich et al. (2014), who compiled a rich data set describing mechanisms of defence (e.g. absent, foul scent, or directed noxious spray), behaviour (solitary or aggregated, daily activity period), geographic range, and habitat type for the carnivores. Predation risk was assessed, using similar data sets for diet and geographic range for mammalian and avian predators. Noxious repellent sprays were strongly associated with predominantly nocturnal activity, where the most likely predatory threat was from close-range mammalian predators that are susceptible to pungent olfactants. In contrast, species with predominantly diurnal activity patterns faced strong threats from avian predators, and here noxious chemical defence was more likely absent, but replaced by defensive social grouping. Stankowich et al. argue that birds are less susceptible to repellent olfactants, hence social grouping with vigilance and mobbing is used as an alternative, more effective defence for species active in the daytime. In carnivorous mammals, therefore, chemical defence is not associated with social living, whereas in arthropods such as the macrolepidopterans it often is (Tullberg & Hunter, 1996), a fact supported by empirically parameterized models of aggregation in chemically defended arthropods (Sillén-Tullberg & Leimar, 1988). An important message from these studies is that defence-ecology patterns cannot easily be generalized from one group to another. There are disappointingly few papers on the scale of the Stankowich et al. paper that examine the ecological correlates of chemical defence, though as we discuss in Chapter 6 some work suggesting that warning coloration of chemical defence is associated with diurnal activity and modified social behaviours.

5.4.2 Ecological and evolutionary consequences of secondary defences

What effect does the acquisition of a secondary defence have for the ecology of a species? Does it change its niche or its life-history characteristics? Does it perhaps have consequences for long-term macroevolutionary effects such as speciation and extinction? The idea that chemical defence may be crucial in macroevolutionary processes originated in the plant literature. In discussing the co-evolution of plants and herbivore enemies, Ehrlich & Raven (1964) implicated chemical defence in adaptive radiation in plant groups, particularly the angiosperms, considering it a 'biochemical shield' that facilitated species persistence and diversification. Ehrlich & Raven's hypothesis is sometimes known as 'escape and radiate' and identifies chemical defence as a so called 'key innovation' promoting adaptive radiation. Ehrlich & Raven's hypothesis is often described at second or third hand, so it is worth quoting the key sentences here: '*a [chemically defended] plant, protected from the attacks of phytophagous animals, would in a sense have entered a new adaptive zone. Evolutionary radiation of the plants might follow, and eventually what began as a chance mutation or recombination might characterize an entire family or group of related families.*' (Ehrlich & Raven, 1964, p. 602).

Though Ehrlich & Raven's paper is key to modern co-evolutionary biology, the paper is somewhat vague as to the mechanisms by which effective defences cause adaptive radiation. A follow-up paper by Berenbaum however provided a more systematic approach (Berenbaum, 1983). Even here the exact mechanism by which diversity is promoted by the removal of predatory threat is not clear in a generalized form, rather Berenbaum suggests exaptation of chemicals that help plants cope with dry conditions into anti-herbivore defences. For a recent viewpoint on ecological opportunity and adaptive radiation, we refer interested readers to the review of Stroud & Losos (2016).

In the animals there are relatively few large-scale evaluations of the escape and radiate prediction that chemical defence increases speciation rates. Arbuckle & Speed (2015) tested the prediction for Amphibia using statistics that measure not just net diversification (as in traditional phylogenetic sister group analyses), but also measure speciation and extinction rates. In keeping with an escape and radiate prediction, they found higher speciation rates in lineages with chemical defence than in those without. However, surprisingly, they also found raised rates of extinction in species with chemical defence. This suggests that on the one hand, chemical defence does open up 'adaptive zones' as Erhlich & Raven suggested for plants; but on the other, the new species that form have a relatively fragile existence. The net effect is that chemical defence actually tends to reduce diversification rates in lineages, because estimated increases in extinction rates were two to three orders higher than the increase in speciation rates. However, across tetrapods Harris & Arbuckle (2016) found an increase in net diversification rates associated with poisonous defence when amphibians were excluded, supporting a general prediction of the escape and radiate hypothesis and suggesting that amphibians are divergent in the macroevolutionary patterns of their toxic defences. Blanchard & Moreau (2017) evaluated the effects of alternative anti-predator defences on diversification in ants, using data for 82 per cent of genera. Sister clade analyses indicated raised diversification associated with larger eyes (greater acuity and threat detection), defensive spines, and colony size, but a negative association with presence of a sting and diversification rates. In plant groups, the evolution of structures that support defence mutualisms has been shown to increase diversification rates (Weber & Agrawal, 2014).

If the acquisition of chemical and other defences does affect macroevolutionary patterns, it is important to determine why this might be. Two explanations are that effective defences allow species to diversify their niche, ultimately causing reproductive isolation and speciation; or to change life-history patterns, making some species more or less

resilient to environmental change. To examine whether chemical defence is associated with niche and life history changes, Arbuckle et al. (2013) examined the musteloid mammals (85 species in all) including the families Mustelidae (badgers, otters, weasels, etc.), Mephitidae (skunks), and Procyonidae (raccoons, coatis, olingos etc.). Most pertinent to the present discussion, chemical defence was associated with increases in the most direct measures of niche space: omnivory, diet diversity, and more diverse activity periods, taking in both night and day times rather than nocturnal or diurnal activity only. There was strong statistical support for a model in which chemical defence preceded the evolution of omnivory, and some support for the more general case that chemical defence preceded changes in other traits. Reduction of the threat from enemies may have allowed members of musteloid species to be relieved of the constraint of hiding (Merilaita & Tullberg, 2005). If competition can be relieved by adopting more varied activity periods, this might require feeding on a wider diversity of food items when there is temporal variation in food availability.

A related possibility is that chemical defence enables prey to make better use of new opportunities if life-history traits are modified. Hence in the musteloid example presence of chemical defence was associated not just with behavioural changes but also with reduced female age at maturity, reduced sexual dimorphism, and reduced longevity, suggesting adaptive change in life history as a consequence of (or requiring) a strong repellent defence. The reduction in longevity is perhaps surprising, since if chemical defence lowers extrinsic mortality we may expect that longevity would increase. Consistent with this prediction, in a phylogenetically corrected study, Hossie et al. (2013) found that the possession of chemical defence in amphibia (in a sample of 106 species) was associated with increased longevity. One possible explanation for the reverse finding in musteloids is that chemical defence does not necessarily reduce extrinsic mortality: though it increases the likelihood that a musteloid prey survives an encounter with a predator, the prey may alter its activities so that it is more behaviourally conspicuous, gaining more resources but exposing itself to more encounters with predators. Perhaps heightened encounter rate explains the earlier female maturation and associated reduced longevity.

Expansion of niche can help explain the diversification effects of chemical defence and hence greater rates at which new species form. But why do chemical defences make extinction more likely in amphibians as Arbuckle & Speed (2015) reported? One explanation is that chemical defence encourages amphibian diversification into habitats which typically have low carrying capacities. Since populations of small size are less resilient to environmental challenges, chemical defence may facilitate short-term gains, but greater long-term likelihood of extinction. A possibility that we think is more likely is that life-history changes, toward greater longevity and later maturation and smaller broods (a slow life-history strategy) again makes the chemically defended population less resilient to adapting to rapid change. 'Slow' organisms, which have longer generation times and fewer (but better quality) individuals, present fewer mutations per unit time and hence have slower reactions to novel environmental challenges.

A final point to make in this section, is that phylogenetic methods increasingly allow us to look at the macroevolutionary patterns of different kinds of defence in large animal groups. Thus, Harris & Arbuckle (2016) were able to show that in the tetrapods, biosynthesis of toxins is less dynamic in terms of gains and losses than sequestration of toxins from diet. Furthermore, in the amphibian clade the rate of gain of defensive poisons is higher than the rate of loss, whereas these patterns are reversed in mammals and reptiles.

5.5 Co-evolutionary considerations

Secondary defences in prey may cause evolutionary change in predators; a potent toxin may for example stimulate the evolution of a detoxification mechanism in predators, which could in turn select for more potent toxins in prey. We consider interspecific co-evolution to occur when there is a cycle of repeated evolutionary effects of one party against the other and vice versa, and a signature time-lag so that one species changes in response to, and therefore after, the other. We may expect co-evolutionary relationships to occur when predator and prey species interact sufficiently and the selective effects from their interaction are strong. Here we discuss likely properties of co-evolutionary interaction, some key examples from the literature, and also the limitations of co-evolutionary explanations.

Antagonistic predator–prey co-evolution can have alternative outcomes. There may be: 1.) an initial period of arms race escalation in which (prey) toxicity and (predator) resistance to toxins increase in relative value across generations. An arms race may be followed by 2.) extinction of one party (for example a prey becomes so toxic it kills off its predator) or by 3.) a stationary equilibrium, in which toxicity and resistance cease to evolve further. Finally an initial arms race could be followed by 4.) fluctuating selection dynamics, in which small-scale arms races alternate with periods of de-escalation in trait values (see detailed discussion of so called 'red queen' co-evolution, in Brockhurst et al., 2014). Here there is continual change, but over a long evolutionary time period there is no change in average values of toxicity and resistance traits.

With no costs in toxin and resistance traits, we may expect stationary equilibria after a period of arms races, unless one side causes the extinction of the other. For arms races to convert into periods of de-escalation (and hence allow fluctuating selection) some kind of cost is required, so that reduced investment is favoured once a toxin or resistance trait reaches a high level but provides little survival benefit. As with costs of defence (section 5.2.2), resistance traits may sometimes confer costs.

5.5.1 Evidence for predator–prey co-evolution

Without a time machine, it is difficult to get evidence of co-evolutionary arms races or fluctuating selection dynamics in natural populations. Hence many studies look instead for evidence of trait correlations across large spatial scales as a sign of co-evolution. If there is fluctuating selection operating over a metapopulation framework, different populations will be at different stages of the co-evolutionary cycle. Hence we should see considerable variation in prey toxicity across populations, but a good match of predator resistance and prey toxicity traits within localities.

In the plant defence literature the classic example is the work of Berenbaum, Zangerl, and colleagues on the co-evolutionary relationships between wild parsnip (*Pastinaca sativa*) and its insect pest, the parsnip webworm (*Depressaria pastinacella*; Berenbaum et al., 1986; Berenbaum & Zangerl, 1992, 1998; Zangerl & Berenbaum, 2005). Wild parsnips defend themselves against herbivory with up to five kinds of furanocoumarin toxin. Levels of these toxins are geographically variable and there is often a good match in traits values between the plant and the webworm. Thus webworms often show heightened capacity to detoxify the locally abundant toxin, and reduced capacity to detoxify locally rare toxins. Berenbaum and colleagues suggest that this system shows evidence of antagonistic fluctuating co-evolution operating at different stages across a large spatial mosaic.

The most thoroughly studied example of potential co-evolution involving animal prey secondary defences is between the rough-skinned newt (*Taricha granulosa*) and its garter snake predator (*Thamnophis sirtalis*), extensively examined in the work of ED Brodie III, ED Brodie Jr, CT Hanifin, and colleagues.

The newt defends itself with a potent neurotoxin known as 'TTX' (tetrodotoxin). TTX blocks sodium channels preventing movement of muscles by interfering with action potentials. In response to TTX, the garter snakes have evolved resistance, but this appears to come at a cost to reduced muscle function, so that more resistant snakes move more slowly (Brodie & Brodie, 1999a).

Newt toxicity is variable across geographic ranges. In some island and mainland areas the newt has very low toxicity levels whereas in other areas toxicity ranges up to very high levels (Brodie & Brodie, 1999a). Toxicity levels higher than that which the snake can resist have been shown to increase survival of newts (Williams et al., 2010), showing that there can be individual selection favouring the trait. On its part, the garter snake seems pre-adapted from ancestral forms to dealing with TTX (Motychak et al., 1999), and often has a good match in its resistance to local newt toxicity over a wide range of toxin levels (Hanifin et al., 2008). However, Hanifin et al. (2008) found numerous places where toxin and resistance levels did not match each other. In all of these sites of phenotype

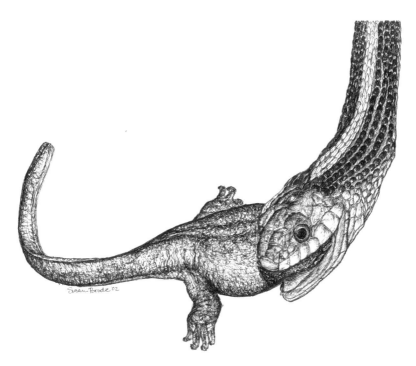

Figure 5.2   A rough-skinned newt (*Taricha granulosa*) is attacked by a garter snake (*Thamnophis sirtalis*).

mismatch, the snake was always resistant to the newt's toxin. As Hanifin et al. point out, this finding runs contrary to our standard expectations of predator–prey co-evolution, that in the 'life–dinner arms race' prey are usually ahead of their predators. Hanifin et al. suggest several explanations for this result, perhaps the most plausible of which is that simple genomic changes in the snake, such as single amino acid substitutions, can lead to high levels of resistance (Feldman et al., 2010), so it can adapt very quickly and escape the evolutionary burden from TTX in prey. This hypothesis may at first glance seem implausible, because TTX resistance in snakes requires modification of several proteins, expressed in different tissues. McGlothlin et al. (2016) have, however, resolved this issue using phylogenetic reconstruction. They found that extreme TTX resistance in snake species has been facilitated by pre-adaptive changes in paralogous genes. The evolutionary history of some snake genomes therefore ensures that some of the necessary genomic changes are already in place, favouring the subsequent, rapid evolution of strong TTX resistance by relatively few further genomic changes.

A complete explanation of these mismatches is difficult, however, because rather less is known about the production of the TTX toxin in the rough-skinned newt. TTX is present as a defensive toxin in a wide diversity of species from bacteria and dinoflagellates to amphibians and fish (Miyazawa & Noguchi, 2001) and its origins in many species are mysterious (Hanifin, 2010). It may frequently be conferred by bacterial symbionts, or perhaps be biosynthesized endogenously. In the rough-skinned newt there is some evidence favouring endogenous production (Lehman et al., 2004; Gall et al., 2012), making the newt's genome the locus of toxin co-evolution.

Co-evolution itself need not be a necessary outcome of predator–prey interactions for at least two reasons. First, the rate of encounter between a predator and a prey species may be so low that the effect of co-evolutionary selection is negligible. Second, phenotypic plasticity in key traits may be adaptively modified without genetical change occurring. Predators may, for example, rapidly learn to avoid certain prey; and they may transmit this information socially, for example from parents to offspring. In some cases predators can neutralize secondary defences behaviourally, as when loggerhead shrikes (*Lanius ludovicianus*) 'exapt' their existing behaviour of impaling prey to hang dead lubber grasshoppers (*Romalea guttata*) on spikes and leave them until the toxin components have degraded (Yosef & Whitman, 1992). On the other side, prey may be able, within their lifetimes, to 'up-regulate' toxicity production mechanisms in the face of heightened threat. Hence greater defence and offence could occur in a correlated, time-lagged fashion, even if no genetic change in predator or prey populations occurs.

5.5.2 Co-evolution—an explanation for defence diversity?

Co-evolutionary considerations may be pertinent to the general questions of why populations are often polymorphic for kinds of toxin; and why individuals often present their enemies with several toxin types.

Rare forms of toxins can frequently be favoured, because predators have little or no recent (co)evolutionary history with them, and hence no resistance. If rarity is always favoured then one consequence can be persistent diversification of toxicity within populations. This form of frequency dependence may help to explain why chemically defended populations often have several forms of toxic compounds present (Speed et al., 2012). Favouring of rare toxins may be co-evolutionary if the predator is constantly evolving alternative resistance and detoxification mechanisms in response to the emergence of new toxin variants. However, again, co-evolution is not necessary. If predators modify their behaviours adaptively by being wary of ingesting prey with rare, novel tastes then toxin diversity will be selectively favoured (Skelhorn & Rowe, 2005).

A second point about toxin diversity is that many defended individuals have more than one toxin. This is widely reported in the chemical ecology literature (Speed et al., 2012). One explanation is that prey need to defend themselves against more than one kind of predator, and hence use alternative defences to do so. Alternatively, dietary variation may lead to variation in chemical defences which protect against different kinds of enemy (Zverev et al., 2017). Where there is only one major kind of enemy, however, co-evolution may provide a promising explanation for diversity in chemical defence. Gilman et al. (2012), for example, modelled the effects of increasing numbers of defense traits on victim–exploiter co-evolution, noting the frequency with which victim species (here the prey) can create defences so effective that they cause the extinction of the (predator) exploiter. The model predicts that evolutionary escape from an enemy becomes more likely as the number of defences increases. Considered in terms of toxin types a key aspect to their model is that as numbers of different toxins within individuals increase, prey only need to 'beat' their predators on one defence trait to repel the predator. In contrast, to gain from an attack a predator must 'beat' all prey defences by having resistance to all toxins. Increasing the number of toxin types therefore increases the chance that a prey has at least one toxin level higher than a predator's corresponding resistance level, so on average prey become increasingly advantaged as the number of different toxins increase (Speed et al., 2015). This is a potentially important co-evolutionary insight, which may explain why prey organisms have multiple secondary defences.

5.6 Unresolved issues and future challenges

The links between chemical defences and other life history and ecological parameters (such as aggregation) and other defences (such as aposematism) were actively and effectively explored by comparative analyses last century, but there have been fewer further investigations of broad cross-species trends since. However, in this chapter and more broadly throughout the book we highlight unresolved questions that may benefit from this approach. The costs of secondary defences is an area that has been actively pursued by cross-species meta-analyses— but these meta-analyses have been limited by lack of available suitable source studies. Hence we would encourage further measurement of aspects of the potential costs of secondary defences: we would particularly encourage those planning such studies to explore how best to design their study to not only be interesting in its own right, but also to contribute to expanding our collective understanding through meta-analyses. The evolution of secondary defences is complex because selection often does not operate through essentially isolated interactions between a predator and prey individuals; there are often also kin-selected effects among prey that potentially share the same predator. However, the situation is even more complex yet, because it is quite common for secondary defensive compounds to be diet-derived and/or to be conferred by symbionts. This means that the interacting network of individuals that impinges on selection for these defences can be complex indeed, and the role of symbionts in particular is deserving of much more attention. In terms of how much organisms should invest in costly secondary defences, we currently have more theory than we have empirical work purpose-designed to test the predictions of that theory—a rebalancing of this would be welcome.

CHAPTER 6

Aposematism

6.1 What is aposematism?

Aposematism is the pairing of two kinds of defensive phenotype: an often repellent secondary defence that typically renders prey unprofitable to predators if they attack them and some evolved warning signal that indicates the presence of that defence. Since secondary defences have been reviewed in the previous chapter we focus here on the functions and evolutionary consequences of aposematic signals.

We could perhaps neatly split these two components of aposematism into two simple categories in which the signal is the primary defence, which is continuously available so that it is always present before attacks begin (following Edmunds, 1974), and the repellent trait that operates as the secondary defence. Life, however, is not so simple because aposematic signals often work to modify the behaviours of predators both before and during attacks. Warning coloration, for example, may increase wariness and hence improve the chances that a chemically defended prey is released unharmed after an attack (Halpin et al., 2008a). An aposematic signal may therefore first tend to reduce the probability that a predator commences attack (a primary defence) and then (as a component of secondary defence) reduce the probability that the prey is injured or killed during any subsequent attack. In this chapter we will therefore consider both the primary and the secondary effects of aposematic signals on prey protection.

Aposematic signals are well known to raise a number of interesting evolutionary issues, including their initial evolution and their integration with sexual and other signals (Rojas et al., 2015). Furthermore, they may have important ecological, co-evolutionary, and macroevolutionary consequences. Before we review each of these properties, however, we begin first by describing the common features of aposematic signals and attempting to show the very wide use to which aposematic signalling is deployed across animals (and perhaps plants too). By way of doing this, we give some indication of the taxonomic distribution of the characteristic.

6.2 Characteristics of aposematic organisms

6.2.1 What characterizes an aposematic organism?

Let's imagine that we have a newly discovered prey species and we want to determine whether it is correct to class it as 'aposematic'. This leads us to the most fundamental question: What criteria would we need to define a prey phenotype as an aposematic signal? The most straightforward answer would be to look for the pairing of conspicuous coloration and a secondary defence, such as a toxin. However, as we explain below, a more nuanced set of diagnostic criteria are needed. To begin, we suggest three minimal criteria, and notice that conspicuousness is itself not one of them.

First, perhaps rather obviously, we need the phenotype in question to be reliably paired with a repellent secondary defence or more generally some trait which can make the prey unprofitable to predators, such as rapid escape or aggressive defence (as in the unusual case of the Norwegian lemming, *Lemmus lemmus*; Andersson, 2015). If the repellent defence is absent, and prey are typically profitable to attack, then the putative signal would be deceitful. Rather than classifying it as 'aposematic', it would be better to consider it as a form of dishonest

Avoiding Attack: The Evolutionary Ecology of Crypsis, Aposematism, and Mimicry. Second Edition. Graeme D. Ruxton, William L. Allen, Thomas N. Sherratt, & Michael P. Speed, Oxford University Press (2018). © Graeme D. Ruxton, William L. Allen, Thomas N. Sherratt, & Michael P. Speed 2018. DOI: 10.1093/oso/9780199688678.001.0001

signal mimicry: either Batesian mimicry of another unprofitable species (Chapter 9) or so called 'auto-mimicry' in which some members of a population of the same species lack repellent defences (Chapter 5). Interestingly, some aposematic species may have target-specific chemical defences, whereby separate defensive compounds are targeted to different predator types (Rojas et al., 2017).

The second criterion is that the phenotype must have evolved as a signal—i.e. been modified by natural selection so that it alters the behaviour of predators, in this case to increase prey survival. It may be helpful to attempt ancestral state reconstructions in phylogenetic trees to ensure that traits that we suppose to be aposematic signals did not precede the evolution of the quality being signalled, here the prey's unprofitability such as a chemical defence. Phenotypes that cannot be demonstrated to have evolved in order to signal are better considered as cues, perhaps informative cues, but not signals per se.

Our third criterion is that the signal must primarily act as a deterrent, causing predators to fail to initiate attack behaviours, where they otherwise would. Hence in our view aposematic signals must necessarily operate as primary defences (as in Edmunds, 1974), even if they sometimes also function as secondary defences, for example reducing the severity of an attack once it has begun (Halpin et al., 2008a). A signal that is only deployed to increase survival once attack has begun, in effect only operating as a secondary defence, such as a flash of colour or a warning sound, could be considered as a combination of startle (diematic) and aposematic display (Umbers & Mappes, 2015). We discuss this in Chapter 10.

We now discuss further diagnostic characteristics that may indicate aposematic signalling.

6.2.2 Aposematic signals are (usually) primarily visual

If we were looking for evidence of aposematism in our newly discovered species, we would likely focus on its visual appearance first, especially if the prey is diurnally active. Although aposematic signals can target any sensory modality in a predator, the fundamentally important one is usually vision.

Indeed the term 'warning coloration' is sometimes used as synonymous with aposematic signalling.

A crucial point is that visual aposematic signals in animals are often constitutive, meaning that they are 'always on' (and hence function as a primary defence, in the sense of Edmunds, 1974). Visual aposematic signals are usually produced by long-lasting pigments in animals, deposited in integuments, scales, or feathers and cannot therefore be switched on and off to match the presence or absence of threats (Cott, 1940). Once an animal has invested in aposematic colour patterning, the phenotype will generally persist. Animals that can change colour rapidly, such as some transiently bioluminescent animals (De Cock & Matthysen, 2003), are of course exceptional to this generalization.

In contrast, smells and sounds are often induced responses to immediate threats. In his classic text Cott (1940, p. 229) made this point succinctly when he wrote that warning odours are 'Often… only released as a last desperate resort, in the face of the aggressor'. Defensive olfactory responses are easy to identify: the famous odours of stinkbugs (Pentatomidae) being perhaps the best-known examples, but ladybirds release well-studied pyrazine odours when disturbed (Marples et al., 1994), as do Neotropical fireflies *Photuris trivittata* and *Bicellonycha amoena* (Vencl et al., 2016) amongst the many examples. Similarly, examples of auditory warning signals are usually only produced when the animal is disturbed, including the 'clicks' of silkmoth caterpillars (Brown et al., 2007) and the famous rattle of the rattlesnake (Rowe & Owings, 1978). Each of these examples could operate as an aposematic signal of unprofitability and/or to induce a startle response and be aversive in its own right (see Chapter 10).

In support of the primacy of vision in aposematism, Rowe & Halpin (2013) reviewed and classified defences in many insect species; in so doing they assumed that colour patterning is present in aposematism as a diagnostic tool, and then found that the distribution of sounds, smells, and tastes is more sporadic.

There are several factors that render the visual modality fundamental in aposematism. First, from the prey's perspective the generation of aposematic sounds may be energetically demanding and similarly generation of odours will use resources including energy. If sounds and smells were to function as primary aposematic signal defences they would need to be 'always on' and could generate a large cumulative resource cost over the lifetime of the individual. Once the energetic and resource costs of pigments have been paid, however, further investment of resources is not necessary (or is not necessary until a new life stage is met, such as metamorphosis in lepidopterans). Many prey species will have to deploy some kind of coloration anyway, for example being cryptic if they are not aposematic. Consequently there might be few additional resource costs to prey that change their coloration between crypsis and aposematism (though there may be trade-offs with thermal regulation; Hegna et al., 2013). In contrast, prey species do not usually make sounds and odours constitutively, hence as aposematic signals they might more clearly represent additional costs.

Second, whereas colour patterns can often be clearly directed at predators perhaps over long distances, olfactory and some auditory signals rapidly degrade or dissipate, especially in complex habitats. This may especially be the case when the prey animal is small, and the costs, for example, of producing odours received over a wide radius are likely high. Consequently odours, and perhaps also sounds and odours, may not be as good as colour patterns at turning predators away well before they get close enough to strike on the prey.

A third point favouring visual aposematic signals is that many aposematic species are diurnal (Merilaita & Tullberg, 2005), a fact that we examine in the section on ecology (section 6.5). They are commonly at risk from visually hunting predators that use daytime illumination to locate prey, and hence they are likely to communicate with them via visual cues. Well-known examples of primarily auditory warning signals are found usually in nocturnal environments, such as many tiger moths (Erebidae: Arctiinae) which signal to bats at night (Dowdy & Conner, 2016), but use visual aposematic signals in the daytime (see comparative analyses in Ratcliffe & Nydam, 2008).

Though consideration of aposematic signals focusses frequently on colour patterning, behaviour itself may be an important component of the visual signals used. Some authors, for example, suggest that chemical defence in insects comes with a syndrome of traits including aposematic coloration and slow, sluggish behaviours (Edmunds, 1974; Hatle & Faragher, 1988; Sherratt et al., 2004a). In support, Srygley (2004) showed that some aposematic lepidopterans fly more slowly and at greater cost than edible non-aposematic and mimetic species. Hence it is important to include modifications to behaviour when evaluating how visual cues are modified to function as aposematic signals (Rojas et al., 2014a).

Though we think it uncommon in animals, the possibility that olfaction is used as a sole, or primary, aposematic signal has been discussed repeatedly in the literature (see discussions of 'olfactory aposematism' by several authors, e.g. Eisner & Grant, 1981; Camazine, 1985; Weldon, 2013). Olfactory cues as primary aposematic traits may be used by fungi and some plants as constitutive aposematic signals (Sherratt et al., 2005). An interesting point is that some olfactory volatiles may be inherently repellent in themselves, and thus be punishing to potential predators, as well as warning of other defences that can cause unprofitability (Rothschild, 1961). Finally here, it is worth mentioning that the volatile odours such as pyrazines (Marples et al., 1994; Rowe & Guilford,1996; Lindström et al., 2001b; Boevé et al., 2009) and the bitter tastes of a prey's chemical defences may affect a predator's response to aposematic colour patterns between capture and ingestion. Smells and tastes may make the predator even more wary of the brightly coloured prey it has picked up, and hence increase the likelihood that the prey is released before it is killed (Skelhorn et al., 2008; Rowland et al., 2013; Halpin & Rowe, 2017).

6.2.3 Visual aposematic signals are often conspicuous, sometimes predictably so

Visual aposematic signals are (very) often conspicuous, in the sense that they tend to attract attention from receivers, including potential predators. Visual aposematic signals often incorporate colours that contrast against the prey's background, and they often include two or more contrasting colours over the organism's body (Poulton, 1890; Cott, 1940; Edmunds, 1974; Stevens & Ruxton, 2012; Aronsson & Gamberale-Stille, 2013). Aposematic signals may have evolved to be effective against a wide range of natural backgrounds, and this may in part explain the success of this strategy (Arenas & Stevens, 2017). Given that aposematism has evolved independently many times, repeated forms of similar pigmentation to achieve visual conspicuousness must be considered a convergently evolved quality in aposematism (Arenas et al., 2015). In many cases it is likely that different species have evolved bright coloration independently in order to stimulate the visual perception of similar predators in similar ways. Hence for a given habitat there may be a limited range of colour patterns which are likely to generate the kinds of conspicuousness that function well as aposematic signals. We often see red, yellow, white, and black pigmentation used, thereby making the form of visual aposematic signals relatively predictable. A study of ladybird coloration 'seen' through the eyes of bird predators bears this out; long-wave red colours are visually conspicuous against green foliage that form prey backgrounds, and they are perceived in a more consistent manner across a range of lighting conditions caused by variation in weather and time of day (Arenas et al., 2014). In a recent study by Preißler & Pröhl (2017), background coloration and dark spot size were both seen to influence predation rate on clay models of the strawberry poison frog (*Oophaga pumilio*); predators avoided the non-local aposematic colour morph models, and the larger a model's spots the lower the attack rate. Though research has focussed on colour patterns, there is growing interest in iridescent and shiny reflections as aposematic signals (Fabricant et al., 2014; Pegram & Rutowski, 2016; Waldron et al. 2017).

It might be tempting to attribute convergence in aposematic patterns wholly to adaptive evolution to match similarities in the receiver characteristics of predators. There are at least two reasons, however, that this is probably too simplistic. A good case study here is the transverse banded patterning seen in many insects, in which typically black alternates with yellow, white, or orange coloration. Examples can be found in caterpillars (e.g. of cinnabar moths, *Tyria jacobaeae,* and of monarchs, *Danaus plexippus*), and in the abdomens of some adult hymenoptera such as bees and wasps. One adaptive reason for this patterning may be to produce high-contrast boundaries within the animal's form, thereby opti-

mally heightening its conspicuousness to predators, hence high-contrast banding patterns could be considered to be adaptively convergent.

However, we can easily propose alternative explanations. One is that the animals' banding patterns follow anatomical segmentation patterns, and hence the form of the actual pattern is determined by underlying anatomy. A second point is that there can be thermal effects from altering colour pattern (Williams, 2007), hence aposematic prey animals could favour high-contrast (light and dark) components within their colour patterns, because this balances conflicting demands from thermal and signalling functions (Hegna et al., 2013).

6.2.4 But aposematic conspicuousness does not always exclude crypsis

Scientists like to classify the phenomena that they encounter, and so we often talk and write about aposematism as if it is a trait that excludes cryptic coloration, and vice versa. However, a growing number of authors speculate, theorize, and occasionally demonstrate that the appearances of many prey can include both components of camouflage and of warning displays. Rothschild (1975), for example, proposed that the yellow and black banding on cinnabar caterpillars would blend at a distance, and provide protective cryptic coloration, but would be distinctive and conspicuous from close up (see also the discussion in Edmunds, 1974). Given that conspicuousness is potentially costly to prey because it attracts the attention of enemies, we might perhaps expect that selection would commonly favour prey forms that include both aposematic and cryptic components in their coloration. Endler & Mappes (2004) proposed that since the acceptability of prey varies between predator species, so the evolved level of conspicuousness may vary with variation in predator community too (and see Mappes et al., 2005). A putative example is the hibiscus harlequin bug (*Tectocoris diophthalmus*), which is both repellant and presumably aposematic to some bird species, but is actually edible to mantid predators (Fabricant & Smith, 2014). The heightened protection which arises from conspicuousness may also be seen as a benefit which can be directly traded-off against the loss of crypsis. It is easy to conceive that the level of aposematic conspicuous-ness is then an optimizable trait in which only in a limited range of risk levels is the best strategy to be maximally conspicuous (Leimar et al., 1986; Ruxton et al., 2009). Varied predator mixes and simple trade-off mechanisms may then provide some general and fundamental reasons for the combination of crypsis and conspicuousness in aposematic colour patterns.

What these two general hypotheses do not say, of course, is how crypsis and conspicuousness can be packaged into the same prey appearance, though Rothschild's idea—blending crypsis at a distance, conspicuousness close up—is a compelling mech-anism. Barnett, Cuthill, and Scott-Samuel have taken a psychophysical approach to testing the Rothschild mechanism, noting that the retina is sensitive to different scales of pattern variation at different detection distances. Narrow bands of con-trasting colours, that they termed 'high-frequency' aposematic patterns, may be highly visible up close, but become increasingly undetectable as distance increases. In one experiment Barnett et al. (2017) created artificial pastry caterpillars showing a range of colour patterns including uniform coloration (green, cryptic; yellow or black, aposematic) and banded aposematic yellow–black patterns that var-ied in the width of the banding from high to low frequency (see Figure 6.1). The pastry caterpillars were pinned to horizontal bramble stems, passerine birds were allowed to feed on them, and prey sur-vival was recorded at census points up to 96 hours. Caterpillars with the high-frequency banding pattern had higher relative survival than the other prey types (Figure 6.1).

To investigate the mechanisms that may facilitate blending-at a-distance further, Barnett & Cuthill (2014) designed an artificial 'dual purpose' prey pat-tern, combining high-frequency striping, consistent with warning coloration, with background-matching coloration consistent with cryptic coloration. When presented pinned to trees in a woodland, human observers detected the dual-purpose pattern at similar distances to patterns with purely cryptic coloration (which lacks striping, 5m), which was less than half the distance for aposematic (high-frequency only) patterning (circa 11m). When pre-sented to wild-foraging bird predators in the woodland, the 'dual-purpose' prey survived longer

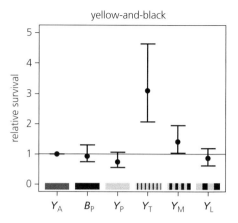

Figure 6.1 Relative survival of artificial prey with different colour patterns, from Barnett et al. (2017). From left to right: Three plain colorations, green, yellow, and black; and three striped patterns, high to low frequency of banding. Relative survival is significantly enhanced by the high-frequency banding pattern (see Plate 4).

than those with just the warning coloration or the cryptic components present. Hence, for these dual-purpose colour patterns, conspicuousness was not appreciably increased over that shown for the cryptic prey, but the presence of the aposematic component increased protection (see extensions to these experi-ments in Barnett et al., 2016a,b).

If many aposematic patterns are in fact of this 'dual-purpose' variety, and if the spatial frequency of patterning is important, as Barnett & Cuthill (2014) proposed, then we may see a common select-ive force that would make aposematic patterning converge on relatively predictable forms. This may help explain the long-made observation that apo-sematic patterns tend to have formal similarities in colour and pattern (e.g. Poulton, 1890; Cott, 1940).

In addition to the work of Barnett and colleagues, there is now a growing set of experiments that test, and generally do demonstrate, that prey patterns can combine crypsis and aposematic conspicuous-ness. With wild-caught predators (great tits, *Parus major*) Honma et al. (2015), for example, show that (presumed) aposematic wing patterns of wood tiger moths (*Arctia plantaginis*; Figure 6.2) may have com-ponents of disruptive coloration in them, making initial detection more difficult. Other researchers use computer games with humans acting as predators, seeking to find images of prey against natural back-grounds. These provide some support for the idea

Figure 6.2 Photo of a wood tiger moth *Arctia plantaginis*, showing warning coloration on the exposed hind wings. The patterning is simple and distinctive. From Bibiana Rojas (see Plate 5).

that natural prey coloration may be cryptic from a distance, and more conspicuous from close up (e.g. Bohlin et al., 2008). Caro et al. (2013) also used human observers in the field to measure detection distances of mounts of (aposematic) striped skunks (*Mephitis mephitis*), (cryptic) bobcats (*Lynx rufus*), and potentially intermediate patterns of spotted skunks (*Spilogale gracilis*, which are chemically defended). They concluded that spotted skunks are of intermediate conspicuousness: less conspicuous than striped skunks, but in some conditions more conspicuous than the bobtails. The human observers did not, however, show effects of viewing distance (between 40m and 140m) on detection rates, but objective colour measurements of animals and their habitats indicated that the striped and spotted skunks' colour patterns do tend to contrast with their immediate backgrounds.

6.2.5 Distinctiveness and simplicity of visual traits may be primary requirements of aposematic colour patterns

An important diagnostic criterion of aposematic signalling may be distinctiveness. A.R. Wallace himself hit on this point when he originated the idea of aposematism. In his text 'Darwinism', for example, he wrote '*Owing to its remarkable power of offence the skunk is rarely attacked by other animals, and its black and white fur, and the bushy white tail carried*

*erect when disturbed, form the danger signals **by which it is easily distinguished in the twilight or moonlight from unprotected animals**'* (Wallace, 1889, p. 233, our emphasis). Prey that are typically unprofitable may benefit by looking and behaving different to those that are edible and profitable. As we have seen in earlier chapters (Chapters 1–4), edible prey are very often cryptic—they have a camouflaged coloration, carry out 'hiding' behaviours, and when disturbed exhibit rapid escape. Aposematic prey may then achieve distinctiveness from edible prey by being visually conspicuous, by not hiding, and by exhibiting relatively slow movements (Sherratt et al., 2004a). However, as Wüster et al., (2004) and others (e.g. Valkonen et al., 2011) showed in their experiments on snake coloration, distinctiveness may be achievable while maintaining crypsis and without imposing strong costs of conspicuousness. Hence, if we are seeking further evidence to determine whether a prey's display is aposematic, then we may look for measurements that show that it is distinctive from the edible species in the habitat (see Merilaita & Ruxton, 2007). A theoretical and empirical investigation by Polnaszek et al. (2017) suggests that increased distinctiveness confers increased predator avoidance to aposematic prey, with the extent of this benefit depending on the prey's relative frequency. Notably they identify conditions in which distinctiveness is not itself beneficial for aposematic individuals.

Cott (1940) added a supplementary point. In addition to distinctiveness, aposematically coloured prey may benefit from simple patterning because this is likely to be reliably recognized, perhaps especially so in poor lighting conditions (Rojas et al., 2014b). Cott's point may explain why naive bird predators are more wary of aposematic prey patterns with large, symmetrical components than with small, asymmetrical components (Forsman & Merilaita, 1999).

6.2.6 Aposematism is primarily a phenomenon in animals, but it may be present in other groups

We focus in this text on animal defences, but aposematism may apply in other groups. Aposematism has thus been variously proposed in plants by Lev-Yadun (2001, 2009; and see Schaefer & Ruxton, 2011; Kavanagh et al., 2016) where, for example, conspicuous coloration on the tips of defensive spines might serve to draw the attention of predators and deter them from attacking. It has been suggested also that aposematism exists in some fungi (Sherratt et al., 2005) and even in the defence of colonies of infectious microbes which undergo olfactory and colour changes, including glowing, during infection (Jones et al., 2015, 2017). An important, but unexplored, question is whether aposematism is intrinsically more effective in animals than, for example, in plants. A key difference may well be that plants may often survive attacks by herbivores, because roots and other tissues are unharmed and tissue regeneration is possible, or because insect herbivores are smaller than the plants that they feed on. Furthermore, plants would lose photosynthetic surface if they replaced chlorophyll with conspicuous pigmentation that signalled aposematically. In contrast, animal prey will be killed by successful attacks by predators. Hence, deterrence of enemies may be more important for animals than it is for plants. Should conspicuousness be an expensive trait, either because of the proximate costs of pigmentation or the heightened costs of conspicuousness, then it may turn up more often in the animals, where deterrence is more valuable.

6.2.7 Aposematism is probably a rare phenomenon in animals

We started this section asking how one could evaluate whether a newly discovered species is in fact aposematic. This brings us to a final characteristic of aposematism, which is that it is probably a comparatively rare defensive strategy. Mappes et al. (2014), for example, reported results of a systematic classification of final instar caterpillars in Finland, finding that about 85 per cent were apparently camouflaged, without signs of aposematic coloration. Only 5 per cent of all species surveyed showed coloration that could be classed as 'strong aposematism' (with the remaining 10 per cent showing signs of aposematic coloration overlaid on camouflage). Similarly, Arbuckle & Speed (2015) estimate that 88 per cent from a sample of 2638 amphibian species are unambiguously cryptic, with 5 per cent showing 'strong aposematism'. It may be the case that colourful aposematic signalling is more common in some locations and habitats (such as tropical forests) than in others (such as temperate habitats, see Adams et al., 2014).

If aposematic coloration is generally rare, however, there can be several explanations, including that its initial evolution is unlikely, or it is evolutionarily unstable and often replaced with crypsis, or that aposematic species may themselves be prone to extinction. We turn next to the initial evolution of aposematism, which it turns out might after all be rather improbable. Subsequently we examine the stability and maintenance of aposematic coloration.

6.3 Evolution of aposematism

6.3.1 Initial evolution of aposematic signals

Aposematism may be relatively rare, and one reason is because its initial stages are inhibited by evolutionary barriers. Even if an aposematic signal is highly advantageous when established and common, it may be maladaptive when new and rare. In particular, the initial evolution of aposematic signalling poses a problem of what we can call 'number-dependent disadvantage'. Here, new, rare forms of aposematic signal are unfamiliar and therefore not associated with unprofitability by predators; organisms bearing this signal are therefore likely to be less well protected than more familiar ancestral forms (Mallet & Barton, 1989; Lindström et al., 2001a; Chouteau et al., 2016). New aposematic forms of unprofitable species are also likely to be more conspicuous than the (likely) more cryptic ancestral form from which they deviate. A new mutant must therefore overcome the joint problems of unfamiliarity to enemies and of heightened apparency. Consequently, and unlike the evolution of cryptic coloration, the initial evolution of aposematic signalling has often been seen as challenging to explain (see Mallet & Singer, 1987). Considerable thought and empirical effort has therefore been put into explaining the initial evolution of aposematic signals, with a strong emphasis being placed on warning coloration.

We present here a brief survey of the now numerous attempts to explain how aposematic signals can originate. Before we do this it is worth mentioning a few general points. First, it is important to clarify the evolutionary (and co-evolutionary) context being used for the initial evolution of aposematic signals. Are we, for example, seeking to explain the very first aposematic signals on the tree of life? In this case we can assume no prior contribution from co-evolution between predators and aposematic prey, and hence no special adaptive preparation by predators to bias responses away from attacks on aposematic prey forms. Alternatively, we may be explaining one of the many later evolutionary events that lead to the evolution of aposematism, in which case it may be sensible to assume the influence of (co)evolved receiver biases. Second, a number of evolutionary treatments here directly address the question *why are aposematic signals so often visually conspicuous?* And so, they provide alternative explanations for the prevalence of conspicuousness in aposematic signalling; this makes them particularly interesting in the context of this chapter. A third point is that many of the explorations of the initial evolution of aposematism assume that the secondary defence is an internally stored toxin. As we shall see, this may be one of the harder classes of aposematic signalling to explain.

1) Absence of predators, relaxed selection, and drift

Perhaps the simplest explanation for the evolution of bright warning coloration is that for some period there was an absence of visually oriented predators, during which bright coloration in prey could evolve by random drift, or in relation to other selection pressures such as mate choice. Subsequently predators arrived in the habitat and had numerous experiences which enabled them to associate bright coloration with toxicity. Perhaps conspicuousness was initially just a cue, and it later evolved into a signal refined, for example, to be more easily learnt and retained. Something like this idea was (to our knowledge) made originally by Beddard (1892) in his considerations of lepidopteran coloration. He wrote, '*In earlier times, when there were no birds—and after all, the chief enemies of caterpillars are birds—brilliant coloration, due to abundant and varied pigment, would be the rule...the advent of bird-life proved a disastrous event for these animals, and compelled them to undergo various modifications, except in the case of those forms which combine brilliant coloration with uneatableness.*' Where most modern accounts of aposematism have crypsis as the ancestral starting point, Beddard, by comparison, assumes a world of bright coloration in which the advent of visually oriented predators causes the evolution of crypsis, away from conspicuous pigmentation, in edible species. In effect, conspicuous, edible prey will be 'filtered off' because they attract attention and cannot survive attacks (a point independently developed in detail by Sherratt, 2002a, and described in point 3 below). It might be possible to begin to test Beddard's hypothesis, if sufficient fossil data on insect coloration could be obtained (e.g. McNamara et al., 2012).

A variant hypothesis could be proposed by assuming that in some locations the chance loss, or reduction in the number, of visually hunting predators for a period causes relaxed selection which allows random drift of alleles for colour patterns (Mallet & Singer, 1987). Some of these colour patterns lead to conspicuous coloration in prey that are also defended. If at the return of the relevant predator species bright coloration were sufficiently common in chemically defended species, then it may be rapidly associated with unpalatability and become an established warning cue.

What both of these hypotheses lack, however, is a compelling explanation for the maintenance of conspicuousness once it has evolved. Even if bright coloration becomes abundantly associated with chemical defence by chance there is nothing obvious to stop the prey coloration reverting to the (less costly) cryptic form over time, perhaps by gradual change. Hence, some other factors that favour conspicuousness per se will be required. One factor discussed earlier is distinctiveness (see section 6.2.5), in which defended prey must maintain some difference in appearance to cryptic, edible prey, and 'relaxed crypsis', or greater conspicuousness is one example (see Merilaita & Ruxton, 2007). We now describe some others.

2) Exploitation of receiver biases

A number of authors have considered that aposematic signals may evolve using conspicuousness because by so doing they exploit biases in perceptual and cognitive properties of predators that enhance protection. Predators may, for example, be especially wary of novel (hence distinctive) and conspicuous traits, and tend to avoid prey that use them as warning signals. Such aversion to novelty (neophobia and dietary conservatism) may be sufficient to protect individual prey and hence enable conspicuous morphs to increase in frequency (Halpin et al., 2008b). This view has been advocated by Marples, Kelly, Thomas, and colleagues, supported by experiments with wild-caught birds as predators and evolving artificial prey (e.g. Thomas et al., 2003; Marples et al., 2005; Marples & Mappes, 2011). Here, bright coloration in populations of artificial prey can 'evolve' across generations, more often than would be expected by chance.

Novelty effects might, however, be short-lived—once a prey form begins to evolve it lacks protection from its novelty. Mallet & Singer (1987), for example, argued that the influence of wariness of new prey coloration would be so brief that its influence would be trivial (*The problem with novelty is that it soon wears off.*', p. 342). However, Lee et al. (2010) showed that in simulation models, novel prey could persist through a dynamic equilibrium, increasing from rarity with benefits from novelty, then decreasing when novelty wears off, but not necessarily to extinction. In this way novel prey may cycle in frequency, maintaining a presence across many generations (and see also Puurtinen & Kaitala, 2006).

In addition to wariness of novel prey items, a number of authors have focussed on the fact that conspicuousness seems to be 'salient', in the sense that it speeds up learning about the aversiveness of prey and perhaps increases the duration of learnt avoidance by making memories resistant to forgetting. These assumptions have been included in a number of theoretical models usually of individual selection, often showing that there are conditions in which receiver biases on learning and memory can promote the evolution of new aposematic forms (e.g. Leimar et al., 1986; Servedio, 2000; Speed, 2001; Merilaita & Kaitala, 2002; Puurtinen & Kaitala, 2006). Guilford (1994) argued that a key factor in the maintenance of aposematic signalling is that it provokes careful handling—he called this 'go slow' behaviour—because this causes predators to take longer during an attack, giving them an opportunity to determine whether a prey is in fact edible (and hence worth eating) or chemically defended (and hence to be rejected; a detailed theoretical treatment is found in Holen & Svennungsen, 2012). 'Go slow' responses of predators to aposematic prey may give the prey a strong individual advantage over chemically defended, but cryptic, conspecifics.

Learning and memory biases may be present even in the absence of co-evolution between prey appearances and prey phenotypes. For example, learning tends to accelerate when events are unexpected or unusual, hence conspicuousness and novelty per se may be inherently 'salient' qualities, in the sense that they tend to increase attention and hence learning. Furthermore there may be predictable physiological reasons that some colour patterns are salient—for example, visual systems often use 'opposed channels' so that cones fire to record e.g. either red or green coloration but never a mixture of both. Consequently, red coloration may be inherently conspicuous in insect prey that rest on green vegetation (Stevens, 2013; Stevens & Ruxton, 2012), and hence inherently salient.

Overall, then, a case can be made that new aposematic forms may increase in frequency because they exploit receiver characteristics, wariness of novelty, and increased learning and memorability by foragers. Conspicuousness as a typically aposematic cue can then be explained by the pre-existing sensory and cognitive biases of predators.

3) Co-evolution of aposematic signals and receiver biases

An alternative (or additional view) is that aposematic signals co-evolve with receiver biases in a self-reinforcing manner. Let's imagine the world as something like Beddard described it: colour patterns of prey evolve in many ways initially without conspicuousness because bird predators have not evolved yet. This leads to a very variable relationship between coloration, edibility, and toxicity across prey species. Subsequently, bird predators do evolve and they exert strong selective pressures. Prey that are highly conspicuous to the birds suffer high attack rates, and those that are edible will likely be killed and eaten. Those that are conspicuous and chemically defended, however, may sometimes survive, because their toxins cause them to be rejected by predators before the prey is killed and ingested.

Hence, as selection proceeds across generations, the predators 'filter off' prey species that are edible and conspicuous, but are more likely to let those that are conspicuous and chemically defended survive. Over generations, conspicuousness and chemical defence become associated in prey populations. This scenario, albeit without Beddard's (1892) speculations about life before birds, has been examined in an analytical model by Sherratt (2002a). Sherratt showed that as conspicuousness becomes associated with defence in prey, so the optimal decisions of predators would increasingly bend toward avoidance and cautious handling of brightly coloured prey. The system becomes then self-reinforcing by co-evolution. Conspicuous coloration in prey then signals that prey have likely survived attacks and are worth avoiding. As conspicuous edible prey are 'filtered off', so predators attack conspicuous prey less often and with less severity, and the spread of aposematic signalling can increase. Sherratt's approach manages not only to explain the evolution of conspicuous signalling in defended prey, but rather importantly it also explains why conspicuousness is so prevalent in aposematism (only defended prey can afford it) and why predators show heightened wariness to aposematic species (they have co-evolved to avoid them).

4) Aggregation and family grouping are causal in the initial evolution of aposematism

Most theoretical and empirical treatments of the initial evolution of aposematism focus on individual selection, evaluating how well rare aposematic individuals can survive compared to more cryptic conspecifics. It is relatively possible, however, to argue that social grouping can help to explain the initial evolution of aposematic coloration. Imagine a family group of chemically defended caterpillars feeding on one host plant. Though all are full siblings, suppose that half have inherited and manifest conspicuous coloration from a parent, the other half are cryptic. When a naive bird attacks and kills one of the group it has an unpleasant experience and will then tend to generalize the aversive experience to the individuals that have the same colour pattern as the victim. If the victim were cryptic, then all the cryptic siblings gain a benefit of a reduced probability of attack, and equivalently if the victim were

conspicuous. The allele for colour pattern in the victim is not directly reproduced, but it is replicated indirectly via the siblings whose survival is increased. Up to this point we can see kin-selected altruism for either colour pattern. However, if the conspicuous colour pattern is more salient (see 1 above) and if the victim is conspicuous, then the predator will avoid the remaining conspicuous siblings more strongly. Across family groups aposematic coloration could increase via kin-selection, raising the frequency of the aposematic morph in a locality (see Harvey et al., 1982; Leimar et al., 1986; Mallet & Singer, 1987). Note though that Guilford (1985) has argued strongly that kin selection should not be invoked in this context. Guilford argues that it is the aggregation of phenotypically similar animals, not relatedness per se, that favours the evolution of aposematism in aggregations (see discussion in Turner et al., 1984).

Empirical researchers have focussed on aggregation and aposematism because in experiments bird predators are often very wary and show low attack rates on aggregated prey. Notably, Tullberg and Gamberale-Stille show strong individual benefits of aggregation in heteropteran and artificial prey from predation by chicks (*Gallus gallus*) in a number of experiments (e.g. Sillén-Tullberg, 1990; Gamberale & Tullberg, 1998). They suggest that aggregations of conspicuous prey present more effective warning signals which, for example, heighten wariness and increase protection. Some empirically focussed researchers do support the idea that aggregation may be causal. Mappes and colleagues (Alatalo & Mappes, 1996; Riipi et al., 2001), for example, used a 'novel world' of artificial prey phenotypes that sought to remove innate, (co)evolved biases. They showed strong benefits of aggregation in part because the conspicuousness costs of groups increase only slowly as the group becomes large, whereas the benefits of rapid predator avoidance are substantial.

5) Gradual evolution of aposematic coloration

Most considerations of the initial evolution of aposematism assume that the aposematic mutant is much more conspicuous than the cryptic conspecifics against which it competes. However, some suggest that aposematism can evolve by small, gradual increases in conspicuousness. Yachi & Higashi (1998), for example, invoked 'peak-shift' to model the gradual evolution of aposematic signals. In peak-shift, receivers learn to discriminate signals, and then respond more strongly to exaggerations of these signals perhaps to avoid misidentifying them. In the Yachi & Higashi model, heightened avoidance of exaggerated warning signals can offset the costs of conspicuousness and lead to gradual increase in this trait. Experimental evidence for a gradual evolution of prey signals is, however, mixed. Gamberale & Tullberg (1996) found that chicks showed exaggerated avoidance of aposematic heteropteran bugs where the size of the bug was increased. This demonstrates that peak-shift-like effects can apply to bird–insect systems. Lindström et al. (1999) used wild-caught great tits as predators, and artificial, black-and-white symbols as signals. In this experiment the birds did not learn to avoid distasteful artificial prey that were slightly more conspicuous than the edible prey available. Lindström et al. argue that there may be a stimulus region close to crypsis in which selection is relaxed. Within this, mutations may accumulate to the point that chemically defended prey cross a threshold in which conspicuous prey are numerous and favoured by selection, especially if peak-shift effects operate.

A different approach to the question of gradual exaggeration of aposematic cues is to use ancestral state reconstructions across phylogenetic trees and evaluate whether aposematic traits tend to increase in magnitude over phylogenetic time. A demonstration of bioluminescence in millipedes supports the idea that aposematic traits increase in this manner (Marek et al., 2011; Marek & Moore, 2015).

6) Initial evolution is more easily explained with physical not chemical defences

It may be easier to explain the initial evolution of aposematic signals when they are associated with physical defences such as sharp, defensive spines of animals and plants (Lev-Yadun, 2001), than for the advertisement of internally stored toxins. We consider defensive spines as an example. These are a taxonomically widespread, likely convergent mode of defence, found in insects (e.g. many lepidopteran larvae), fish (species of pufferfish), and mammals

(such as the porcupine). Defensive spines are often externally situated on a prey and can therefore be seen by predators and recognized as a threat before damaging attacks take place. Physical defences such as spines can therefore be considered to have self-advertising properties, in a way that chemical defences cannot. Aposematic colour patterns that add visual conspicuousness to structures like a porcupine's spines, then, arguably function to better draw a predator's attention to the threat before contact is made. If predators are familiar with the threat that the physical defences pose then new conspicuous colour patterns can add protection without the need for additional and costly learning events by drawing attention to the threat. The frequency-dependent 'learning barrier' to initial evolution of aposematic signals may not, therefore, pertain in these cases. Speed & Ruxton (2005a) showed in a simple evolutionary model that aposematic signalling of this nature could easily evolve, in a wide range of conditions, from a cryptic ancestral population. Some physical defences also incorporate chemical defences, such as when defensive spines puncture the skin of enemies and release poisons into them (e.g. Murphy et al., 2009). Speed & Ruxton suggest a scenario in which visual aposematic signals evolve initially to draw attention to physical defences, but can become associated with a chemical defence of stinging spines subsequently.

7) Ecological conditions lower the costs of initial aposematic coloration

There are a number of ways that local conditions can lower the survival costs imposed on novel aposematic signals. First, conspicuous coloration may be one of a set of alternative developmental outcomes, and only deployed when the benefits of crypsis are small, and hence the relative costs of aposematic coloration are low. In the desert locust *Schistocerca gregaria* (Orthoptera: Acrididae), for example, there is a density-dependent switch from cryptic and edible to conspicuous and toxic phenotypes as prey density exceeds a threshold (Sword et al., 2000).

A second way that ecological conditions can favour the initial evolution of aposematism is if the prey species is already behaviourally conspicuous or grows to a size that 'hiding' as a defensive strategy is not effective. Animals, for example, that move around freely for foraging opportunities, or that engage in showy sexual displays, will add a smaller marginal increase in conspicuousness when they gain new conspicuous colour patterns than those that are, by their natural history, originally hidden and camouflaged (Guilford, 1988; Merilaita & Tullberg, 2005; Speed & Ruxton, 2005b; Speed et al., 2010). In the case of prey with colourful sexual displays, it may be that they are subsequently and easily 'exapted' as aposematic signals after a protective secondary defence evolves.

A general point that we can make here, then, is that many theoretical (and some empirical) treatments of the initial evolution of aposematic signalling take a pessimistic starting point, choosing highly cryptic ancestral species and therefore imposing high additional costs of encounter rate when aposematic signals evolve. What seems more likely (to us at least) is that the initial evolution of aposematic forms is more likely to occur in prey species that are not very cryptic, by reason of high density, foraging, or mating demands.

6.4 Maintenance of aposematic signalling

Focussing on barriers to initial evolution, the last section provided reasons to expect aposematic signalling to be rare. However, it is perhaps also plausible that aposematism is rare because, even when established, it is typically not maintained over long evolutionary periods. Aposematic signalling may, for example, be unstable over evolutionary time and frequently replaced by crypsis within lineages. Alternatively, there could be macro-evolutionary explanations, such as relatively reduced diversification or heightened extinction rates in lineages with aposematism, compared to lineages which contain crypsis (as we saw for chemical defence, Chapter 5).

To examine these possibilities, Arbuckle & Speed (2015) analysed the relationship between variation in coloration and patterns of diversification in the amphibians (using >2500 species). They found heightened speciation rates in lineages with aposematic coloration compared to those with cryptic coloration, but no increase in extinction rates (see also Santos et al., 2014; Wang & Shaffer, 2008). The

relative rarity of aposematic coloration in amphibians was instead explained in this study by the intrinsic instability of the trait over long evolutionary timescales. Aposematic coloration was estimated to switch to cryptic coloration twice as often as crypsis gave way to aposematism. Aposematism in amphibians is, then, on the one hand associated with periods of species diversification, but on the other hand the trait itself is typically unstable over macroevolutionary timescales in this group.

One explanation for this instability is that selection by predators for or against conspicuousness may be very variable across habitats and across time, because the composition of predator communities is itself variable (Mappes et al., 2014; Nokelainen et al., 2014). In addition, the chemical defences on which aposematic signalling is often based may also be highly variable within and between populations (Speed et al., 2012). If aposematic signalling is underpinned by variable, and sometimes weak, secondary defences, selection may favour reversion to crypsis (Speed & Franks, 2014). Aposematism may then be favoured in a relatively narrow range of ecological parameters which are prone to variability, rendering the trait unstable over long timescales. We return to this point in section 6.5 below.

6.5 Ecology of aposematism

We restrict our discussion of ecology to three key ecological components: first, variability in selection regimes imposed by ecological variation in predators; second, the effect of aposematism on prey niche; and third, the explanations for signal variation, including whether aposematic signals are 'honest'.

6.5.1 Ecology 1: Effects of predictable variation in predator communities

We will begin this section by considering how threats from predation vary in habitats subjected to seasonal changes. Every Spring starts with a set of predators educated from experience in previous seasons who know that conspicuously coloured prey are often chemically defended, and tend to avoid them. When young, naive predators emerge in the middle of the early Summer period however, then

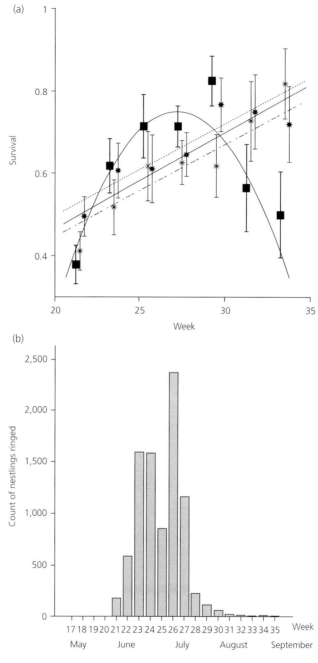

Figure 6.3 Summary of results from Mappes et al. (2014). (a) Mean observed survival of three artificial larval types (*N* =1171, ±s.e.m.) over a season (non-warning-coloured: blue squares, small signals: orange stars, large signals: red bold stars) and the predictions for two best models (models D and E). The seasonal relationship for non-warning-coloured prey (blue curve) is identical in both models but differs for warning-coloured prey. Model D predicts an identical increase for small and large signals (red solid line) and model E a higher survival for large signals (red dotted line) than for small signals (orange line). (b) The estimated nest-leaving dates of juvenile passerine birds in Finland (see Plate 6).

there can be many naive, inexperienced predators who may attack brightly coloured prey at high rates. Aposematic individuals may be especially vulnerable in this period. Toward the end of the season this vulnerability may be reversed as the new predators are increasingly experienced and tend to avoid aposematic signalling. The consequence is that aposematic conspicuousness may be excessively costly in the middle of the season, but beneficial either side.

This prediction was confirmed in a field experiment in which Mappes et al. (2014) investigated the survival of three forms of the caterpillar of the wood tiger moth—cryptic (black), highly conspicuous (large signal: black with large portions of orange coloration) and intermediate conspicuousness (small signal: smaller areas of orange pigmentation). They recorded prey survival over the Spring to Autumn months and simultaneously the age-structure of the avian predator community. As Figure 6.3a shows, the cryptic forms (squares) were advantaged when the young fledglings started to forage (mid-season, week 22, Figure 6.3b), but in the other periods conspicuous coloration was superior.

An analysis of the phenology of macrolepidopteran caterpillars in Finland (in the same paper, Mappes et al. 2014) showed that conspicuous species are least prevalent in this middle part of the year. Hence, temporal variation in predator abundance appears to drive temporal patterns of variation in the abundance of aposematic forms in a habitat. This study gives us considerable insight into the vulnerability of aposematic coloration to variation in ecological conditions. Each season aposematic signalling may become unprofitable in the 'fledging period'. If that period increased in duration for sufficient seasons, because for example of increased reproductive periods in predators, we could reasonably imagine that aposematic signalling could be destabilized and replaced by crypsis. Note though, that not all the evidence supports this particular pattern; Langham (2005), for example, found that older jacamar birds (*Galbula ruficauda*) were more likely to sample novel, conspicuous prey than younger birds, who were more hesitant. It is worth mentioning here also that seasonality as a cause of rapidly changing selection on aposematic phenotypes is unlikely to apply in tropical ecosystems, in which the seasonality is less pronounced and the breeding season of birds tends to be longer.

Aluthwattha et al. (2017) have proposed that spatial variation in predation pressure may modulate selection for, and maintain variation in conspicuousness of, aposematism. Their study of aposematic Danainae (Nymphalidae) butterflies and their mimics found that highly conspicuous individuals experienced fewer attacks from avian predators when background predation pressure was low, but attack rates increased rapidly as background predation pressure increased. The least conspicuous individuals, conversely, experienced higher attack rates at low predation pressures, but reduced predation with high predation pressures, seemingly benefitting from crypsis.

Selection can, of course, vary across space too, and a growing number of ambitious field experiments demonstrate this. Nokelainen et al. (2014), for example, showed that yellow vs white polymorphisms for warning coloration on the wings of the adult wood tiger moth provide protection against alternative bird species (white appears most protective against accentors, Prunellidae; yellow against Paridae, tits, chickadees etc.). Hence, relative variation in the distribution of these two groups helps predict the variation in these colour patterns across large areas. Willink et al. (2014) report alternative selective forces across large areas in Costa Rica on artificial models of cryptic and conspicuous dendrobatid frogs (*Oophaga granulifera*) by birds, lizards, and crabs. Lizard predation, for example, would select for cryptic coloration in prey, whereas bird attacks were more often deterred by conspicuousness. Variation in the relative frequency of different kinds of predator can therefore affect the net value of aposematic signalling. Similarly, variation in raptor populations may affect survival of aposematic and cryptic forms of snake coloration. Valkonen et al. (2012) showed that predation on model vipers by raptors was affected by the presence of buzzard species, which unlike the other raptors in the study would attack warningly coloured models. Hence, again, an established aposematic signal may become vulnerable and be replaced by cryptic coloration if there is a change in the suite of predators in a habitat so that sufficient numbers of them ignore the signal.

6.5.2 Ecology 2: Aposematic coloration and its relationship to niche and behaviour

Since chemical defence is often paired with aposematic signalling, we may expect that there are similar effects on niche, often to widen it, and hence we refer the reader to Chapter 5, section 5.4.2.

However, coloration mediates many of an animal's interactions with its environment, so it is not surprising that we may find additional, or different, effects of aposematic coloration on niche and life history than we would by chemical defence alone. For example, where Arbuckle et al. (2013) found that chemical defence has some effect of increasing diet range in the musteloid group (see Chapter 5, section 5.4.2), they found a stronger association between omnivory and aposematic coloration than with chemical defence alone. Aposematic coloration may, in a sense, amplify the protective effects of chemical (or other) defences, preventing many attacks rather than merely deterring predators once attack takes place. In this way, aposematic signalling may open a wider set of opportunities within a habitat (Speed et al., 2010). Furthermore, warning coloration may itself be strongly associated with niche change: Merilaita & Tullberg (2005) found a strong phylogenetic association between the presence of aposematic coloration and transition to diurnal from nocturnal in selected macrolepidopteran groups, mainly Arctiidae.

Warning coloration may then be a result, or a cause, of quite fundamental changes in the biology of the prey organism. Two important case studies are worth describing here, though of course others exist. First, in a phylogenetically corrected study of the Carnivora, Stankowich et al. (2011) demonstrated that 'salient' (highly contrasting, presumed aposematic) colour patterns evolved in a manner positively correlated with more 'sophisticated' anal spraying methods of secondary defence, and with body length, movement into open habitats, and increased burrowing capacities. The data available at the time of their analysis did not allow Stankowich et al. to reveal the sequences of events in Carnivora coloration, so causality remains unresolved. Did, for example, aposematism precede or succeed initial movement to more open habitats? In a study of similar biological range, Santos & Cannatella (2011) found that the presence of aposematic coloration in

dendrobatids was correlated with active metabolism and 'aerobic range' (a measure of physical activity), as well as with body size. They ask whether metabolic change for greater activity resulted from, or caused, the evolution of aposematic coloration and chemical defence.

Warning coloration may also be associated with variation in behaviour. If aposematism frees animals from behavioural constraints, we may find them modifying and diversifying their behaviours in response to diversifying sexual selection or to take advantage of new opportunities (Guilford, 1988; Merilaita & Tullberg, 2005). Comparative methods are useful here, showing for example the associations and perhaps the sequence by which new behaviours and aposematism come into being. Santos et al. (2014) notably showed association between the presence of alkaloid sequestration and conspicuous coloration with acoustic characteristics of male calls in dendrobatid frogs. Variation in a principal component that included the temporal pattern of sound production significantly explained the presence of alkaloids; specifically, short-pulsed sounds predicted an increased likelihood of sequestering alkaloids (by 1.4 times) and of using conspicuous coloration (by 1.2 times; similar effects associated with low pitch calls, but this is associated with larger body size). Species in which males use exposed calling sites, rather than being hidden, were also more likely to be chemically defended and conspicuously coloured. Santos et al. suggest that aposematism enables males to use exposed positions in which their calls are likely to travel further. A key point is that estimations of ancestral states suggest that modification to vocal characteristics followed as a consequence of, rather than preceded, aposematic coloration.

Other studies on dendrobatids explore the general prediction that aposematism is associated with behaviours that enable greater exploitation of social or ecological opportunity, but would cause increases in visibility to enemies. Evidence from dendrobatid frogs is currently supportive, but mixed (see review in Rojas, 2017). Rudh et al. (2012, with *Oophaga [Dendrobates] pumilio*), Willink et al. (2013, *O. granulifera*) and Galeano & Harms (2016, *O. pumilio*) all reported positive correlations, especially of coloration and aggressive (male–male) behaviours, whereas Dugas et al. (2015, with *O. pumilio*) found no simple correlation. Crothers & Cummings (2015, *O. pumilio* again) found that conspicuousness did correlate with individuals' speed of initiating aggressive territorial defence, and furthermore that individuals used brightness of a competitor as a social cue, such that the difference in brightness tended to predict the outcome of an interaction. Hence, variation in conspicuous coloration in (*O. pumilio*) provides intraspecific social information, whether or not it is informative to predators.

In summary, there is a growing body of research which links the evolution of aposematic signalling with changes in the biology of prey species. Sometimes these changes are profound, moving activity periods from night to day, or reversing the dietary range of the species. Aposematic signals also seem very often to interact with sexual selection, opening the door to diversification of traits, if species have not found ways to use colour patterns so that they work independently on different audiences (Cummings & Crothers, 2013).

6.5.3 Ecology 3: Variation and honesty in warning coloration

On the face of it, aposematism within a species is a mutualistic enterprise. Individuals that have secondary defences evolve a consensus phenotype to warn enemies about their unprofitability, an evolved and widely used warning signal (in Chapter 7, we see that this mutualism can be extended across species in Müllerian mimicry). In a simplified view, the more common the aposematic signal, the more effective it will be, since predators will have many opportunities to learn, and many reminders about the unpleasant consequences of attack. Hence, we should expect a rule of signal uniformity within aposematic signals (Mallet & Singer, 1987). Indeed there is evidence from field-based experiments (Mallet & Barton, 1989; Noonan & Comeault, 2009) and the preferences of wild-caught predators (Langham, 2005) that novel variation in warning coloration is disproportionately punished.

Though many aposematic colour patterns are similar within species, variation is nonetheless frequently reported. We deal with such variation in this section. Note we consider here only that variation in pigmentation that will be sufficient to be noticeable to predators (Arenas & Stevens, 2017), leaving aside that which is too minor to affect predator decision-making. We are also careful to discriminate such variation that occurs within populations, which might indicate varied levels of signal honesty or dishonesty within a prey population, from that which occurs between species or between populations which reflect broader selective regimes, as we explain later.

1) Relaxed and disruptive selection and variable costs

First, there are some relatively straightforward explanations that involve relaxed selection. It may be, for example, that predators perceive variation in warning signals, but choose to categorize the variant phenotypes together into one class that they treat similarly, mainly by avoidance. Here the variation in phenotypes may be selectively almost neutral. There is some suggestion of this in work examining predation on models of dendrobatid frogs (*O. histrionica*) by Amézquita et al. (2013), including that such generalization is more likely when prey are at high abundances (see Mallet & Joron, 1999). Reports that predators focus on selected, highly salient features such as overall coloration, and pay less attention to less salient components of prey appearance (Aronsson & Gamberale-Stille, 2008; Rönkä et al., 2018), suggests that selection may be relaxed on some components of warning coloration, but not others, and account for variation in those traits.

Note that in studies of geographical variation in colour patterns, it may be possible to test whether drift alone can account for diversification in warning coloration. A notable study by Brown et al. (2010) showed that the wide range of colour variation in the strawberry poison dart frog (*O. pumilio*, which is distributed over the Bocas del Toro archipelago in Panama) is not caused by neutral processes alone. Rather, diversifying selection must be included to explain the range of colour variation (see related study on the wood tiger moth, by Galarza et al., 2014).

Varied forms of disruptive selection make for plausible explanations of signal variation. Rojas et al. (2014b), for example, propose that variation in lighting conditions within habitats can maintain simultaneous selective advantage on varied colour forms (and recently, that variation in colour forms correlate with variation in movement; Rojas et al., 2014a). Temporal changes in selection and population size may often be important (Galarza et al., 2014). Alternatively, varied levels of sexual selection combined with variation in the effectiveness of anti-predator coloration may cause disruptive selection across habitats and lead to variation in conspicuousness of prey colour forms. Endler & Rojas (2009) examine the disruptive effects of variation in predator abundance and mate availability, showing how variation in both will lead to variation in coloration and behaviour. Taking the wood tiger moth as their case study, Gordon et al. (2015) propose that a combination of migration, positive frequency-dependent mating advantage, and variation in predator communities all combine to cause the persistence of variation in colour forms, which affect fitness through anti-predator defence and sexual selection. Kang et al. (2017) have recently suggested that differential predation is influential in the divergence in multiple traits of the aposematic oriental fire-bellied toad (*Bombina orientalis*) between South Korean mainland populations and populations on Jeju Island. Jeju Island populations apparently experience higher predation pressure than mainland toads, and exhibit less bright and less chromatic aposematic ventral coloration, lower activity overall, and a tendency to rest underwater. Increasingly, researchers are taking this kind of broad geographical approach to warning colour variation, so we invite interested readers to follow the growing set of papers which evaluate this further (some excellent examples are: Comeault & Noonan, 2011; Mochida, 2011; Rojas & Endler, 2013; Hegna et al., 2015).

Another relevant point to make here is that genetic correlations, as arise from linkage or pleiotropy, between signalling and other traits may lead to quite complex determinants of colour variation. This point is well made by Lindstedt et al. (2016) who demonstrate a negative genetic correlation between larval and female adult coloration in the wood tiger moth (*Arctia plantaginis*), so that larvae with large warning signals turned into adults with smaller signals. Lindstedt et al. propose that this correlation can contribute to additive genetic variation in these traits, especially as both are subject to multiple and, likely, variable selection pressures. This leads us to a final point, that trade-offs between coloration and other biological functions such as mating (as demonstrated by Nokelainen et al., 2012) may be a proximate cause of variation in prey species, and these may not always be apparent to empirical researchers. An interesting question is: how many important trade-offs are obscure to us, but important in causing variation in aposematic coloration?

2) Honest signalling as an explanation of signal variation

One attractive hypothesis is that the observable variation in aposematic signalling is maintained by its role in the signalling system itself. More conspicuous phenotypes may, for example, predict higher levels of toxicity. If this were the case, aposematism may be a 'quantitatively' honest signalling system, in which variation is essential to its functioning.

A paper by Summers & Clough (2001) arguably initiated much of the interest in this idea, though it was earlier discussed in theoretical terms by Guilford and Dawkins (1993). Using phylogenetically corrected regression, Summers & Clough showed that across dendrobatid species there is quantitative variation in both toxicity and conspicuousness in colour patterns and, furthermore, there is a positive correlation between these traits. Subsequently, a number of papers report such positive conspicuousness–toxicity correlations between species (for example: marine opisthobranchs, Cortesi & Cheney, 2010; ladybird beetles, Arenas et al., 2015: see reviews in Summers et al., 2015; Crothers et al., 2016) or between geographical populations of different species (e.g. Maan & Cummings, 2011; for the frog species *Oophaga pumilio*). However, because the correlations being reported are across genetically or geographically independent entities, it is very hard to see these as evidence for evolved signalling honesty, in which variation in conspicuousness between species has a signalling function to predators. It would take some feat of migration and cognition for predators to impose honesty across geographically separate populations, and some rather non-intuitive selection mechanisms by which the conspicuousness levels of several species is driven by mechanisms that enforce mutual signal honesty.

Furthermore, it may be that fitting regression or correlations through cross-species data sets leads to the appearance of continuous correlations, whereas in fact we have two groups of prey: a mostly cryptic group (those poorly protected by toxicity) and an aposematic group (high toxicity). If the values for these are placed on the same axes and a regression line plotted, we might gain a significant effect whereas the real state may be closer to categorical.

Accounts of cross-species variation in conspicuousness can, though, tell us about the processes that cause exaggeration of conspicuous coloration within species of different levels of defence. It may be, for example, that in a group like the dendrobatids the processes that cause aposematism tend to lead to higher equilibrium levels of conspicuous phenotype when chemical protection is greater. Sexual selection is a possible explanation, in which better protected species have freer range to exaggerate conspicuous coloration, so that the driver is now courtship advantage, rather than anti-predator deterrence. It would be instructive to find case studies where we can compare cross-species correlations in larval vs adult forms, for only in the latter would sexual selection drive brightness. We note also that at least one amphibian study found the reverse of the more common relationship—a negative toxin–conspicuous correlation (Wang, 2011; readers interested in explanations for negative correlations are referred to the review by Summers et al., 2015 and the experimental work of Darst et al., 2006). This exception suggests that we need a larger set of evaluations to understand cross-species patterns in nature.

Although we do not see cross-species or cross-population correlations as evidence of honest signalling per se, there are a few empirical studies that have demonstrated conspicuousness–toxicity relationships within species; positive associations, including in paper wasps (venom sac vs colour pattern in *Polistes dominula*, Vidal-Cordero et al., 2012), frogs (*O. pumilio*, Crothers et al., 2016), and beetles (Blount et al., 2012; Arenas et al., 2015); but negative in the wood tiger moth (Reudler et al., 2015) so there is some (mixed) evidence for honest signalling

within aposematic species. There are several theoretical attempts to explain honest signalling (positive correlations) within aposematic populations (see review in Summers et al., 2015). Some explanations rely on the idea that coloration is a condition-dependent trait (Blount et al., 2009; Holen & Svennungsen, 2012). Notably, Blount et al. (2009) proposed that toxin generation and storage invoked oxidative stress, which required anti-oxidant molecules to mitigate the damage. However, pigment molecules (e.g. carotenoids, melanins, and pteridines, among others) may themselves have a primary use as anti-oxidants. Pigmentation may then mechanistically trade-off against coloration. Prey with good anti-oxidant resources can be both more toxic and have more pigmentation than those with poorer anti-oxidant resources. An experiment that manipulated the availability of food (aphids) to larval ladybirds (seven-spot, *Adalia septempunctata*), showed the predicted positive toxin–pigment correlations (between precoccinelline, a precursor toxin molecule, and elytra carotenoid concentrations; Blount et al., 2009; Winters et al., 2014; and see Crothers et al., 2016).

A major issue in all of these studies is the measurement of toxicity. Some studies measure the concentrations of the toxic chemicals directly (e.g. Blount et al., 2012). Other studies use lab assays with model species as bio-indicators of toxicity (effects on mice, e.g. Summers & Clough, 2001, or *Daphnia* in Arenas et al., 2015). Results from experiments based on lab models are not straightforward to evaluate, however, because the toxin in question could affect the model system differently to ecologically relevant predators (Weldon, 2017). The alkaloid of the two-spot ladybird is, for example, not harmful at least to one likely bird predator species (blue tits, *Cyanistes caeruleus*, Marples et al., 1989), and is only a mild deterrent to ant predators (Marples, 1993). However, it turns out that the compounds in this species are actually quite toxic to *Daphnia* used in lab assays (Arenas et al., 2015). This discrepancy calls for caution in interpretation. Bolton et al. (2017) have recently taken steps to verify that dendrobatid toxicity characteristics gathered from lab mice can be reproduced in more ecologically appropriate arthropod species (and we recommend interested readers look at the very interesting discussion in Crothers et al., 2016). The gold standard is, we would argue, first to measure the variation of concentrations of relevant chemical compounds within the prey population, so that there is an objective basis for evaluating variation, and second to know how natural predators respond to this variation. When lab models (such as *Daphnia*) are used as a proxy, they need to be calibrated against real predators for us to really understand the nature of toxicity in natural populations.

6.6 Co-evolution of aposematic signals and receiver psychology

In the preceding chapter, we spent some time discussing the evidence for co-evolution between chemically defended prey and their predators. Traits of toxicity (prey) and resistance to toxicity (predators) may be linked in tight co-evolutionary relationships, with evolutionary change limited to few genomic sites. Researchers can then look for evidence of correlation in traits as they vary over wide spatial scales. Co-evolution in aposematic signals is less tractable, and has a more limited evidence base. We do know that there is often 'something special' about the effects of aposematic coloration on the cognition of predators. Aposematic coloration may increase wariness, speed up learning, increase the likelihood that prey will be rejected before consumption, and perhaps reduce forgetting rates. These may be evolved, special responses to aposematic coloration, and aposematic coloration may evolve to match the perceptual-cognitive biases of their predators. The evidence for repeated co-evolution of these traits in relation to each other is to the best of our knowledge not present. There are signs of interest in the empirical literature, for example when researchers look at variation in predator diet choice from geographically distant sites (Exnerová et al., 2015).

6.7 Future work and conclusions

We focussed in this chapter on the prey side of aposematism, but as a way to conclude, we briefly point out that aposematic signalling is an important experimental tool in the study of the predator side. As an experimental tool, aposematism has been important in the discovery and investigation of dietary conservatism, which has widespread implications for our understanding of diet choice in animals (Marples & Kelly, 1999). Aposematism has also proven useful in the study of cognition. Thorogood et al. (2018) for example use aposematism as a tool to show the importance of social learning in the evolution of animal signalling. In their 2016 review 'Learning about aposematic prey', Skelhorn et al. draw attention to the fact that predators do not simply learn to avoid prey utilizing aposematic signals, but instead make adaptive decisions regarding when to gather information about such prey and when to include them in their diets despite their defences. Skelhorn, Rowe, Halpin, and colleagues have used aposematism as a laboratory model with which they have demonstrated that predators are physiological strategists in their approach to diet, taking into account toxin burdens, concentration of toxins, and the size of prey in their decision-making (Smith et al., 2014; Barnett et al., 2012). In their review, Skelhorn et al. (2016a) argue that current knowledge about what predators learn about aposematic prey is lacking. We fully support this 'call to arms' and would encourage future studies investigating how predators gather and use information about aposematic, defended prey.

CHAPTER 7

The evolution and maintenance of Müllerian mimicry

7.1 Introduction

In his classic paper, Bates (1862) reflected on the similar appearances of a number of Neotropical butterfly species and proposed that the phenomenon could be explained by natural selection on palat-able species to resemble unpalatable (or otherwise unprofitable) species ('Batesian mimicry', Chapter 9). Palatable species experience this selection simply because would-be predators will tend to refrain from attacking them when they resemble something harmful. However, in the same monograph Bates also noted that some common and apparently distasteful butterfly species also resembled one another. While Bates speculated that this resemblance could have arisen due to shared physical conditions, it was Johannes Friedrich ('Fritz') Müller who proposed the explanation that continues to be favoured today. Indeed, reflecting the honour bestowed on Bates, mimicry among unpalatable or otherwise unprofitable species has since become known as 'Müllerian mimicry'. Müller's idea was very simple: predators often have to learn to avoid unpalatable (or otherwise unprofitable) prey, and in these instances there will be strong selection on unpalatable prey to adopt a similar appearance, because it serves to share the mortality cost of educating inexperienced predators. In effect, they will face selection to evolve the same set of warning signals.

To formalize Müller's arguments, we begin by outlining the model he presented to illustrate his theory (Müller, 1878; translation in Müller, 1879). Of course, Müller invoked his quantitative arguments

for illustration only and the real world is inevitably more complicated than that. In section 7.3 we revisit the model and reframe it, showing that the optimal sampling strategy when predators are faced with unfamiliar prey can explain both Müllerian mimicry and neophobia (a reluctance to attack novel prey). Next we survey some of the best-known examples of shared resemblances among unprofitable species that cannot be explained simply on the basis of shared ancestry. The central part of this chapter is devoted to considering the evidence that the shared resemblance has arisen for the reason that Müller hypothesized. Finally we consider some of the properties of Müllerian mimicry systems in more depth, including ecological and co-evolutionary phenomena, and raise some fundamental questions that have only been partly resolved.

7.1.1 Müller's theory

As described in our opening remarks, Müller's argument was based on 'strength in numbers': two or more unpalatable species may face selection to adopt a common appearance because it helps reduce the mortality cost involved in teaching naive predators to avoid them. Following Müller's (1878, 1879) original argument, let a_1 and a_2 be the numbers of two approximately equally unpalatable (or otherwise defended) species in some definite district during one summer, and let n be the number of individuals of each distinct, unpalatable species that are killed by predators during a season before their distastefulness is generally known. If the species are distinct in appearance, then each species

Avoiding Attack: The Evolutionary Ecology of Crypsis, Aposematism, and Mimicry. Second Edition. Graeme D. Ruxton, William L. Allen, Thomas N. Sherratt, & Michael P. Speed, Oxford University Press (2018). © Graeme D. Ruxton, William L. Allen, Thomas N. Sherratt, & Michael P. Speed 2018. DOI: 10.1093/oso/9780199688678.001.0001

would lose n individuals in the course of educating predators to avoid them. If, however, the two species were exactly alike in appearance, then the first species would lose an average of only $a_1n/(a_1+a_2)$ individuals and the second would lose only $a_2n/(a_1+a_2)$ individuals. Under these conditions, a mimetic mutant of species 1 that perfectly resembled species 2 would tend to spread from extreme rarity (in that it will have a higher mean survivorship) so long as $a_2>a_1$. So, all else being equal, the rarer unpalatable species would face selection to resemble the more common unpalatable species. We revisit this model from an opti-mization perspective in section 7.3, but first we consider some of the examples of resemblance among unprofitable prey that motivated Müller, and many others that have since come to light.

7.2 Examples

Here we present a range of instances of mimicry among unprofitable species. The list is not intended to be exhaustive (we have left out excellent cases of Müllerian mimicry in snakes, burnet moths, and cotton stainer bugs, to name a few). However, in each of the cases below we include evidence that: (i) the species involved are unprofitable to attack by would-be predators, (ii) the species are similar in visual appearance to their would-be predators (or

at least human observers), and (iii) the shared appearance is not simply the consequence of shared ancestry. We note two features of these examples at the outset.

First, although Müllerian mimicry is often presented as involving resemblance between just two species, it frequently involves larger collections of similar-looking species—'mimicry rings' (Mallet & Gilbert, 1995). Linsley et al. (1961) for example, describe a series of 'lycid complexes' that include collections of unpalatable lycid beetles, arctiid moths, parasitic hymenoptera, and flies, all of which are orange in coloration with black tips (Figure 7.1). Similarly, tarantula hawk wasps in the genera *Pepsis* and *Hemipepsis* have some of the most painful stings known to man and form Müllerian (and Batesian) mimicry complexes with many other species of stinging tarantula hawks, as well as numerous flies, beetles, and moths (Schmidt, 2004).

Second, one might expect that selection by predators on unprofitable prey species to adopt a common warning signal would promote near-universal uniformity in appearance. As we show, this is indeed often the case at a local scale, although in some unusual cases a single unprofitable species can be locally polymorphic with different forms participating in different mimicry rings (section 7.5.5). However, despite strong localized selection for uniformity, species engaged in Müllerian mim-

Figure 7.1 (a) A lycid beetle and (b) an arctiid moth, found together in southern Ontario. These species are both unpalatable to birds and their colour pattern similarity suggests they are Müllerian mimics. Photo courtesy of Dr Henri Goulet (see Plate 7).

icry often have surprisingly high geographical diversity in colour patterns, sometimes referred to as 'spatial mosaics'. Although this polymorphism at the geographical scale is somewhat counter-intuitive given the nature of selection, it is now readily understood. Moreover, the matching of these colour patterns between the same pair (or more) of unprofitable species across geographical scales serves to reaffirm that mimicry has not arisen through shared ancestry. We consider the phenomenon of geographical polymorphisms in Müllerian mimics in section 7.5.4.

7.2.1 Neotropical butterflies

Following the original observations of Bates and Müller, Neotropical butterflies have provided some of the most celebrated and intensively studied examples of Müllerian mimicry. The best known examples derive from the genus *Heliconius* and nine smaller genera from the Neotropical tribe Heliconiini (Nymphalidae: Heliconiinae), along with butterflies from the tribe Ithomiini. Indeed, Neotropical butterflies are the poster child for Müllerian mimicry and we will return to them on multiple occasions throughout this chapter.

The genus *Heliconius* comprises 46 described species, all of which have conspicuous (and often shared) wing patterns as adults (Figure 7.2). *Heliconius* butterflies are widely considered unpalatable to vertebrate predators due to cyanogenic glycosides which are sequestered from their *Passiflora* host plants as larvae (Cardoso & Gilbert, 2007) or newly synthesized in larvae or adults (Engler-Chaouat & Gilbert, 2007). This unpalatability has been confirmed in a number of feeding trials. For example, Brower et al. (1963) presented wild-caught silver-beaked tanagers (*Ramphocelus carbo*) with alternating sequences of lab-reared *Heliconius* butterflies and palatable satyrid butterflies. While the palatable butterflies were always attacked when offered, the majority of the heliconids were either rejected at the outset or consistently rejected after only 1–3 attacks by the birds. Intriguingly, the nature of the responses of birds (ranging from not touching to pecking, killing, and consuming) differed dependent on the heliconid species offered, although there was also substantial variation among individual birds. More-

Figure 7.2 An unpalatable *Heliconius numata* butterfly, shown here sitting on a Heliconia flower. This Müllerian mimic is unusual because it exhibits local polymorphism in which several different forms of the same species co-exist in the same locality, with different forms participating in different mimicry rings (see section 7.5.5). Photo credit: Mathieu Joron (see Plate 8).

over, a very high proportion of tanagers avoided touching a heliconid mimic from another species after they had learned to avoid their original heliconid model, providing early evidence for a protective effect of mimicry.

The Neotropical tribe Ithomiini (Nymphalidae: Danainae) comprises circa 380 species drawn from 50 genera (Elias & Joron, 2015). Many ithomiine species have partly or entirely transparent wings ('glasswings') although their wing patterns are still considered conspicuous. Like *Heliconius* butterflies, many ithomiine species are involved in mimetic complexes and are often the most abundant species in these complexes (Elias & Joron, 2015). Adult male ithomiines are thought to acquire their chemical defences, pyrrolizidine alkaloids, from decaying flowers or stems of Apocynaceae, Boraginaceae, and Asteraceae (Trigo, 2011). Intriguingly, these pyrrolizidines are then transferred to females during copulation as a nuptial gift (Elias & Joron, 2015; see Conner et al., 2000 for a similar example in an arctiid moth). A number of experiments have confirmed the unpalatability of adult ithomiines to avian and arthropod predators. For example, Arias et al. (2016b) recently found that wild-caught great tits (*Parus major*) consumed proportionately less crushed-up ithomiine *Mechanitis polymnia* (wings removed) than similar palatable mealworms.

7.2.2 Wasps

Velvet ants are wingless wasps (Hymenoptera: Mutillidae) that are ectoparasitoids of the larvae of solitary bees and wasps. Not only are egg-laying females protected by a painful sting, chemical secre-

tions, and a hard cuticle but they also tend to be conspicuously coloured. Wilson et al. (2012) quantitatively compared the appearances of 65 species in the velvet ant genus *Dasymutilla* and placed them into six morphologically distinct and geographically delimited mimicry rings (Figure 7.3). Phylogenetic analysis suggested that these shared colour patterns are primarily the result of independent evolution rather than a consequence of their phylogenetic history. In follow-up work, comparing the appearances of 351 species from 21 velvet ant genera, only 15 species did not fit into any mimicry ring (J. Wilson et al., 2015). Yet the complex may be even bigger than that. Rodriguez et al. (2014), for example, argued that another group of solitary parasitic wasps, spider wasps (*Psorthaspis* spp), showed parallel geographical variation in colour pattern, making it

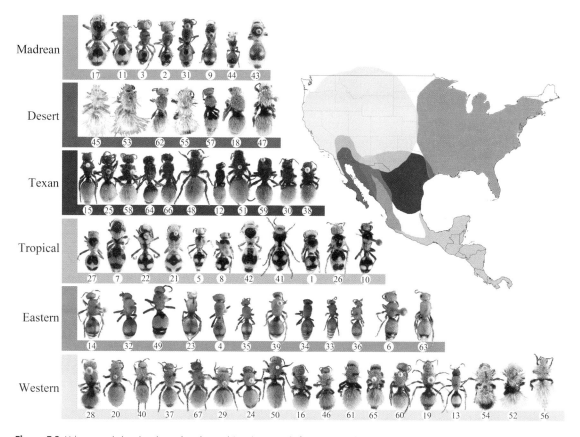

Figure 7.3 Velvet ant mimicry rings in north and central America. A total of 65 *Dasymutilla species* (and three populations of *D. bioculata*) were classified by Wilson et al. (2012) into six mimicry rings on the basis of their shared colour, pattern and geographic location. The indices under each species are voucher numbers, also used to highlight their position on the phylogenetic tree. Reprinted from Wilson et al. (2012) with permissions (see Plate 9).

one of the largest Müllerian complexes described to date. Not surprisingly, the complex has also attracted putative Batesian mimics, including beetles, antlions, and spiders. Intriguingly, male velvet ants lack the sting and are considered relatively harmless compared to females. However, while males are often conspicuously coloured, these putative Batesian mimics do not always resemble or even participate in the same mimicry ring as their conspecific females.

7.2.3 Millipedes

Apheloriine millipedes are endemic to the temperate forests of North America, and very well protected from predators: each individual can secrete 18-fold the amount of hydrogen cyanide necessary to kill a pigeon-sized bird (Marek & Bond, 2009). The millipedes also exhibit conspicuous colour patterns that vary in hue, including yellow, red, orange, and pink—traits that cannot be used for conspecific signalling since the millipedes are blind. Although it has long been postulated that Appalachian apheloriine millipedes are engaged in Müllerian mimicry, Marek & Bond (2009) formally analysed the colour patterns and spatial distribution of seven species of cyanide-generating millipede endemic to the mountains of Virginia, Kentucky, and Tennessee. They showed that while individual species vary in colour and

pattern over large spatial scales they share very similar appearances where they co-occur, an outcome that could not arise simply due to shared ancestry, making this a compelling example of Müllerian mimicry in a group not well known for it (Figure 7.4).

7.2.4 Catfish

Named for their barbels that resemble cat's whiskers, many catfish are defended by spines and bony plates. In a particularly comprehensive study, Alexandrou et al. (2011) described colour pattern similarity among over 400 catfish specimens from the subfamily Corydoradinae. These fish swim in multispecies shoals in streams, rivers, and floodplains throughout South America and have retractable venomous spines and bony plates that serve to defend them against a range of fish and avian predators. As might be expected from a diverse group, the catfish species exhibit a range of colour patterns, including potentially cryptic and disruptive patterns but also highly contrasting patterns (black and white, orange and black, and colours that highlight their spines). After fitting a phylogeny to help understand the historical basis of this colour pattern variation, then placing this variation in a geographical context, Alexandrou et al. (2011) identified 52 species belonging to 24 different Müllerian mimicry rings, each composed of two or

Figure 7.4 Millipedes of the Apheloria clade (top row) and their co-mimics in the Brachoria clade (bottom row). Photo by Paul Marek. Reprinted from Merrill & Jiggins (2009) with permissions (see Plate 10).

three relatively unrelated (hence reproductively iso-lated) co-existing species. The convergent colour patterns among distantly related species is a clear indication of natural selection for colour pattern similarity, but their phylogenetic distinctiveness may well be more than fortuitous. Indeed, in this case the vast majority of co-mimics in the same ring dif-fer from one another in snout length and stable nitrogen isotope content, indicating that they do not compete for benthic resources. This finding contrasts with what we know about the case of Müllerian mimicry reported in the venomous catfishes in Lake Tanganyika in the same year (Wright, 2011), with co-mimics being closely related and occupying similar feeding niches. We leave further discussion of this intriguing example to section 7.5.3.

7.2.5 Bumblebees

It is widely appreciated that female bumblebees (*Bombus* spp) sting and that certain palatable fly species such as hoverflies have evolved to mimic them (see Chapter 9).

Although there are only about 248 recognized species worldwide, bumblebees are highly poly-morphic, which has led to taxonomic confusion and many different names (Williams, 1998). Plowright & Owen (1980) analysed the morphological appear-ances of different species of bumblebee within England and North America and concluded that co-occurring species were much more similar in colour patterns than their taxonomic affinities would pre-dict. In a tour de force, Williams (2007) greatly extended this work by rigorously classifying the multiple (about three per species) visible colour patterns of 219 non-parasitic bumblebee species distributed worldwide, and again confirmed spa-tially aggregated colour-pattern groups comprising a number of different bumblebee species.

In a more directed analysis, Hines & Williams (2012) documented the colour patterns and distri-butions of polymorphic *Bombus trifasciatus*, *B. haem-orrhoidalis*, and *B. breviceps*. Collectively these species exhibit approximately 14 distinct colour patterns, but once again they exhibit similar colour patterns in the same geographical location, despite their phylogenetic distances.

7.2.6 Poison frogs

Poison frogs (including species from the family Dendrobatidae) are well known for their toxicity as well as their highly conspicuous colour patterns. Their toxicity derives from a range of alkaloids, which are obtained from their arthropod diet of mites, ants, beetles, and millipedes (Saporito et al., 2009). As might be anticipated, several potential incidences of Müllerian mimicry among these frogs have been reported. In particular, the dendrobatid poison frog *Ranitomeya* (formerly *Dendrobates*) *imitator* has four distinct morphs in north-central Peru, each of which resembles a distinct co-occurring dendrobatid species from the genus *Ranitomeya* (Stuckert et al., 2014b; Symula et al., 2001). These distributions collectively form a spatial mosaic with narrow transition zones (Twomey et al., 2016). Intriguingly, *Ranitomeya imi-tator* is generally considered more toxic than the spe-cies it resembles in that it has a greater number of distinct alkaloid types and more total alkaloids (Stuckert et al., 2014a), and it currently occurs at higher abundances (Twomey et al., 2013). However, it is noteworthy that all the mimetic morphs in this sys-tem appear to belong to this single species. While genetic analysis of the variability and genetic differ-entiability of populations raises important questions (Chouteau et al., 2011), phylogeographic arguments support the contention that *Ranitomeya imitator* has rapidly adverged to resemble different model species as it expanded its range (Twomey et al., 2016). So its 'imitator' species name may be well deserved.

7.3 Müller's theory revisited

Having been introduced to Muller's theory and surveyed some candidate examples, it is now time to re-visit Müller's theory with a more sophisti-cated representation of predator decision making and evaluate its implications for the evolution of shared warning signals. In particular, the idea of a light suddenly going on in the predator's brain after attacking exactly n distasteful prey is clearly a convenient abstraction. Moreover, even if this were the case, one might expect the number of prey attacked before compete rejection would be based on a number of factors, including the unprofitabil-ity of the prey in question and their abundance.

Indeed, researchers are beginning to recognize that decision making by predators is not simply a matter of learning but is often strategic—balancing the benefits of correct decisions against the costs of incorrect ones (Skelhorn et al., 2016a).

It turns out that the optimal sampling strategy for unfamiliar prey can be deduced from first principles, albeit with certain assumptions. As Sherratt (2011) noted, predators encountering an unfamiliar prey item are effectively faced with an exploration–exploitation trade-off. Do they take the risk and attack the unfamiliar prey item (for future informational rewards, but also uncertain immediate benefit or cost), or do they exploit their available information (which may suggest it is unprofitable to attack) and ignore it altogether? In this case, the optimal decision for the predator can be identified by comparing the known future payoff to a predator

if the prey item is rejected (i.e., a payoff of 0) with the estimated future payoff from attacking it, assuming the continued use of its optimal decision rules as it learns more about the properties of the prey type (Box 7.1).

If predators adopt this optimal sampling strategy (or use rules of thumb that approximate the optimal strategy) then the number of unprofitable prey that should be sampled before complete rejection should rise with the density of these unprofitable prey. This is because the value of obtaining information about common prey phenotypes is higher than that of obtaining information about rare prey phenotypes. Despite the fact that the number sampled before complete rejection (n) can increase with the density of unprofitable prey (a), the rate of increase of the former is less than the latter. Thus, the per capita attack rate on unprofitable prey is predicted to

---

Box 7.1 The optimal sampling strategy for unfamiliar prey

Let N be the total number of an unfamiliar prey type with a given appearance that a predator can possibly sample in its lifetime. This unfamiliar prey type could potentially belong to one or more of up to k profitability classes. Let f_i be the immediate cost or benefit of attacking an individual prey item of profitability class i. The information currently available to a predator is the number of prey items of the unfamiliar prey type that have been attacked that belong to profitability class i (i.e. r_i for $i = 1 \ldots k$ where $\sum_1^k r_i = n$). A predator should continue to attack an unfamiliar prey type if its estimated future payoff from attacking the prey type exceeds its expected future payoff from not attacking it. However, in making this calculation it needs to take into account the fact that its future estimates of payoffs (and hence its optimal decisions) from attacking the unfamiliar prey type are liable to change as it learns more about them.

The future payoff from not attacking the unfamiliar prey type in the remaining ($N - n$) trials on the basis of its current information ($r_1, r_2 \ldots r_k$) can be labeled $S_D(r_1, r_2 \ldots r_k)$ with subscript D representing deferral. Since no beneficial or costly prey are attacked, then the future payoff from deferral, $S_D(r_1, r_2 \ldots r_k) = 0$. By contrast, the future expected payoff from attacking (S_A) an individual of the unfamiliar prey type under the same conditions is:

$$S_A(r_1, r_2 \ldots r_k) = \sum_{i=1}^k \pi_i [S(r_1, \ldots, r_i + 1, \ldots, r_k) + f_i] \quad (7.1)$$

where π_i is the expected probability that the prey item is of profitability class i. In effect, this estimate of future payoff represents the immediate expected payoff from attacking the prey item, weighted by its likelihood i.e. ($\sum_{i=1}^k \pi_i f_i$) plus the future payoff for all its potential future encounters (assuming it makes optimal decisions) weighted by their likelihoods, i.e. $\sum_{i=1}^k \pi_i S(r_1, \ldots, r_i + 1, \ldots, r_k)$. Clearly the estimated probabilities of belonging to a given profitability class (π) will depend on its current information ($r_1, r_2 \ldots r_k$), but these beliefs will be updated as more information is gained about the profitability of the unfamiliar prey type. A convenient way to represent updated beliefs is Bayesian learning—in which prior beliefs are converted to posterior beliefs as new information is gained. If we employ a Dirichlet distribution ($D[\alpha_1, \alpha_2 \ldots \alpha_k]$) to represent the priors (and consequently the posteriors) then at any stage

$$\pi_i = \frac{(\alpha_i + r_i)}{\sum_{i=1}^k (\alpha_i + r_i)}$$

where α_i is a concentration parameter of the initial Dirichlet prior, reflecting the initial relative confidence that the prey type is of profitability class i.

By working backwards from the maximum number of trials N, we can deduce the optimal decision for a predator at each informational state ($r_1, r_2 \ldots r_k$) and hence its optimal sampling strategy. The above approach assumes that the number of unfamiliar prey N is known, but the model is readily generalized to identify the optimal sampling strategy when the density of unfamiliar prey also has to be estimated (see Sherratt, 2011; Aubier et al., 2017 for more details).

diminish (and indeed reach an asymptote) the more prey there are with this appearance, just as Müller had argued (Sherratt, 2011).

What is the optimal sampling strategy for a predator when the prospective mimics differ in unprofitability? In general, predators should sample fewer individuals of an unprofitable prey type before complete rejection the more costly they are to attack (Sherratt, 2011). Indeed, if an unfamiliar prey was sufficiently rare and the costs of attacking prey should they prove to be unprofitable sufficiently high, predators should refrain from attacking them altogether. This is because both the future value of information and the immediate payoff from attack are estimated to be low. While differences in unprofitability and abundance among species can collectively combine to influence which of the species faces the strongest selection for mimicry, the basic tenets of Müller's model in terms of selection for a shared warning signal remain (for a formal demonstration, see Aubier et al., 2017). We return to consider experimental data on the sampling strategies of predators in section 7.5.1. We also consider how Müller's theory applies when co-mimics to differ in unprofitability in section 7.5.6.

7.4 Evidence for Müller's hypothesis

Here we discuss evidence for Müller's hypothesis. We begin by considering field-based evidence for the general form of positive frequency-dependent selection anticipated by Müller. We then consider more focussed studies on predators themselves that have attempted to elucidate the cognitive basis of selection for warning signal uniformity.

7.4.1 Field assessments of the benefits of adopting a common warning signal

There has been a range of field experiments to assess the benefits of adopting a common warning signal. *Heliconius* butterflies have been the predominant study species, but there have also been a number of recent studies on other groups, notably on poison frogs.

One of the earliest field experiments to test the advantages of Müllerian mimicry was described by Benson (1972)—see Figure 7.5a. Working in Costa Rica, Benson (1972) stained part of the forewings of a group of *Heliconius erato* butterflies, creating a non-mimetic form (red band stained black) and a mock-treated control (similar area of black stained black, in order to control for the effects of painting). The sample sizes of individuals marked were relatively low, but by re-sighting individuals at roosting sites it was possible to estimate two indices of selective (dis)advantage: mortality and beak marks. In the first year of study (1968), the non-mimetic forms had lower estimated longevity than their mock-treated controls, while in both years (1968 and 1969) the wing damage was higher in non-mimetic forms than the mock-treated controls and unaltered *H. erato* combined. Although these results are entirely consistent with Müllerian mimicry, it should be borne in mind that beak-marked butterflies have (by definition) escaped and remain alive.

In a subsequent study, Mallet & Barton (1989) reciprocally transferred local races of *H. erato* ('postman' and 'rayed') across their narrow hybrid zone in northern Peru (see Figure 7.5b) so that they could evaluate the relative success of forms that were rare (individuals transferred across the hybrid zone) with forms that were common (controls caught, marked, and released on the same side of the hybrid zone). Overall, proportionately fewer experimental (i.e. novel rare morph) butterflies were recaptured than controls (i.e. established common morphs). Among recaptures, a higher proportion of the rare morph than the common control morphs of *H. erato* carried beak marks, suggesting that predation (most likely by jacamars) played some role in generating differences in establishment rates. Indeed, the differences in survival rate between experimental and control releases were individually significant at two sites where jacamars were common, and not significant at the two sites where jacamars were perceived rare or absent.

In an even more ambitious field study, Kapan (2001) made use of the unusual polymorphism exhibited by *Heliconius cydno* in western Ecuador (Figure 7.5c). Two colour morphs (yellow and white) of this species resemble two different species: *H. eleuchia* (yellow) and *H. sapho* (white) respectively. Where *H. eleuchia* and *H. sapho* occur in the same locality, the yellow and white colour morphs of *H. cydno* vary in frequency. Yellow and white *H. cydno* were caught at two source sites and released in other sites that were dominated by a particular co-mimic. Treatments involved the introduction of a morph that differed from the dominant co-mimic while in the controls the forms were similar. Overall a lower proportion of treatment (rare morph) individuals were re-sighted than control individuals, and these individuals had both lower initial establishment rates and higher subsequent disappearance rates. If predators quickly learn to avoid morphs, then one might expect that there would be little difference in survivorship of controls and treatments when they were released at high density. Indeed, release density also affected survival differences: while there was a significant difference in the estimated life expectancy of treated and control butterflies that were released at low density, there was no such effect at high density. Of course, we still need to understand how polymorphism in *H. cydno* can persist when one might expect that distasteful butterflies should converge on a single shared warning coloration. Kapan (2001) suggested that the solution may lie in geographically variable selection, or relaxed selection at higher density, but we leave consideration of polymorphism in Müllerian mimicry systems to section 7.5.5.

Researchers have increasingly been using artificial models to compare rates of predation between variants of mimetic phenotypes. In a recent experiment,

Chouteau et al. (2016) placed artificial wax models of butterflies with coloured paper wings in transects at six different locations in northern Peru (Figure 7.6a,b). These artificial models (over 100 of each colour form) were designed to match natural colour variants of *H. numata* that varied geographically (in some locations certain colour patterns were rare, in other locations the same colour pattern was common). Models of the common palatable brown butterfly *Pierella hyceta* were also distributed to assess the overall intensity of predation in each area. In several locations the models with exotic (rare) colour patterns showed signs of attack more frequently than the intermediate and common colour patterns for that area. Moreover, after fitting a generalized linear mixed model to their data (available in their Supplementary Information), it is evident that while relative frequency of the phenotype in the area was a highly significant predictor of attack rates on the various artificial models and attack rates varied between locations, the specific colour pattern phenotype had no detectable influence per se. Thus, it is clear that there is strong localized selection imposed by predators for colour pattern uniformity, even in this species that is known to exhibit local colour polymorphism.

Chouteau et al. (2016) note that even individuals with the appearance of the common morph are not

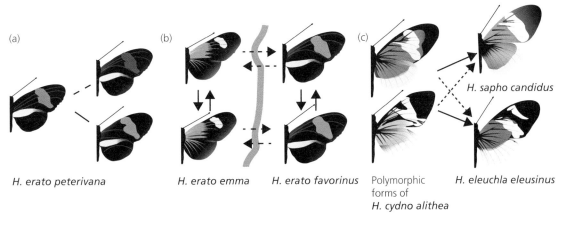

H. erato peterivana   H. erato emma   H. erato favorinus   Polymorphic   H. eleuchla eleusinus
                                                                  forms of
                                                                  H. cydno alithea

Figure 7.5 Field tests of Müllerian mimicry. (a) In Benson's (1972) experiment the establishment success of altered and mock-treated *H. erato* butterflies were compared. (b) In Mallet & Barton's (1989) study, different forms of *H. erato* were reciprocally transferred across their hybrid zone and their establishment success compared with controls moved within each area. (c) In Kapan's (2001) experiment, white and yellow forms of *H. cydno* were separately transferred to sites dominated by white co-model (*H.sapho*) and the yellow co-model (*H. eleuchia*) and their establishment success compared. From Kapan (2001), re-drawn from *Nature* with permission (see Plate 11).

Figure 7.6 A collection of artificial butterflies (coloured paper wings with wax bodies) with patterns resembling variants of *H. numata*, placed out in the field by Chouteau et al. (2016) to quantify the local susceptibility of each colour form to predation (b) an example of one such butterfly deployed in the field. Photos courtesy of Mathieu Chouteau (see Plate 12).

completely protected from predation (an observation seen in other experiments such as those with poison frogs), leading the authors to suggest (in contrast to earlier mark-recapture studies indicating rapid learning) that predator knowledge is saturated only for the most common warning signals. However, there are many reasons why artificial prey with the common appearance might experience some predation. Of course, any influx of naive predators in the area would also explain why morphs are not completely protected. Moreover, given their resting positions and lack of movement, some birds might suspect that the butterflies are not what they appear to be. Perhaps most importantly, in adding 100 additional individuals of a given colour pattern then the perceived local density of this phenotype will be augmented, leading predators to sample a few more of them to confidently establish their (un)profitability (Sherratt, 2011). In this sense, there may be a Heisenberg uncertainty principle operating—in conducting well-replicated field experiments to assess the nature of frequency dependence in Müllerian mimicry systems, the underlying conditions are also altered.

Not all experiments establishing positive frequency dependence have been conducted with Neotropical butterflies. The leaf beetle *Oreina gloriosa* is a relatively sedentary metallic-coloured beetle that is chemically defended by cardenolides and found in isolated populations throughout the mountains of Europe. Its potential predators include

European robins (*Erithacus rubecula*) and Eurasian wrens (*Troglodytes troglodytes*). Other leaf beetle species of the genus *Oreina* tend to share the same local colour patterns and chemical defences through sequestration/synthesis. Borer et al. (2010) was able to crudely classify the leaf beetle community in 20 separate high-altitude locations as being dominated by the metallic green phenotype or metallic blue phenotype. After tethering 10 green and 10 blue *O. gloriosa* at these locations, the authors found that beetles that matched the locally predominant colour morph had a significantly higher survivorship (by a factor of 1.67) compared to those beetles that did not match, once again indicating the potential for localized positive frequency dependence.

Analogous experiments have also been carried out with dendrobatid frogs. For example, *Dendrobates tinctorius* exhibits significant geographical variation in colour pattern in French Guyana. To understand the factors that might maintain this variation, Noonan & Comeault (2009) placed out hand-painted clay frogs that resembled the local phenotype (yellow dorsum, black legs), along with clay frogs that resembled those from a coastal population (blue legs, black dorsum, yellow stripes) and a cryptic brown form. They found that brightly coloured novel forms were more likely to suffer an attack from avian predators (deduced from characteristic peck marks) than both the local aposematic form and cryptic forms. Nevertheless, this study was conducted in only one locality and (like

Figure 7.7 Clay models of frogs (bottom row) hand-crafted by Chouteau & Angers (2011) to resemble two aposematic phenotypes (striped and reticulated) of *Ranitomeya imitator* and a non-conspicuous frog (top row). Images courtesy of Mathieu Chouteau (see Plate 13).

Benson's classical study) could potentially have arisen from frequency-independent rather than frequency-dependent selection, with models displaying the yellow dorsum simply attacked less. In a more complete reciprocal 'transfer' experiment, Chouteau & Angers (2011) noted that the dendrobatid frog *Ranitomeya imitator* had a vivid green reticulated appearance in a high-elevation valley in northern Peru, yet exhibited a distinct gold-striped pattern at a lowland site just 15km away. Motivated by these observations, they placed 900 hand-painted clay frog models at each of the two sites. The clay models comprised 300 frogs with the locally dominant colour pattern, 300 of the locally novel pattern, and 300 resembling a non-aposematic (brown) frog (see Figure 7.7). Although attack rates were relatively low in each of the sites (10–14 per cent), in each locality the exotic morph experienced up to four times more attacks by avian predators (identifiable through their characteristic peck marks) than the local one, and two times more than the non-conspicuous phenotype. This form of positive frequency-dependent selection readily explains selection for common warning signals in the manner that Müller had anticipated.

There has been a range of field-based experiments investigating the nature of frequency dependence using artificial baits that do not resemble any particular organism. While unnatural, they do have the benefit that they are easier to replicate and manipulate. In an early experiment, Greenwood et al. (1981) presented wild birds with yellow and red pastry baits (in a 1:9 ratio and in a 9:1 ratio), both of which had been made unpalatable using quinine sulfate. While birds showed a frequency-independent preference for red baits they showed no tendency to consume disproportionately more of the rarer form of unpalatable prey. In a subsequent experiment, Greenwood et al. (1989) presented yellow and red pastry baits (made equally distasteful with quinine hemisulfate) at different ratios (1:9 and 9:1) to garden birds, this time with separate control populations involving similar pastry baits without quinine. In this experiment disproportionately more of the rare forms of distasteful baits were indeed attacked by birds. Intriguingly, more of the more common distasteful form were taken in the process, a result we return to in section 7.5.1. No such frequency dependence was observed in the palatable control populations (which were consumed at far higher rates), suggesting that the frequency dependence arose due to aversion learning.

7.4.2 Direct observations of predators reacting to warningly coloured unpalatable prey

Much of the fieldwork on Müllerian mimicry has understandably focussed on the relative rates of survival of different prey phenotypes at different relative frequencies, but we can also understand the processes that generate positive frequency-dependent selection by directly observing predator behaviour. For example, one of the most basic assumptions of Müller's hypothesis is that there is a degree of learned avoidance of unpalatable prey—if avoidance were always innate then there would be no predators to educate and no burden to share (Müller, 1879). It has long been recognized that some predators have evolved innate avoidance for highly dangerous prey. For example, naive avian predators will tend to avoid coral snake patterns (Smith, 1975, 1977). Likewise, naive predators frequently show heightened wariness when presented with novel and/or conspicuously patterned prey (e.g. Schuler & Roper, 1992). However, there are now numerous studies (albeit often with artificial prey) demonstrating that predators often need to learn to associate colour patterns with unprofitable experiences. These experiments typically show that predators gradually reduce their attack rates on conspicuously coloured unprofitable prey over a series of test sessions (Alatalo & Mappes, 1996; Ham et al., 2006; Kazemi et al., 2014; Skelhorn et al., 2016a). For example, Dell'Aglio et al. (2016) placed out chromatically and achromatically coloured artificial heliconid butterflies with novel warning patterns (found only in the Amazon Basin) along three trails in Panama in two separate 4-day trials, 5 days apart. These artificial prey were rendered unpalatable using quinine monohydrochloride dehydrate. While the attack rates on the achromatic butterflies was higher than the chromatic butterflies, the proportion of both heliconid-like prey types attacked declined in successive trials. Likewise, Chouteau & Angers (2011) found that the frequency of new predation attempts on their clay frog models declined over time.

Field and semi-field experiments with relevant natural predators of heliconid butterflies have also provided evidence of learned avoidance. For example, jacamars specialize in preying on fast-flying insects such as dragonflies and butterflies and have long been implicated as selective agents for Batesian and Müllerian mimicry. When exposed to Neotropical butterflies for the first time, naive young jacamars tend to attack the unpalatable butterflies without inhibition yet subsequently reject them, showing learned aversion (Chai, 1996). By contrast, Langham (2004) presented 80 wild-caught (hence somewhat experienced) caged adult rufous-tailed jacamars (*Galbula ruficauda*) with individuals of three morphs of *Heliconius* butterflies—the local morph of *H. erato*, and two other morphs of *H. erato* (red and black) that were simply artificial morphs created with permanent ink. Overall, the novel (red or black) *Heliconius* was attacked by 29 of 80 jacamars (36 per cent) in these first trials, while the local morph was completely avoided. Clearly for adults of this predator species, at least, its avoidance of butterflies exhibiting the familiar warning signals was complete. More importantly, it also shows that the jacamar uses subtle differences in colour patterns when making discriminative decisions and is capable of selecting out novel variants. In follow-up work, 46 of the same birds were re-caught from between 4 to 429 days since original testing and tested again in the same manner. A total of 39 of the birds showed the same attack predilections in terms of complete avoidance or attacking one or the other novel morphs, suggesting that they had highly repeatable behavioural responses.

In an earlier study, Pinheiro (2003) released live mimetic Neotropical butterflies (all participants in Müllerian mimicry rings) and non-mimetic butterflies close to predatory tropical kingbirds (*Tyrannus melancholicus*) and cliff flycatchers (*Hirundinea ferruginea*). To evaluate the role of the likely prior experience of these predators, he conducted his experiments in three different Amazonian habitats (rain forest, city, and low growth 'canga' vegetation), where only particular species of butterflies co-occur. The highest proportion of sight-rejections by birds occurred at the forest site where cliff flycatchers selectively attacked the cryptic and palatable species and avoided almost all the Müllerian mimics on sight. By contrast, the cliff flycatchers in the canga did not discriminate among butterflies and attacked most mimetic and non-mimetic species. Tropical kingbirds in the city also attacked most non-mimetic species, but did tend to sight-reject some species of Müllerian mimic.

Following Chouteau & Angers' (2011) field demonstration of frequency-dependent selection for common forms of the dendrobatid frog *Ranitomeya imitator* (see section 7.4.1), Stuckert et al. (2014b) conducted trials in which they presented young chickens (*Gallus domesticus*) with either the spotted morph of *R. variabilis*, the spotted morph of *R. imitator* that resembles it, or the striped morph of *R. imitator* that does not have this appearance. All chicks initially expressed a degree of interest in the frogs, and a number picked them up, but some chicks expressed innate neophobia and did not taste the frogs they were presented with. Comparing hesitation times in the pre-learning and post-learning trials it was clear that chicks learned to avoid the brightly coloured frogs. However, both striped and spotted phenotypes were subsequently avoided at similar rates, independent of the training environment. So, while the work demonstrates that avian predators can learn to avoid this type of prey it does not directly explain the localized frequency dependence that had previously been observed in this system. Of course domestic chickens are not wild predators and they have been bred to put on weight. Moreover, as the authors note, this high degree of generalization may have arisen as a consequence of the experimental design (which re-used chicks, presumably for ethical reasons, so that the predators had previously been exposed to frogs with both phenotypes). We return to stimulus generalization in section 7.5.3.

7.5 Questions and controversies

7.5.1 Do predators take a fixed *n* in aversion learning?

As noted in section 7.1.1, Müller speculated that predators would take approximately a fixed number of prey with a given appearance before completely avoiding them, when making his case for positive frequency-dependent selection on warning signals. However, this was simply for illustration. As reported above, Greenwood et al. (1989) found that while disproportionately more of a rarer form of distasteful baits were attacked by garden birds, more of the common form of baits were attacked, suggesting that the number sampled (*n*) before rejection increases with prey density. Lindström et al. (2001) likewise found that while the proportion of unpalatable artificial prey attacked by captive

great tits declined with increasing density available, the total number of unpalatable prey consumed increased with absolute density. Analogous phenomena have since been found in a variety of systems. For example, Beatty et al. (2004) conducted experiments in which humans searched for artificial computer-generated prey, some of which were costly to attack. Once again, it was found that the per capita 'mortality' of the unprofitable forms declined as their frequency increased, but that the human predators attacked more of the common form of the unprofitable prey available. To explore this phenomenon further, Rowland et al. (2010c) presented male chicks with mixtures of green, palatable (water-coated) and red, unpalatable (quinine-coated) crumbs. The starting density of edible crumbs was fixed at 10 or 20, but the density of unpalatable crumbs was varied from 5, to 10, to 30 for each edible prey density (6 treatments overall). Chicks in all treatment groups readily learned to discriminate between quinine-coated red prey and water-coated green prey. Moreover, the mean per capita attack rates (total attacked/total presented) of unpalatable prey decreased as the number of unpalatable prey in the population increased. Once again, however, chicks attacked significantly more unpalatable red crumbs the more that were available.

Why would predators sample more unprofitable prey before the experiment ends? Greenwood et al. (1989) suggested that it may have resulted from predators leaving the study area before they were fully educated to avoid the rare form, but after they were educated about the common form (see also Chouteau et al., 2016 above). Mallet (2001) similarly suggested that a possible explanation for the Lindström et al. (2001) results was that learning was not complete. However, Beatty et al. (2004) examined the encounter and attack sequences of the human volunteers and noted that even rare unprofitable prey tended to be rejected by the end of the experiment, indicating that avoidance learning was complete. An alternative explanation for the phenomenon may be that predators occasionally return to attacking the more common unprofitable type since the opportunity cost of not attacking them, should some prove to be palatable, is higher (Beatty et al., 2004). Sherratt (2011) formalized this argument, showing that as the density of an unfamiliar prey type increases then information about their properties increases in value, so that predators should sample more of them.

7.5.2 Is one signal better than two?

If Müller's theory was correct then the positive frequency dependence generated by predators ('strength in numbers') should lead to one signal in a given number of unprofitable prey incurring lower overall per capita mortality than if the same number of prey had two signals. So, while the per capita mortality on a given unprofitable prey type should decline with increasing density, another comparison can be made—unprofitable prey that exhibit two distinct signals (some with one signal, the other with another) should experience higher overall mortality than prey with a single signal. Note at the outset, however, that learning to associate two separate signals with unprofitability may not be particularly challenging; by contrast, learning to associate many different signals with profitability and unprofitability is likely to be much more demanding. Therefore one might expect the intensity of selection for Müllerian mimicry to increase with community complexity (Beatty et al., 2004; see section 7.5.3).

The above 'one signal is better than two' prediction has been tested on a number occasions, with mixed results. For example, in a 'novel world' experiment, Rowe et al. (2004) presented individual wild-caught great tits with palatable and unpalatable artificial prey (slices of almond, rendered unpalatable by soaking in chloroquinine phosphate) hidden under cards carrying particular symbols. Either 20 palatable and 20 unpalatable prey were presented, or just 40 unpalatable prey alone. The unpalatable prey were presented either with a single signal, or in two forms (50 per cent of each type that were judged similar or more distinct in appearance). The authors found that the extent of pattern similarity between co-mimics (identical, similar, or dissimilar) had no detectable influence on their overall survivorship. Thus, two patterns did not appear any more difficult to learn to avoid than just one (although in the last two trials there was a trend of indicating that increasing pattern similarity

reduced mortality). Related experiments conducted as part of more complex designs, in which co-mimics vary in unpalatability, have likewise reported that the predation pressure on unpalatable prey with two signals is not always significantly higher than on the same number of unpalatable prey with just one signal (Ihalainen et al., 2007; Lindström et al., 2006).

Nevertheless, Rowland et al. (2010a) argued that the signals of each prey type used in the Rowe et al. (2004) experiment had shared features, so the study may be more appropriately considered a comparison of the success of perfect and imperfect mimics. Likewise, simultaneously varying both the level of unpalatability and signal design (of uncertain conspicuousness and discriminability) can make inferences difficult. Motivated by the above challenges, Rowland et al. (2010a) conducted a 'novel world' experiment involving wild-caught great tits foraging for cryptic palatable and conspicuous unpalatable prey (almonds with or without chloroquine phosphate) hidden under cards with one of two signals (a black-filled square and a symbol resembling a spoked wheel). These symbols were confirmed at the outset to be approximately equally visible when palatable. The symbols also generated similar avoidance learning patterns when unpalatable, and they were shown to be discriminable in that the birds did not completely generalize their learned avoidances to prey bearing the alternative symbol. Rowland et al. (2010a) then presented great tits with 50 cryptic palatable prey, and 50 unpalatable models, with or without an additional 50 unpalatable mimics that had either the same symbol as the model (generating 100 unpalatable prey with one signal) or a different symbol (generating 50 unpalatable prey with one symbol and another 50 with the alternate symbol). The overall per capita mortality of unpalatable prey declined as their density was increased from 50 to 100, a result readily understood in terms of an overall dilution effect (trials were always ended after the birds killed 40 prey items in total). More importantly, however, the authors found that the model had significantly lower mortality when presented alongside prey sharing the same symbol, compared to when presented with alternative prey with a different symbol.

7.5.3 How does Müllerian mimicry evolve?

How does Müllerian mimicry evolve from a putative starting point of two or more species with distinct appearances? Assuming the simplest case of just two species, does one species evolve to mimic the other ('advergence'), or do both species evolve towards one another in appearance ('convergence')? As we will argue, to fully address this question we also need to consider the phenomenon of generalization, since the extent of generalization will influence the degree to which intermediate phenotypes are protected from predation and hence the nature of the adaptive landscape.

Müller (1878) argued that mimicry would frequently involve convergence (English translation): *'the question of which one of two species is the original and which one is the copy is an irrelevant question; each had an advantage from becoming similar to the other; they could have converged to each other'*. Dixey (1920) shared this view, labelling such reciprocal advantages of mimicry as 'diaposematism'. In effect, the early proponents of diaposematism argued that a mutant of a common species that resembled a rarer species may not lose all of its protection from resembling the common species—on the contrary, it may gain as a 'jack of all trades' by looking like both. Intuitively, one might expect convergence of this nature would be more likely to arise when the species are already reasonably similar in appearance, so that intermediates would survive at high rates. Moreover, if there are two stimuli that a predator is learning to avoid, then there may be a form of 'peak shift', in which a predator's maximum aversive response arises not around any one of the stimuli, but at some intermediate point (Balogh & Leimar, 2005).

Ironically, however, at face value Müller's simple model beginning with readily differentiable (i.e. no intermediate) phenotypes suggests that selection for mimicry is a one-way street, i.e. advergent (Marshall, 1908). In particular, it has already been argued (section 7.1.1) that if there are two unprofitable species (1 and 2) then species 1 should be selected as the mimic (and species 2 as the model, but not vice versa) if $(n_1/a_1) > (n_2/a_2)$ where n_1 and n_2 refer to the number of each type consumed through avoidance learning. As more mimetic mutants arise and are favoured,

the selective pressure for advergent mimicry only increases.

So far, the empirical evidence available has tended to support advergence (see Mallet, 1999 for a detailed discussion), although such inferences are not always clear-cut. For example, despite being more abundant and having more toxins than its models (Stuckert et al., 2014a) the poison arrow frog *Ranitomeya imitator* appears to have adverged to resemble different model species in different geographical areas (Symula et al., 2001; Yeager et al., 2012) most likely as it spread into new areas from rarity. Likewise, Marek & Bond (2009) have argued that Brachoria millipedes likely evolved the colour patterns of the more established and widespread species (but see Merrill & Jiggins, 2009). Similarly, ecological and genetic arguments (Mallet, 1999; Flanagan et al., 2004) suggest that Müllerian mimicry among *H. melpomene* and *H. erato* may have arisen mostly via advergence of *H. melpomene* towards *H. erato*. However, historical reconstruction based on genetic analyses by Hines et al. (2011) suggest that while *H. melpomene* adverged on the red-banded patterns of *H. erato*, both species acquired the Amazonian rayed mimetic pattern around the same time, making it harder to discern the leader and/or follower.

Despite continuing empirical and theoretical debates, the advergent–convergent dichotomy may well be too simplistic. For example, it is possible that even when two species experience selection to resemble one another, differences in the mutational space available and intensity of selection produces an outcome which is predominantly, but not exclusively, advergent in nature (Balogh & Leimar, 2005). Likewise, Sheppard et al. (1985) noted that while Müllerian mimicry might involve an initial stage of advergence (which is particularly likely if one of the species was cryptic and the other conspicuous), mutual convergence in both species might subsequently be selected for as the species come to resemble one another (Franks & Sherratt, 2007). Note also that while harder to prove, a shared appearance among closely related Müllerian mimics may arise simply through conservation of the ancestral state. This is precisely what Machado et al. (2004) found in their analysis of the yellow-black pattern in chemically defended soldier beetles (Cantharidae). More recently, Wright (2011) made a case for the conservative co-evolution of colour patterns among the venomous *Synodontis* catfish species of Lake Tanganyika, noting that those species confined to Lake Tanganyika were more similar in appearance than related *Synodontis* species occurring outside of the lake. In this way, predation maintains the similarity, while shared phylogeny readily explains its origins.

Since Müllerian mimicry can in theory evolve even by small mutational steps when the extent of generalization (the phenomenon of attributing common properties to distinguishable objects) across colour patterns is wide (Balogh & Leimar, 2005; Ruxton et al., 2008), then it is important to consider how broad generalization tends to be in natural systems, and when and why predators would tend to generalize broadly. One insight as to the breadth of generalization exhibited by natural predators comes from considering co-occurring mimicry rings, and comparing the success of intermediate phenotypes. Indeed, the very persistence of multiple co-existing mimicry rings (see section 7.5.5) serves to illustrate the fact that predators do not simply generalize across all conspicuous forms. In most *Heliconius* species, a single mimetic colour pattern is maintained within populations by positive frequency-dependent selection (see section 7.4.1). However, *H. numata* is unusual in having a high number of morphs present in each locality, with each morph a close Müllerian mimic of a coexisting ithomiine species (Joron et al., 1999). The presence of these multiple co-existing forms allowed researchers the opportunity to investigate the relative success of crosses that are intermediate in phenotype.

To understand why such intermediates are so rare, and the types of experiment that have been conducted, we first need to understand more about the fascinating genetics of *H. numata*. Genetic variation in wing colour pattern morphs in the species is controlled by about 20 genes clustered in a single genomic region (hence low recombination) referred to as the 'supergene P', with a range of 'superalleles' (Joron et al., 2006). With so many morphs in the same locality, and plenty of opportunity for hybridization, one might wonder how the polymorphic mimicry of different species is ever maintained. It turns out that the majority of crosses between local phenotypes resembling different models continue to

resemble one model or another because of the strong hierarchical dominance that has evolved among sympatric (but not parapatric) superalleles (Joron et al., 2006). Nevertheless, intermediate heterozygous phenotypes (the products of rare recombination or incomplete dominance) can occasionally be generated. In an ambitious experiment, Arias et al. (2016a) deployed over 5000 artificial butterflies (comprising paper wings with wax bodies) that displayed colour patterns exhibited by *H. numata* at eight sites in northern Peru. Included in the colour pattern variants were mimetic forms, intermediate phenotypes, exotic aposematic morphs of the species (absent from the local community but found elsewhere), and cryptic phenotypes resembling a palatable butterfly species. After 72 hours exposure in the field, it was found that both the non-mimetic exotic morphs and intermediates were attacked more than the local mimetic forms. These results suggest clear adaptive peaks associated with locally abundant mimetic morphs, surrounded by valleys in which intermediate phenotypes are purged. Not only do such results help explain why the unusual genetic dominance may have evolved in this species, but they also indicate that the agents of selection were relatively picky with relatively narrow generalization around the ithomiine models they were familiar with. Indeed, simply having aposematic coloration provided no clear benefit to the exotic morph since it was attacked more frequently than the cryptic morph.

The phenomenon of generalization has been widely researched by psychologists but it is of increasing interest to evolutionary biologists interested in signal evolution. Müller did not consider the appearance of profitable prey at all when making his arguments, and in effect he assumed that predators learn to recognize the characteristic features of each and every distinct unpalatable prey type independently. Yet, as Fisher (1930) observed, '*being recognized as unpalatable is equivalent to avoiding confusion with palatable prey*', so categorization of prey types will almost inevitably involve seeking rules to distinguish profitable from unprofitable prey (MacDougall & Dawkins, 1998; Sherratt & Beatty, 2003). For example, if most prey with a yellow stripe are generally found to be unpalatable then a predator may think twice about attacking an unfamiliar prey item with a yellow stripe, even if it is readily distinguishable from other such prey with yellow stripes. In this way, even unprofitable prey species that are readily distinguishable may mutually reduce one another's attack rate, so long as they share a common appearance property (Chittka & Osorio, 2007).

While theoretical treatments (Balogh & Leimar, 2005; Ruxton et al., 2008) have tended to consider generalization curves immutable, it is clear predators should be prepared to generalize widely when the relative cost of attacking an unprofitable prey item is high. Indeed, Kikuchi & Sherratt (2015) recently showed that it would sometimes pay predators to lump unfamiliar prey that had common traits together—even if they could be phenotypically distinguished—if learning more about them separately was potentially expensive. In effect, while learning to avoid moderately distasteful prey may involve relatively narrow generalization curves, it is likely that learning to avoid lethal prey would involve very broad generalization—so that any unfamiliar prey sharing some of the characteristics of prey known to be highly unprofitable would be given a wide berth.

Other factors may also affect the breadth of generalization. In particular, predators only have a finite capacity for processing information (Sherratt & Peet-Pare, 2017), so one might expect that simple (hence broadly applicable) discriminative rules are likely to be employed when the phenotypic diversity of profitable and unprofitable prey phenotypes is high (MacDougall & Dawkins, 1998). In support, Beatty et al. (2004) conducted experiments on humans foraging for artificial computer-generated prey and concluded that imperfect mimetic forms are more likely to spread in complex communities compared with simple communities, because of the need for predators to adopt simple rules of thumb to deal with the high signal diversity. Likewise, in laboratory trials, Ihalainen et al. (2012) found that only those great tits that had foraged for prey in communities of artificial prey with little signal diversity actively selected against inaccurate mimics. There is also gathering comparative evidence that community complexity can influence the nature of selection for Müllerian mimicry. For example, Wilson et al. (2013) found that the extent of mimetic similarity in velvet ants varied between Müllerian mimicry rings,

and suggested that this variation likely arose from there being wider generalization in more diverse communities.

In sum, Müller's model predicts advergence, and empirical data tend to support this view. However, when the unprofitable species are relatively similar in appearance and/or the predators generalize widely (an outcome particularly likely when the prey are highly unprofitable and/or the prey community is particularly diverse), then it is possible that there is ultimately selection favouring convergence of both forms towards one another. The observation of mimicry rings is a reflection of the adaptive landscape generated by predators, and the fact that there are distinct peaks clearly shows the converse: there are also deep valleys where intermediates are selected against.

7.5.4 Why are mimetic species variable in form between areas?

We have already described examples of Müllerian mimics, such as *Heliconius erato* and *H. melpomene*, which, while locally monomorphic, vary considerably every 100km or so *'as if by touch of an enchanter's wand'* (reported commentary of Bates in Müller (1879), page xxix). Analogous spatial variation has been reported in a number of other Müllerian mimics, such as the burnet moth *Zygaena ephialtes*, which exists in four different colour forms in different areas throughout Europe, with each form a member of a different Müllerian mimicry complex (Turner, 1971), cotton stainer bugs (genus *Dysdercus*) which show widespread coincident intraspecific variation in colour pattern (Zrzavy & Nedved, 1999)) and *Eulaema* bees which likewise co-vary in colour pattern over South America (Dressler, 1979)– see also our description of Müllerian mimicry in bumblebees, velvet ants, and frogs (section 7.2). Although spatial mosaics are seen as something of a paradox in Müllerian mimicry systems (e.g. Langham, 2004), so long as there are mechanisms to generate the geographical diversity, their subsequent maintenance is a direct consequence of the frequency dependence favouring common forms, imposed at limited spatial scales (Joron & Iwasa, 2005).

A range of theories, not mutually exclusive, have been put forward to explain the underlying source of the geographical diversification, particularly in Neotropical butterflies. For instance, one possibility is that local populations simply diversified from one another when in isolation due to Pleistocene ice ages. However, a variety of historical and geographical arguments have since been raised that render this theory unlikely (see Jiggins, 2017 for a review). Instead, some researchers place more emphasis on Wright's 'shifting balance' theory (Wright, 1932) to explain the geographical diversification of Müllerian mimics (Mallet & Singer, 1987). This argument maintains that stochastic processes generated by genetic drift can help explain how novel variants might cross adaptive valleys and ultimately be selected upon. Naturally, this mechanism would be expected to be more powerful in small populations and in areas temporarily devoid of predators, such as species spreading into new areas. For example, Sherratt (2006) explored stochastic simulations which showed that a spatial mosaic can form under local positive frequency-dependence generated by Müller's 'fixed n' rule, but only so long as there were chance effects, such as the temporary absence of predators from certain areas, that could help generate the initial spatial diversity. In support, Chouteau & Angers (2012) extracted genetic samples and conducted predation experiments either side of a transition zone in the poison frog *Ranitomeya imitator* in northern Peru. Although transition zones may be special for a number of reasons, the authors concluded that the frog population in this zone had both high phenotypic diversity and relatively relaxed predation (even clay models of palatable frogs were less frequently attacked than elsewhere). In this way the hybrid zones, or simply areas with relaxed predation, may be important engines of phenotypic diversity, but precisely how these new forms might succeed elsewhere (or spread within the zone) is unclear.

In a memorably titled 'Shift happens…' opinion piece, Mallet (2010) makes a strong case for how shifting balance can help explain warning signal diversity in *Heliconius* butterflies. Likewise, Jiggins (2017) contrasts the shifting balance and refugia explanations, favoring the former. Nevertheless, both authors suggest that selection may play some role in explaining the generation of phenotypic diversity necessary for mosaics. This selection could be frequency-independent and arise through local

adaptation to particular abiotic (e.g. light environment) or biotic (e.g. social mating preferences, different predator species) environments not related to mimicry. However, another means through which novel forms may at least be initially protected is neophobia. Indeed, Aubier & Sherratt (2015) showed that the optimal sampling strategy for unfamiliar prey could generate spatial mosaics of co-varying colour patterns far more readily than Müller's fixed n rule, simply as a consequence of the cautiousness expressed by predators when novel forms were rare and the fact that attacking unprofitable prey is relatively costly (Figure 7.8). This adaptive neophobia allows particularly rare forms the chance to persist for long enough to increase in density, but once they are sufficiently common to be attacked then they will only thrive if they have shared warning signals. In this way, neophobia can help generate diversity by allowing rare morphs an opportunity to persist, but it does not obliterate the positive frequency dependence maintaining monomorphism at higher densities. Ironically then, the very same (optimal) sampling behaviour that leads to Müllerian mimicry can also help explain the generation and maintenance of spatial polymorphisms.

Intriguingly, while the zones at which different phenotypic forms meet often coincide with a barrier to dispersal such as a major river, one can gain an insight as to the nature of selection from the width of the zone in which morphs from neighbouring areas intermix. As one might expect, the more movement, the wider the zone—but the more intense selection against hybrids, the narrower the zone. Indeed, the width of the zone is approximately proportional to the rate of dispersal (measured in terms of gene flow σ) divided by the strength of selection (measured in terms of $\sqrt{s}$) (Jiggins, 2017; Mallet & Barton, 1989). Simulations of Aubier & Sherratt (2015) confirm that low rates of dispersal are more likely to generate highly patchy mosaics, which helps explains why transitions in *Heliconius* arise in the order of 100km, whereas the transitions in poison frogs arise in the order of 10km (Mallet, 2014). Several mathematical and simulation models suggest that the hybrid zones between different phenotypic forms can form even without physical barriers, in which case they can move in a manner influenced by the curvature of the boundary (Kawaguchi &

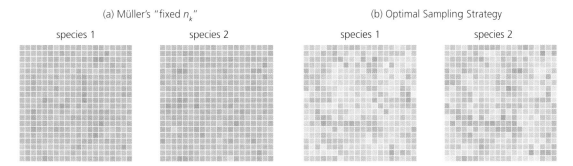

Figure 7.8 Examples of prey community composition in the spatial model (a 20 x 20 regular lattice, with two unprofitable species) after 500,000 generations when (a) Müller's "fixed n" model and (b) the optimal predator sampling strategy is implemented. The same nine distinct morphs of each unprofitable species were assumed possible. The local population morph composition is shown for both species, with the colour of any given cell indicating the dominant morph phenotype of that species in that area. Initially, all individuals were represented by the red morph. When local predators sample a fixed n in each local population then this monomorphic mimicry is maintained across the lattice, because rare mutant morphs are always attacked. However, when predators employ the optimal sampling strategy then they will often avoid attacking rare morphs, only attacking them when they reach a density at which they becomes worthwhile to sample. Under these conditions a diverse spatial mosaic readily arises, with the forms of each species matching one another in any given area but with the dominant morph varying between areas. See Aubier & Sherratt (2015) for model details (see Plate 14).

Sasaki, 2006; Sasaki et al., 2002; Sherratt, 2006). Indeed, a well-known juncture between two forms of the 'postman' pattern of *H. erato* in Panama has been documented to move about 47 km in 17 years (Blum, 2002). Another intensively studied hybrid zone is found near Tarapoto in northern Peru where two forms of *H. erato* (and two corresponding forms of *H. melpomene*) meet. In this case, however, both the width and position of the hybrid zone has remained remarkably stable in the past quarter century (Mallet et al., 1990; Rosser et al., 2014). Here high levels of precipitation at the edge of the easternmost Andes may generate a population density trough for butterflies, effectively creating a barrier that traps the hybrid zones at the foot of the mountains. Overall, it is clear that local positive frequency dependence serves to promote local homogeneity, but it can also preserve regional heterogeneity once local differences have arisen by some other means. Hybrid zones where two forms meet are have understandably attracted considerable interest since they highlight an intriguing discontinuity.

Warning colour patterns may also be used as mate recognition cues, so mate choice (by males and/or females) may serve to reinforce or undermine the localized frequency dependence imposed by predators. In particular, several studies have shown that male *Heliconius* butterflies exhibit preferences for mates that have similar wing colour patterns as themselves (Jiggins et al., 2001, 2004; Kronforst et al., 2007; Merrill et al., 2011a; Finkbeiner et al., 2014). It turns out that male mating preferences and their colour patterns tend to be inherited together in *Heliconius* butterflies (Chamberlain et al., 2009), while linkage mapping studies have confirmed that traits associated with mate preference loci map very closely to loci underlying colour pattern variation (Merrill et al., 2011b). So, colour and preference for colour are closely associated in these species. Since the local patterns within *Heliconius* that are preferred in mating trials are also the ones that experience less predation (e.g. Finkbeiner et al., 2014), then the mating preferences observed in *Heliconius* butterflies may well have evolved as a consequence of the fact that inter-morph hybrid offspring with unusual colour patterns suffer high predation. In this way, selection for mimicry may generate mating preferences that further promote phenotypic uniformity.

A fascinating counter-example has recently been reported in the locally polymorphic *H. numata*. Here females of all morphs exhibit a strong aversion toward males displaying their own phenotype, that is they exhibit disassortative mating preferences (Chouteau et al., 2017). The ultimate reasons for these diversity-generating mate preferences are unclear, but one possibility is that it helps promote genetic heterozygosity. Whatever its selective benefits, just as assortative mating can promote uniformity, disassortative mating preferences may in part help explain the maintenance of the unusual local polymorphism in this species (section 7.5.5).

In a more preliminary study, Twomey et al. (2014) quantified mate choice in the poison frog *Ranitomeya imitator* (see section 7.2.6). Both sexes of *R. imitator* frogs are able to express mate preference, with males generally initiating courtship and females choosing to follow up or take no interest. Working at a contact zone between the striped and a Varadero (blotched) form of *R. imitator*, Twomey et al. assayed mutual mate choice by presenting a single male of a given form with a striped and a blotched female and quantifying the proportion of total courtship time that was for the striped morph. Striped *R. imitator* from near the contact zone preferred to court fellow striped morphs over blotched morphs. However, blotched morphs, as well as striped morphs from well outside the contact zone, do not exhibit a clear preference for either morph. This asymmetric courtship preference parallels other studies of incipient speciation which have found greater selection against hybridization in the contact zones and greater selection to avoid hybridization in some morphs (notably, rare ones) than others (e.g. Hoskin et al., 2005).

7.5.5 How can multiple Müllerian mimicry rings co-exist?

While spatial polymorphisms can be understood, it is much harder to understand how members of the same taxonomic group can co-exist in several different rings all in the same general location. For example, ithomiine butterflies from a single site have been classified as members of 18 distinct co-existing mimicry rings (Beccaloni, 1997). Likewise, different bumblebee species may belong to a different mimicry rings in the same location (see section 7.2.5). Even males and females of the same species may belong to distinct mimicry rings, as seen in certain sexually dimorphic ithomiine species (Willmott & Mallet, 2004). Although examples of polymorphic mimicry in Müllerian mimics are relatively few and far between, it has been reported in a few species

namely *H. numata*, *H. ismenius*, *H. doris*, and some populations of *H. cydno* (Jiggins, 2017; Kapan, 2001). *H. numata* provides one of the most celebrated examples of polymorphic mimicry at its most extreme, in that it co-occurs in a number of distinct forms with each form a member of a different mimicry ring, dominated by a tiger-patterned ithomiine species. Outside Neotropical butterflies, we see additional examples of polymorphisms in Müllerian mimics. For example, the bumblebee *Bombus rufocinctus* is dimorphic with forms that resemble two distinct earlier emerging *Bombus* species (Plowright & Owen, 1980). All of these examples are something of a puzzle, given that we have previously presented multiple lines of evidence showing that unprofitable species are under strong selection to evolve the same warning signals, thereby reducing the costs of predator education.

There are several general explanations for this paradoxical diversity, which are not mutually exclusive. One explanation is based on habitat segregation. Many sites may actually comprise a range of different microhabitats ranging from open forest to closed canopies, and the different mimicry rings may be separable in space and/or time. So while mimics in ostensibly different rings may be occasionally seen together, members of the mimicry rings and their predators spend most of their time sufficiently separated to be considered distinct communities. Gompert et al. (2011) presented a model showing that mimetic diversity could indeed be maintained in this way, analogous to the manner in which spatial polymorphisms are maintained over a much larger scale. Many ithomiine butterflies show a considerable degree of vertical stratification. For instance, species with tiger patterns tend to fly relatively high above ground, whereas species with transparent wings fly in the understory and low above ground (Elias & Joron, 2015). Differences in the heights of their host plants are likely causes of this stratification (e.g. Beccaloni, 1997; Papageorgis, 1975). Indeed, Willmott & Mallet (2004) found that ithomiines in the same mimicry ring utilize the same larval host-plant species significantly more often than expected in two out of five communities surveyed. Likewise, Hill (2010) reported that there was habitat segregation among ithomiine mimicry complexes which were driven largely by the abundant species in each complex, although there was a degree of overlap among the complexes. Given that predators such as birds also partition forest microhabitats (Walther, 2002), it is clear that this spatial separation could contribute to the stable co-existence of multiple mimicry rings in this group. Although one might anticipate that the microhabitat preferences which separate the different colour-pattern rings of ithomiine butterflies could be explained by their shared ancestry, it turns out that their preferences are more similar than one would predict based on their shared ancestry (Elias et al., 2008). Thus, either colour-pattern mimicry has been selected in species that share the same niche, or the selection for colour-pattern mimicry has also led to the convergence of the ecological preferences of co-mimics. The evidence for microhabitat partitioning is less clear-cut for taxonomic groups such as *Heliconius* (Mallet & Gilbert, 1995). However, it is possible that *Heliconius* at least partially exploit the segregation seen in ithomiines, and there is some evidence that co-mimics tend to share similar nocturnal roosting heights (Mallet & Gilbert, 1995; Mallet & Joron, 1999).

The sympatric intraspecific polymorphism in *H. numata* is perhaps the hardest to explain of all, with up to 11 of its 38 known mimetic colour pattern forms occurring in a single population (Joron et al., 1999). It has been suggested (Charlesworth & Charlesworth, 2011) that *H. numata* could act as a parasitic ('quasi-Batesian') mimic on its ithomiine models, with the decline in warning signal effectiveness with increasing mimetic load generating selection for polymorphism. We leave a discussion of the nature of Müllerian mimicry when species differ in their unprofitability to section 7.5.6. However, given that *H. numata* is particularly unpalatable (Arias et al., 2016b) and often relatively common (Charlesworth & Charlesworth, 2011) this explanation now appears unlikely. Instead, Joron & Iwasa (2005) developed a model to show that microhabitat segregation of ithomiine models could generate stable polymorphism in *H. numata* so long as the spatial segregation was strong and/or the mimic was not too distasteful (as noted above, this remains questionable for *H. numata*). Naturally, if ring formation was all due to microhabitat segregation one has to wonder why so few Müllerian mimics are genuinely polymorphic in a given location. It could be that the spatial segregation of its particular ithomiine models encourages it, but *H. numata* has also evolved some remarkable genetic architecture which appears to limit the formation of non-mimetic intermediate morphs that would have relatively low survivorship (Arias et al., 2016a; Le Poul et al., 2014). As noted earlier, the dissassortive mating preferences of female *H. numata* (Chouteau et al., 2017) may also play a role in maintaining the local diversity.

Once established, how are rings maintained? We have already discussed the phenomenon of generalization when discussing how Müllerian mimicry might evolve (section 7.5.3), noting that when predators only have narrow generalization, then intermediates may be always at a disadvantage. In a seminal paper, Turner (1984) drew an analogy of an astronomical attractor—the more species that join any given ring, the more powerful is its 'gravitational attraction'. Following up on these arguments, Franks & Noble (2004) presented an individual-based model showing that Müllerian mimicry rings can readily evolve and be maintained for precisely the reasons that Turner (1984) had in mind, with predators generalizing their learned aversion over a limited region of phenotypic space, thereby selecting out individuals that differ from the most common warning signals. Nevertheless, it is unclear precisely how permanent these configurations are and how they get started (Franks & Noble, 2004 simply assumed random starting conditions and observed what happened). Extending the gravitational analogy, it is possible that any given ring could ultimately act as an all-consuming black hole. However, in contrast to single species of Müllerian mimics, when different rings are all well populated with co-mimics, then the per capita mortality experienced in the course of educating predators is likely to be very low so there will be little or no advantage in switching rings, at least in terms of predation, and the long-term stability of the rings will be maintained.

7.5.6 What is the nature of selection when the species differ in unprofitability?

It has long been appreciated that prey vary in their palatability to would-be predators.

Dixey (1920, p. 564), for example, asserted that *'distastefulness is relative; that it exists, like other means of defense, in degrees that may vary indefinitely from species to species.'* Nicholson (1927, p. 34) likewise noted that *'every intergrade appears to exist between the most distasteful species and those which are palatable to all predators.'* Given this inevitable variation, there has been a great deal of debate as to the nature of the relationship between Müllerian co-mimics when one species appears much better defended than the other. In particular, does the less unpalatable species undermine the effectiveness of the more unpalatable species, thereby parasitizing it in a manner analogous to Batesian mimicry (Chapter 9), or are the signals mutually reinforcing in the manner envisaged by Müller? This question is important, not least because if moderately defended prey acted as parasites eroding the deterrent of the common warning signal then it could help explain the puzzling cases of polymorphism observed in unpalatable species (see section 7.5.5).

A wide range of models of predator behaviour have been developed to determine the nature of mimicry between prey with unequal defences. Some of these models have been based exclusively on learning, others have been based on identifying optimal state-dependent decisions (e.g. hunger) in well-informed predators, and a few have combined learning and state dependence. Here we briefly review some of their key findings.

In the most systematic evaluation of the consequences of aversion learning to date, Speed & Turner (1999) quantified the probability of the less unpalatable prey species parasitizing the effectiveness of the signal of a more unpalatable species in 29 different learning models. Parasitism in these cases was judged to occur if (i) the probability of attacking the model–mimic phenotype on encounter was higher than a distinct model control (held at equal density) or (ii) the attack rates of the model–mimic phenotype increased as the probability of encountering a mimic (at the expense of encountering no prey at all) increased. This second criterion implicitly assumes there is no other form of density dependence (for example, the predators do not exhibit a saturating functional response in which an increase in food availability reduces the per capita mortality), which we return to below. A surprising

27 of these 29 models were found to be capable of generating a form of parasitism in which the moderately unpalatable prey eroded the protection afforded to the more unpalatable prey type. Given the fact that the mimics in question are unpalatable and yet harm their more unpalatable models, they have been dubbed 'quasi-Batesian' mimics (Speed, 1993; Speed & Turner, 1999).

A study by Balogh et al. (2008) explored a similar set of learning models to those investigated above and likewise reported the potential for quasi-Batesian mimicry involving a long-term non-zero attack probability for unpalatable models. Intriguingly, they argued that if variation in prey unpalatability enhances learning through a surprise effect (i.e. the outcome was different from predicted in the recent past) then the model can benefit even more from being mimicked by a less unpalatable mimic because the signal is that much more effective. The authors dubbed this phenomenon 'super-Müllerian', and argued that the mutualism could arise even if the mimic was palatable ('quasi-Müllerian').

In the Speed & Turner (1999) model of conditioned aversion, unpalatable prey had a characteristic asymptotic probability of attack on encounter, with prey of neutral palatability considered to have an asymptotic attack rate of 0.5. Likewise, Balogh et al. (2008) assumed that prey could be charac-terized by a particular asymptotic attack rate. Moreover, in all of the models the (un)palatability of the prey types were considered fixed, which effectively assumes fixed predator hunger levels (Speed & Turner, 1999). Clearly a 50 per cent probability of attack on encounter is an arbitrary threshold, and one might wonder why a predator that found a given prey type unpalatable would not learn to permanently reject it. While many researchers continue to use the terms 'unpalatable', 'distasteful', 'inedible', or 'defended' when discussing Müller's hypothesis, his theory is better at explaining mimicry when predators sample and ultimately reject prey. For this to happen, the prey species must not simply be unpleasant, but generally sufficiently unprofitable to attack to dissuade further sampling.

Since associative learning is a means to an end (maximizing payoff) then to understand predator behaviour from an adaptive perspective, it helps to take a cost–benefit approach. Such an approach is particularly appropriate once one considers the fact that predators tend to be more prepared to attack unpalatable prey items in times of nutritive need. For example, European starlings (*Sturnus vulgaris*) increased their attack rates on quinine-injected mealworms when their body masses and fat stores were experimentally reduced, while choice trials clearly indicated the response was not due to any hunger-based reduction in the discriminatory abilities of the birds (Barnett et al., 2007). In these cases, a given prey type such as a quinine-injected mealworm may continue to be unpalatable, yet hunger makes it profitable for predators to attack these prey. Likewise, Skelhorn & Rowe (2007b) revealed that the dose of quinine sulfate administered to European starlings affected the number of chem-ically defended prey that they were subsequently prepared to consume, with birds receiving high toxin doses less inclined to eat chemically defended prey.

To address the state-dependent nature of (un)profitability, a number of dynamic programming models have been developed and explored (e.g. Kokko et al., 2003; Sherratt, 2003; Sherratt et al., 2004b). Each of these models have assumed that the properties of the prey types are already known to the predator and the predator simply seeks to make decisions that balances its toxin and energy levels to ensure maximum long-term survivorship (thus the predator dies with too low an energy level, or with too high a toxin dose). For example, Sherratt et al. (2004b) found that when distinct palatable prey are rare then those prey types which contain relatively low doses of toxins become profitable to consume by (hungry) predators with low energy levels. Faced with hungry predators, then the weakly toxic prey could gain selective advantage by mimicking prey with a higher dose of toxins. Whether the more toxic model gains from this mimicry depends on how its success is measured. First and foremost, the model controls were always attacked at lower rates on encounter than the model–mimic phenotype. This indicates that when predators are well-informed about the nature of prey types then a highly toxic model would experience greater protection from predators if it had a unique and recognizable appearance compared to a case in which it were confused with weakly unpalatable prey (which are occasionally eaten by hungry predators). Nevertheless,

increasing the mimic density sometimes increased and sometimes decreased the attack rates of predators on the model–mimic phenotype. The relationship tended to be negative if the increase in mimic density increased the total prey density in the system (i.e. the additional mimics represent a new supply of potential food) but positive (increasingly parasitic) if the increase in mimics came at the expense of well-recognized mimic controls.

In fact, the possibility that weakly defended prey could act as Batesian mimics of better-defended models in periods of nutritive need has long been discussed. de Ruiter (1959, p. 353) felt that it was common, arguing that the Müllerian mechanism '*is very unlikely to be realized except when predators live in the presence of such a superabundance of food that they never have to resort to relatively distasteful prey.*' Wallace (1882) made a similar argument when he proposed many '*difficult cases of mimicry*' could be understood in terms of Batesian mimicry, so long as some predators find the unpalatable species profitable to attack. Are weakly toxic prey subject to predation by hungry predators therefore quasi-Batesian mimics? We prefer to treat such prey simply as temporary Batesian mimics since they are after all profitable to eat by hungry predators. Thus, despite their distastefulness, if a prey species is profitable to consume then they are Batesian mimics for the period of time they remain profitable to consume—however temporary that relationship may be. Indeed, the possibility that relationships between co-mimics can sometimes switch from mutualistic (Müllerian) to parasitic (Batesian) should not come as too much of a surprise to ecologists. For example, plant–mycorrhizal relationships frequently switch from parasitism to mutualism dependent on the underlying nutritive conditions (Johnson et al., 1997). Likewise, the mutualism between cleaner fish and their clients in coral reef fish can degrade into parasitism if ectoparasites are rare (Cheney & Côté, 2005).

The above state-dependent models assumed complete knowledge and did not assume any form of learning at all, which is so central to Müller's hypothesis. Ironically, the distinct model never gains from the mimicry per se because, in contrast to Müller's arguments based on associative learning, the predator was assumed to have complete information about the nature of the various phenotypes at the outset. In a recent paper, Aubier et al. (2017) sought to bridge this gap by developing a state-dependent optimization model in which predators have to learn the attributes of prey types and yet make optimal decisions based on this incomplete information as to whether to attack the prey they have encountered. Here the introduction of a moderately unprofitable mimic (derived from a discriminable non-mimic population) of a more unprofitable model caused an increase in overall consumption of the model–mimic phenotype because there were simply more such prey with this appearance (so information about them is more valuable), and because the mean cost of attacking unpalatable prey is lower, warranting more sampling. However, this increase in consumption was more than offset by the increase in overall density of prey sharing the model appearance, generating a decline in the per capita attack rates and classical Müllerian mutualism.

It is important to note that parasitism readily arose in the above model when the payoff from attacking moderately defended mimics varied among individual predators, so that they are profitable to attack by some predators but not others. Under these cases then the moderately defended mimics could be parasitic on their models, just as classical Batesian mimics are. Note, however, that this is not a new phenomenon, but the very same mechanism that has always been invoked to explain Batesian mimicry. Indeed, the fact that the very same optimal sampling model invoked to explain Müllerian mimicry also readily predicts Batesian mimicry strongly suggests that the two phenomena are intimately related. An earlier attempt by Honma et al. (2008) to combine a model of learning with the classical optimal diet model came to a similar conclusion. Thus, they found that a parasitic relationship could emerge only under very restricted conditions in which the availability of alternative prey was so low that the predator must either starve or eat unpalatable prey. Under these conditions, such unpalatable prey are (by definition) profitable and therefore the relationship is classical Batesian mimicry.

Of course, one could argue the above state-dependent mechanisms, while broadly correct in theory, never operate in practice because: (a) all natural distasteful prey are so highly noxious that, once

learning is complete, they are never attacked by predators and/or that (b) alternative palatable prey are always so abundant that predators are never hungry enough to need to eat these prey. Cyanide-producing millipedes and poison frogs may well be simply too unprofitable to attack to ever warrant re-sampling by reasonably informed predators. It has been rumoured among field biologists that *Heliconius* butterflies 'taste like dog shit' but it is reasonable to assume that some predators under some conditions would find them profitable to eat, particularly in times of food scarcity. Moreover, given the fact that specialist insectivores are capable of handling prey in such a way as to reduce their ingestion of toxins (Brower & Calvert, 1985) then we suspect that it is likely that natural predators will occasionally have recourse to attacking chem-ically defended prey for the food they can provide. Naturally, if attacks were too common then the warning signal would provide no benefit at all—indeed it would be detrimental to the prey, since they would be highly visible to predators.

So far, experimental evidence for quasi-Batesian mimicry (which we prefer to consider classical Batesian mimicry when the defended mimics are profitable to attack due to hunger or an ineffective defence) has been mixed. Speed et al. (2000) conducted a field experiment with pastry baits containing a range of doses of quinine and found that increasing the density of moderately unpalatable mimics at the expense of distinct (non)mimetic controls increased the mortality of the more toxic model. Rowland et al. (2007a) conducted a 'novel world' experiment with wild-caught great tits, and found that when moderately unpalatable mimics were added to a system with highly unprofitable models, then great tits attacked more prey in the model–mimic mixture than one would expect if all the prey were highly unprofitable. However, despite this increase in overall consumption, the presence of moderately unprofitable mimics actually decreased the mortality on the models through a simple dilution effect, suggesting that the net effect of moderately unprofitable mimics was mutualistic in the way Müller had anticipated rather than parasitic. More recently, Rowland et al. (2010b) conducted an analogous experiment, this time keeping the total density of prey constant by reducing the density of distinct (non)mimetic controls as the number of moderately unprofitable mimics increased. In contrast to their earlier work, they found that the per capita mortality of the moderately unpalatable mimics and more unpalatable models went up as the proportion of mimics increased, suggesting evidence for quasi-Batesian mimicry. Rowland et al. (2010b) suggested that the differences in results of their two studies arose as a consequence of whether the total density of prey was fixed. However, you might expect predators to treat prey with distinct appearances entirely independently from models and mimics, and in both experiments the combined densities of models and mimics were increased in comparable ways. An alternate possibility is that some of the moderately unpalatable prey were profitable to consume by the birds, especially those with low toxin doses (indeed, the authors also showed evidence that toxin content influenced foraging decisions). So, once again, we see that moderately toxic prey can occasionally be profitable to consume, making them genuine Batesian mimics.

7.6 Overview

The subject of Müllerian mimicry has seen significant advances over the past couple of decades. Not only are more examples of Müllerian mimicry being uncovered in a range of new taxonomic groups from millipedes to catfish, but with the advent of genomics we now know a great deal about the genetics underlying colour-pattern evolution in the group best known for it, namely *Heliconius* butterflies. We now have a far more nuanced understanding of the strategy that predators should employ when learning to avoid unfamiliar unprofitable prey and there have been a number of field studies establishing positive frequency-dependent selection by predators in the manner that Müller had envisaged. For instance, deployment of artificial models—including clay frogs and wax butterflies—has revealed the phenotypes most vulnerable to predation under field conditions. Following some targeted experimentation, we now know that prey with low doses of toxins may, under some circumstances, act as parasitic Batesian mimics of better defended models. However, the reasons for this may be simply that such prey are profitable to attack rather than through some demonstrably sub-optimal psychological feedback mechanism.

Future work should focus on just how easily Müllerian mimics switch to being Batesian and vice versa. We also need to know a lot more about mimicry ring formation—are these rings simply human constructs or do they reveal the limits of predator generalization? Are they fixed or can rings effectively merge? What factors influence the fidelity of Müllerian mimicry? We have learned a lot over the past decade, but there is still much more to do.

CHAPTER 8

Advertising elusiveness

8.1 Introduction and definition

In Chapter 6 we discussed *aposematic signals* that act to inform the predator that while it might easily be able to capture the prey, consuming the captured prey might be unattractive to the predator because it is chemically or otherwise defended. The signals considered in this chapter aim to inform the predator that an attempt to catch the prey is likely to be unsuccessful. That is, they warn that prey will be difficult to capture, rather than somehow being unappealing to attempt to consume post-capture. Necessarily, these signals will be restricted (unlike aposematic signals) to mobile prey that can mount an active response to impending attack, generally involving fleeing. Also unlike aposematic signals, they will normally not be displayed continuously, but triggered by the perception by the prey that they are under imminent risk of predatory attack. We differentiate between two different types of such *elusiveness signals*: *pursuit deterrent* and *perception advertisement*. These two types of signal are used against coursing and ambushing predators respectively. In *pursuit deterrence*, the prey signals to a coursing predator that it is a particularly fleet individual, that the predator will struggle to close on, and/or a particularly strong individual, that will be difficult to subdue if the predator does succeed in closing on it. Certainly, predators select prey individuals on traits such as size, sex, age, and individual behaviour—such as activity during high risk periods (Heurich et al., 2016)—with smaller and poorer quality individuals, for example, being targeted in particular (Tucker et al., 2016); and so, an individual signalling fitness and escape ability may dissuade a predator from targeting it. *Perception advertisement* is directed at a stalking or ambushing predator that relies on being able to gain close proximity to the prey prior to detection; the prey signals that it has detected the predator prior to it coming sufficiently close to mount a successful attack. In both these cases, the signal can only be effective if mounting attacks is expensive to predators in some way such that they have an interest in biasing their attacks towards situations where they are more likely to be successful. In such a situation, the signal could be selected providing it offers sufficiently reliable information to predators to make their attack strategy more cost-effective.

Aposematic signals are effective because predators generalize from aversive experiences; that is, the prey signal that they have a similarity (in terms of unpalatability) to previous prey encountered by the predator. There is also an element of generalization required for elusiveness signals to be effective but, by signalling, the prey also seek to differentiate themselves from other potential prey. At least some individuals of the prey species (or group of similar-looking species) concerned must be of interest to the predator, or else the type of signalling discussed here would not be required. Hence, with perception advertisement the particular prey individual is signalling that, on this particular occasion, it presents an unattractive option for the predator to try and capture because it has already detected the presence of the predator. Similarly, for pursuit deterrent signals the prey may be signalling that it is currently particularly fleet by comparison to alternative prey the predator might target. Such fleetness can change between encounters through factors such as exhaustion or injury. For this reason, the types of signals described in this chapter tend to be displayed only in response to the prey detecting a specific potential predatory threat. Such prey are signalling something

Avoiding Attack: The Evolutionary Ecology of Crypsis, Aposematism, and Mimicry. Second Edition. Graeme D. Ruxton, William L. Allen, Thomas N. Sherratt, & Michael P. Speed, Oxford University Press (2018). © Graeme D. Ruxton, William L. Allen, Thomas N. Sherratt, & Michael P. Speed 2018. DOI: 10.1093/oso/9780199688678.001.0001

about active defence through flight that must be initiated after the predator has been detected, which again makes it natural for such signals to be triggered by a predator's detection by the prey. In comparison, because aposematic coloration generally describes signals that can be used after capture, these signals are generally (but not always) continuously displayed. The general expectation that signals of elusiveness will not be continuously displayed should not be seen an absolute requirement. We can imagine that individuals of a particularly fleet species might modify their appearance to differentiate themselves from otherwise similar-looking prey species that are easier to capture. Such a continuously displayed signal of elusiveness has been suggested in butterflies (Pinheiro et al., 2016), but currently no strong evidence exists.

The borderline between the types of signalling described here and aposematism is difficult to draw definitively. The hornet *Vespa velutina* is a predator of bees, including the Asian hive bee *Apis cerana*. When a hornet approaches a colony of these bees, guard bees simultaneously vibrate their abdomens, producing a characteristic visual display and buzzing noise. This signal is not produced by the close approach of harmless insects to the colony, and the signal seems to act to cause the hornet to abort its approach toward the colony. However, since a mass of bees can kill a hornet by suffocating and/or overheating it in a ball of bees, we see this as more akin to an aposematic display, unlike the perception advertisement signals discussed in this chapter. Essentially, where the signal indicates defences that might simply produce an aversive experience but might in some cases (like this) be lethal, we consider the signal to be aposematic. However, in describing the hornet–bee interaction, Tan et al. (2012) link it with the type of signals discussed in this chapter, because it is triggered by close approach of the predator. However, as discussed in Chapter 6, this can be true of aposematic signals too.

Hence, the types of signals considered in this chapter are signals given by prey during the close approach of a predator, that act to inform the predator that the prey would be difficult to catch and subdue. Other displays may be given by undefended prey that act to simply startle the predator, and these are fully discussed in Chapter 10. Since the

key informational aspect of the signals discussed in this chapter is how easy it will be for the predator to capture the signaller, we have introduced the umbrella term *elusiveness signalling*. In the next section we will document current empirical evidence for the existence of such signalling.

8.2 Empirical evidence of elusiveness signals

The empirical study of elusiveness signalling is challenging since, by definition, such studies require observation of encounters between potential predators and mobile prey. Further, it must be demonstrated that the predator (rather than conspecifics) is the target perceiver of the signal, and that the signal does not function exclusively through aposematism, startle, or deflection—mechanisms discussed in Chapters 6, 10, and 11 respectively. Nevertheless, there are now a number of case studies where valuable evidence has been collected, and we summarize some of what we consider the most interesting ones in the following sections. However, because of the challenges discussed in section 8.1 above, we have to expect that these cases potentially represent only a fraction of the prevalence of elusiveness signalling in the natural world.

8.2.1 Stotting by gazelle

The behaviour most commonly cited as a signal to predators of individual prey's escape ability is stotting (jumping with all four legs held stiff, straight, and simultaneously off the ground). This has been observed in several mammalian species but has been investigated most fully in the Thomson's gazelle (*Gazella thomsonii*). Caro (1986a,b) identified eleven non-exclusive potential functions of stotting. Two of these involved signalling to the predator: either that it has been detected or that the prey individual is particularly fleet. Caro suggested that cheetah need to approach within 20m of gazelle in order to have a chance of a successful attack. Stotting during a chase occurred when the gazelle was generally further than 60m from the cheetah, and occurred mostly towards the end of unsuccessful chases. These observations are compatible with stotting being a signal of prey athleticism to the predator. The outcomes of 31 cheetah hunts are summarized in Table 8.1. These data may suggest that stotting is related to both failed chases and the cheetah giving up a hunt without giving chase (although sample sizes are too small to be certain). Again, this interpretation is consistent with stotting being a signal of fleetness. Caro initially favoured the 'predator detection' hypothesis over the 'high quality' hypothesis (although both could operate). He realized that if stotting only functions to indicate to the cheetah that the gazelle has detected it, then explaining its occurrence during chases is chal-len-ging. He suggested that 'the act of fleeing itself may be insufficiently unambiguous to inform the predator that it has been detected'. This might be because gazelle sometimes make quick dashes to gain respite from biting insects (Tim Caro, personal communication). Crucially, Caro notes that the 'high quality' hypothesis 'is challenged primarily because so few fleeing members of a group (less than 20%) stott, and stotters do not get captured after pursuit any less than non-stotters (see Table 8.2). The fraction of individuals stotting does not seem critical to us as this might be what is expected when only high-quality individuals in a group can afford to send a costly signal that (for example) reduces running speed. Further, the results shown in Table 8.2 do not have the sample sizes to persuade us that it is safe to assume that stotting during a chase has no influence on the eventual outcome of that chase. However, it is true that they provide no evidence in support of such an association.

FitzGibbon & Fanshawe (1988) demonstrated that the individual Thomson's gazelles that African wild dogs (*Lycaon pictus*) selected to chase from a group stotted at lower rates than those that dogs did not select to chase. However, cause and effect are unclear—it could be that gazelles react to being chased by stotting less, rather than dogs selecting those that stott less. Fitzgibbon & Fanshawe argued against this alternative explanation as follows:

'In the eight cases when gazelles were observed both before they were selected and while being chased, four increased their stotting rate when selected and four decreased it. This implies that the gazelles are as likely to decrease their stotting on being hunted as they are to increase it, and that the dogs are selecting on the basis of stotting rate. In addition, wild dogs were seen to change the focus of a hunt from one gazelle to another on five occasions, and on four of these the gazelle preferred was stotting at a lower rate.'

Hasson (1991) pointed out that Fitzgibbon & Fanshawe's data also shows that the one occasion where the dogs switched to a faster stotting individual is also the case where the two gazelles' stotting rates were most similar. Fitzgibbon & Fanshawe also report that those gazelle that were selected but which outran the wild dogs were more likely to stott and stotted for a longer duration than those that were successfully captured. Although this is suggestive of stotting being a signal of ability to escape, there is an alternative explanation: animals that escape may not be more able to stott but simply have more oppor-tunity to do so, as they are not as closely pursued. Together, these studies of stotting behaviour come closer to convincing evidence of signalling of ability to evade attack, rather than awareness of the predator's presence. However that stotting is a signal (of any kind) to predators has not been demonstrated to the exclusion of all other

Table 8.1 Outcomes of 31 cheetah hunts—redrawn from Caro (1986b).

Outcome	Chase successful	Chase failed	Hunt abandoned
Gazelle stotted	0	2	7
No stotting	5	7	24

Table 8.2 Outcomes of 10 cheetah hunts on neonate gazelles—redrawn from Caro (1986b).

Outcome	Chase successful	Chase failed	Hunt abandoned
Gazelle stotted	1	0	0
No stotting	7	2	0

candidate explanations, although it certainly remains highly plausible.

8.2.2 Upright stance by hares

Holley (1993) reported on observations of naturally occurring interactions between brown hares (*Lepus europaeus*) and foxes (*Vulpes vulpes*) on English farmland. He observed 32 occasions in which feeding adult hares were approached by a fox. On 31 of these occasions, as the fox closed to within 20–50m of the hare, the hare stood bipedally, ears erect, directly facing the fox, turning its body so as to remain facing the fox as it moved, continuing until the fox moved away. This behaviour was not seen consistently in other contexts during 5000 hours of observation of hares. Hence, Holley suggested that this behaviour was a signal to the fox that it had been detected. In support of this interpretation, Holley pointed out that none of the 31 'signalling' hares was attacked. He further argued that such signalling is plausible because adult hares are sufficiently fleet that foxes are only successful in catching fit hares when catching them unawares, allowing capture before the hare is able to accelerate to its maximum speed. On five occasions, a fox appeared near hares as a result of using cover to approach undetected. In these five cases, the hare did not show this 'signalling' behaviour, but either moved away from the fox or adopted a crouched stance that Holley suggested allows fast acceleration if flight is required. This was interpreted as the hares not signalling in circumstances where signalling of detection may not deter the fox, because it has already managed to get near to the hare. Holley also reported that in 28 encounters between hares and domestic dogs, this 'signalling' behaviour was not shown. He interpreted this by suggesting that domestic dogs are coursers, willing to pursue hares over long distances in order to exhaust them, are thus not reliant on surprise like foxes, and so are less likely than foxes to be concerned about being detected by a hare. Hence, such signalling would not be expected to deter dogs from attacking.

Holley considered alternative explanations for this bipedal behaviour. He argued that the bipedal behaviour is unlikely to serve solely to improve the hare's view of the fox, because it is not used against

dogs. Further, he suggested that in low vegetation the fox is visible to the hare at much greater distances than the distance at which this behaviour is induced (never more than 50m). He also argued that bipedal behaviour is unlikely to serve solely as a signal to conspecifics because it was not given to dogs, because the hare turns its body so as to constantly face the fox, and because on 13 of the 31 occasions when it was used the signaller was solitary.

We consider this study as strongly persuasive that the hare's behaviour serves at least in part as a signal to the fox that it has been detected. The turning of the body to face the moving fox is strongly suggestive of signalling to the fox. A follow-up study, using video analyses, that allowed quantification of this aspect of behaviour would be useful. Such a study could also search for evidence that hares detect foxes before they have closed to 50m; perhaps detection can be identified by head turning or increased vigilance. However, given that the original study involved only 32 encounters between foxes and hares in 5000 hours of observations, such a follow-up study would be very challenging.

8.2.3 Vervet monkey alarm calls to leopards

Isbell & Bidner (2016) report on a powerful use of technology to study the responses of leopards (*Panthera pardus*) to the alarm calls of vervet monkeys (*Chlorocebus pygerythrus*). The researchers deployed acoustic recorders, camera traps, and GPS collars on both species. They found that leopards often quickly changed direction and moved away from the monkeys, or broke off approaches to monkeys after the monkeys gave alarm calls (see Figure 8.1). Since leopards typically hunt by ambush, there appears to be no benefit to them in lingering once they have been detected. Vervet monkey alarm calls in response to leopards are well known for their function as a warning call to conspecifics. Their predator-deterrent function was recognized only because the devices allowed the researchers to min-imize their presence (leopards tend to be wary of humans) and remotely monitor the behaviour of both predators and prey. This study very much deserves follow-up investigation. Firstly, it would be valuable to study leopard responses to recordings in order to demonstrate that abandoned attacks are not just

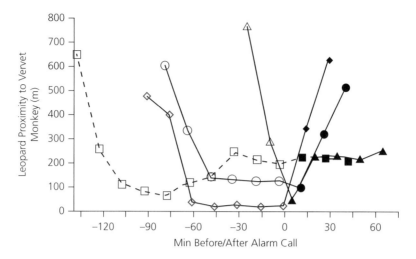

Figure 8.1 Modified by Lynne Isbell from results presented in Isbell & Bidner (2016). Proximity of collared leopards to collared vervets on the three nights when a leopard approached the vervets' sleeping site, and on the one evening when vervets approached the leopard, before and after vervets gave alarm calls. Open symbols: proximity before alarm calls began; closed symbols: proximity after alarm calls began. Solid lines: nights leopard moved away >200 m from vervets after alarm calls; dashed line: night vervets moved >200 m away from leopard before giving alarm calls. Distance between consecutive symbols indicates speed of travel.

related to alarm calling but actually triggered by leopards detecting such calls. Secondly, other functions of the calls deserve exploration, given that monkeys used these calls mostly at dawn and dusk, but leopards approached vervets most closely in middle of the night (Isbell & Bidner, 2016).

8.2.4 Other potential signals by mammals

Another very commonly cited putative elusiveness signal is tail-flagging in white-tailed deer (*Odocoileus virginianus*). Caro et al. (1995) provided a useful synthesis of the available data. They concluded that tail-flagging may indicate individual ability to escape, on the grounds that flagging animals ran faster than those that did not. However, as no studies have investigated the responses of natural predators to this behaviour, and Caro et al. suggested that alternative explanations based on crypsis exist, this remains only conjecture.

Observation of the behaviour of predators in response to putative signals can be challenging but, when successful, can be insightful. Clark (2005) implanted radio trackers in a number of timber rattlesnakes (*Crotalus horridus*) and was thus able to set video cameras on snakes when they were basking or sitting in ambush beside runways used by

small mammals. This allowed him to observe harassment displays by chipmunks (*Tamias striatus*), eastern grey squirrels (*Sciurus carolinensis*) and by a wood thrush (*Hylocichla mustelina*). Foraging snakes were over 4 times more likely to abandon sites after harassment than foraging snakes that were not subjected to displays; whereas displays had no effect on movement rates of basking snakes. Foraging snakes also moved further when moving soon after harassment than when moving for other reasons, and again this difference was not seen in basking snakes. In a follow-up study, Barbour & Clark (2012) found that Californian ground squirrels (*Spermophilus beecheyi*) frequently gave tail-flagging displays to northern Pacific rattlesnakes (*Crotalus oreganus oreganus*). They found that the probability of ambush-foraging rattlesnakes striking at a tail-flagging squirrel decreased with spatial separation between the two, but remained high for all distances in non-flagging squirrels. Flagging was associated with squirrel vigilance, as measured by readiness to dodge a snake strike. Tail-flagging, again, was seen to increase the rattlesnake's likelihood of leaving the ambush site. Interestingly, a study by Putman et al. (2015) found that adults were both more likely to detect snakes than pups and perform more tail-flagging toward snakes. Snake detection and

Plate 1 Examples of countershading, counterillumination, and reverse countershading (see page 42 for details).

(a)   (b)   (c)   (d)   (e)   (f)   (g)

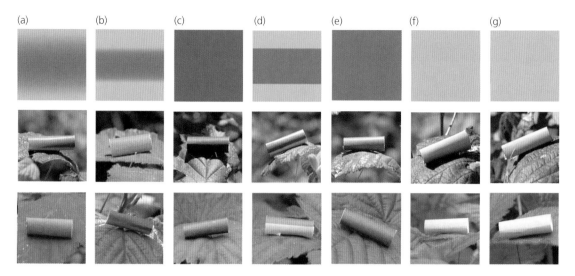

Plate 2 Treatment groups for experiment on countershading (see page 48 for details).

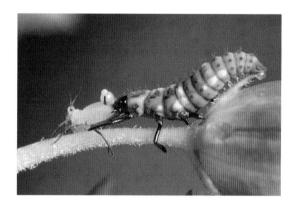

Plate 3 An aphid employing defensive secretions against a ladybird larva (see page 73 for details).

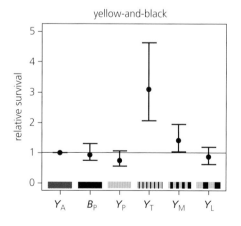

Plate 4 Relative survival of different types of artificial prey from an experiment to investigate aposematism (see page 88 for details).

Plate 5 A moth displaying conspicuous warning signals (see page 89 for details).

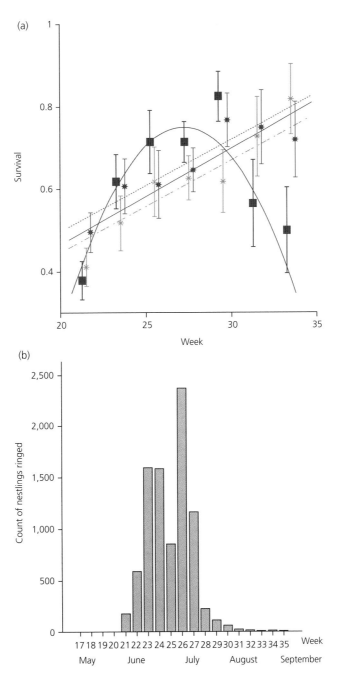

Plate 6 Seasonal variation in emergence of avian fledglings from nests and survival of different types of artificial prey in an experiment to explore how naive individuals might affect selection for aposematism (see page 96 for details).

Plate 7 A beetle and a moth that may be Müllerian mimics (see page 104 for detail).

Plate 8 A Heliconius butterfly sitting on a Heliconia flower (see page 105 for more on these butterflies).

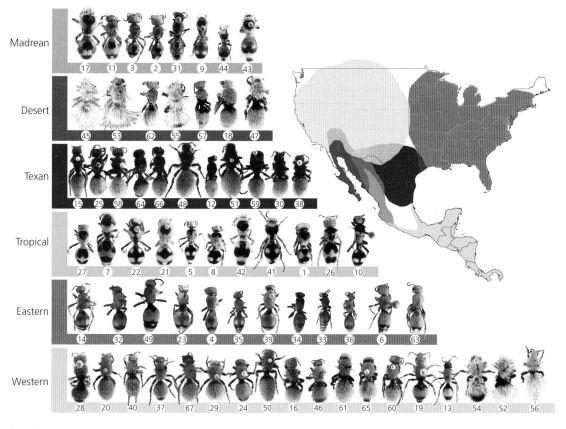

Madrean
17 11 3 2 31 9 44 43

Desert
45 53 62 55 57 18 47

Texan
15 25 58 64 66 48 12 51 59 30 38

Tropical
27 7 22 21 5 8 42 41 1 26 10

Eastern
14 32 49 23 4 35 39 34 33 36 6 63

Western
28 20 40 37 67 29 24 50 16 46 61 65 60 19 13 54 52 56

Plate 9 Velvet ant mimicry rings (see page 106 for detail).

Plate 10 Above are three millipedes of the Apheloria clade; below are three co-mimics (see page 107).

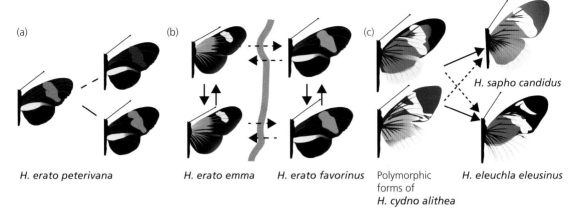

(a)

H. erato peterivana

(b)

H. erato emma   H. erato favorinus

(c)

H. sapho candidus

Polymorphic
forms of
H. cydno alithea

H. eleuchla eleusinus

Plate 11 Three field experiments to test Müllerian mimicry in butterflies (see page 111).

(a)   (b)

Plate 12 Right are the different morphs of artificial butterfly used in an experiment on Müllerian mimicry; left is an example of one as they were deployed in the field (see page 112).

Plate 13 Below are hand-crafted clay models designed to resemble two conspicuous and one inconspicuous frogs shown above (see page 113 for detail).

(a) Müller's "fixed n_k"   (b) Optimal Sampling Strategy

species 1   species 2   species 1   species 2

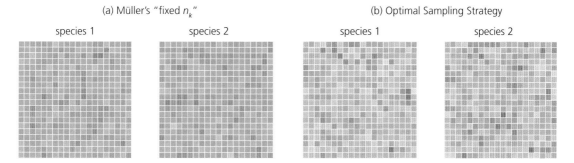

Plate 14 Predictions of the spatial distribution of different prey colour morphs from models employing different assumptions about predator behaviour (see page 121 for details).

Plate 15 A moth whose wing patterning is reminiscent of two flies feeding on a bird dropping (see page 151 for detail).

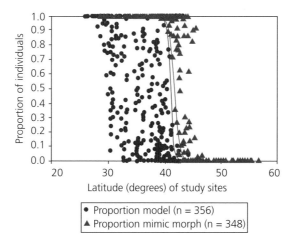

Plate 16 A transition zone between mimic and non-mimetic forms of a butterfly (see page 156 for detail).

Plate 17 Left is a flesh-fly mimicking weevil; right is a potential flesh-fly model (see page 160 for details).

♀
f. cyrus
♂

♀
f. polytes
♂♀
Pach. aristolochiae

♀
f. theseus
♂♀
Pach. aristolochiae
black from

♀
f. romulus
♂♀
Pachliopta hector

***Papilio polytes* (Batesian mimic)**

Plate 18 Females of the non-mimic form of a butterfly look like the males (top line); whereas three mimetic forms (first column) look like distantly related species (second column – see page 168 for details).

(a) Original frame

(b) Motion map

Plate 19 Above is a frame from a video of several zebra moving in the same direction; below is the output of a computational model of motion detection applied to this image that implies a patchwork of motion signals in different directions (see page 210 for details).

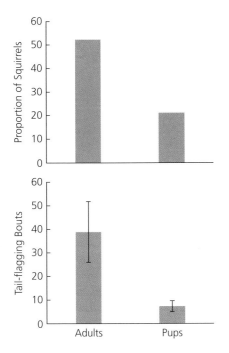

Figure 8.2 Adapted by Bree Putman from results presented in Putman et al. (2015). Age differences in the proportion of squirrels that detected a hunting rattlesnake when coming within 1 m of one, and the number of tail-flagging bouts that were directed at a snake when one was detected.

appropriate anti-predator signalling, therefore, seems to improve with age (see Figure 8.2). Together these studies are very highly suggestive of elusiveness signalling towards these snakes.

Striking the ground with one or both feet to produce an audible thump is a near-universal response to predators in macropodid marsupials. This has often been suggested as an elusiveness signal. However, as reviewed by Rose et al. (2006), exploration of the behaviour's function remains very limited. A great many putative elusiveness signals to predators have been suggested within the artiodactyls (including different calls, tail movement, foot stamping, and changes of gait); in general, the function of these is difficult to evaluate in the absence of studies involving responses of predators. Caro et al. (2004) presented comparative analyses that sought to link the presence or absence of these different potential signals to morphological, ecological, and behavioural variables across 200 species of artiodactyls. This study did not yield any strong trends that could be interpreted as suggestive of signalling to predators

specifically. Indeed the main function of this work may be to highlight the limits of the approach to the study of this animal group at present, where behaviour patterns are poorly documented, the formation of clear hypotheses is difficult, independent variables are correlated, and species may show a given behaviour in some ecological conditions but not others. Logistically challenging as it undoubtedly is, consideration of this study suggests that in evaluating the impact of such putative signals to predators, we really do need to observe interactions with predators.

8.2.5 Singing and distress calling by birds

Cresswell (1994) reported on extensive and carefully recorded naturally occurring predation events by a raptorial bird (the merlin, *Falco columbarius*) attacking a songbird (the skylark, *Alauda arvensis*). The merlin clearly selected a skylark for pursuit before any song by the skylark was heard. If singing by the skylark was heard (by Cresswell), then it started very soon after the pursuit began. For pursuits where the merlin gave up without capture, merlins chased non- or weakly singing skylarks for longer periods compared to skylarks that sang strongly. These chases often exceeded five minutes in duration over several kilometres, so that the costs of not signalling or responding to the signal were high for the skylark and merlin respectively. Merlins were more likely to catch a non-singing skylark than a singing one. The study also showed that skylarks that did not sing on attack (and therefore probably could not sing) were more likely to attempt to hide from the merlin rather than outrun it. The author himself admitted that one drawback to this study is that he does not demonstrate that singing whilst being pursued is costly, although other studies have shown that singing in skylarks in the absence of pursuit is energetically demanding. However, in all other respects this study comes very close to a convincing demonstration of pursuit deterrence signalling.

Laiolo et al. (2004) caught a number of lesser short-toed larks (*Calandrella rufescens*) and recorded the calls they made on flying away after being released from a bag. The authors found variation in aspects of the call structure to be related (separately) to variation in body condition and immunocompetence. They argue that these latter two factors are likely to be associated with ease of capture, and thus the calls may afford predators an ability to detect and preferentially select those that would be most profitable to chase.

8.2.6 Visual signalling of contrastingly coloured birds' tails

One of the earliest suggestions of quantitative evidence for elusiveness signalling comes from observation of tail-flicking behaviour by eastern swamphens (*Porphyrio melanotus*: a large bird of the rail family) when approached by a human (Woodland et al., 1980). Flicking rates were highest in the birds nearest the approaching human; birds' tails were orientated towards the human and flicking rate increased with decreasing separation between human and bird. Woodland et al. interpreted these observations as evidence that tail-flicking was a signal directed towards predators. Craig (1982) challenged this interpretation, suggesting that tail-flicking was more likely to be a submissive signal directed at dominant conspecifics. He suggested that the reason that birds nearest the approaching predator signal most was that these birds were likely to be low-quality birds at the periphery of groups. These peripheral subdominants have increased need to signal submission to dominants when the human 'predator' drives birds closer together. The closer the human gets, the more signalling is required as the birds are driven closer together when they bolt for nearby water. Craig also suggested that the apparent orientation of the tail towards the human was simply a necessary consequence of the bird walking away or preparing to do so.

Craig's alternative interpretation certainly complicates the situation, but further work on this system may be worthwhile since there are other observations in Woodland et al. (1980) that the alternative explanation does not easily accommodate. For example, when birds on reed beds were approached by boat, 'birds would remain virtually on one spot, standing erect, looking back by rotating their body so that the rump flash was directed towards the intruder'. Again the flicking rate increased as the intruder approached.' This observation seems easier to reconcile with tail-flicking being a signal directed to potential predators rather than one directed

to conspecifics. Further, Randler (2007) reported observations of tail-flicking in the Eurasian moorhen (*Gallinula choropus*) that seem difficult to reconcile with signalling to conspecifics. He found a positive relationship between flicking and both vigilance and nearest neighbour distance. He found a negative relationship between flicking and flock size. In playback experiments, moorhens increased both vigilance and tail-flick rate in response to calls of potential predators but decreased in response to calls of conspecifics. Further, flicking rates were no different between adults and (socially submissive) juveniles.

One puzzle about tail-flicking is what keeps the signal honest, since no obvious time or energy cost is apparent; birds may well be able to feed and signal at the same time. Alvarez (1993) suggested that the cost is an added risk of predation from secondary predators, as envisaged by the theory of Bergstrom & Lachmann (2001) discussed later in this chapter. This hypothesis could be tested with flicking and non-flicking mechanical model birds. Alternatively, honesty might be caused by anatomical constraint; it may be that balance considerations preclude feeding and flicking, but this has not been explored. Alvarez et al. (2006) observed moorhens and found a positive correlation across individuals between flicking rate and various measures of condition. If it is the case that feeding and tail-flicking are mutually incompatible, then tail-flicking rate might be an honest measure of condition, since birds in better condition might be more able to forgo feeding opportunities in order to signal. However, as we stated above, such incompatibility has yet to be demonstrated. Alternatively, if conspicuousness to predators increases with flicking rate, then only individuals in good condition could risk signalling in this way. In a quite different species (a ground feeding insectivore, the white wagtail *Motacilla alba*), Randler (2006) found that wagging was positively related to vigilance and negatively correlated with pecking for food; however, the constraint or cost that might drive these correlations remains to be identified.

Both sexes of the turquoise-browed motmot (*Eumomota superciliosa*) have an exaggerated pendulum-like tail display. These displays were experimentally triggered by a potential predator (a human emerging from a blind) near to nesting colonies, in experiments reported by Murphy (2006). The signal

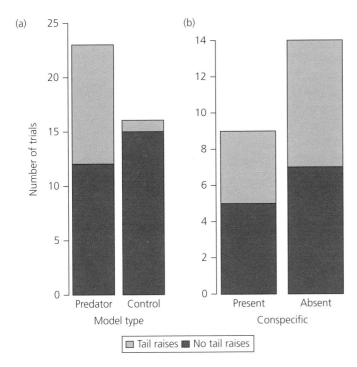

Figure 8.3 Created by Pierre-Paul Bitton from data presented in Bitton & Doucet (2013) on the factors affecting a tail-raising signal in the elegant trogon.

was given by both sexes, and was given by unpaired individuals, paired individuals out of visual contact with their mate, and by individuals out of visual contact with any conspecific. The displays were generally (see section 8.4) not observed in the absence of this simulated predator, were triggered immediately on the appearances of the human, and stopped coincident with the removal of this stimulus. From this, Murphy suggests that the signal is likely to be directed at predators to inform them that they have been detected. Bitton & Doucet (2013) explored a tail-raising signal in a woodland bird, the elegant trogon (*Trogon elegans*). They found that the signal was more likely to be induced by a model of a potential predator than a suitable control model, and that the presence of conspecifics along with the model did not influence propensity to signal (see Figure 8.3). Effects of this signal on predators await exploration, and an alternative explanation where the predator induces the display as a nonfunctional fear-response that the predator does not respond to cannot currently be ruled out.

Intriguingly, it might be that some birds' prey themselves use pursuit deterrence signalling. Pin-

heiro et al. (2016) suggest that signalling difficulty of capture to predators may be widespread in butterflies, and that this ability may not be limited to palatable butterflies. Further, they point out that the possibility that the evolution of pursuit deterrence signalling mimicry preceded that of unpalatability should be further investigated within the context of classical Müllerian and Batesian mimicry. This is an interesting idea worthy of further investigation.

8.2.7 Willow tit alarm calls and attack preferences by pygmy owl

In a simple but insightful experiment by Kareksela et al. (2013), a pygmy owl (*Glaucidium passerinum*) was exposed to several dummy willow tits (*Poecile montanus*) in an aviary. A small loudspeaker emitting characteristic willow tit alarm calls was associated with one of the dummies, and the owl invariably attacked one of the 'non-calling' dummies. A very valuable follow-up would be to explore if this selection by predators is specific to alarm calling.

8.2.8 Visual displays by lizards

Leal & Rodríguez-Robles (1997) reported that a lizard (*Anolis cristatellus*) performed conspicuous 'pushup' displays, in which the body is moved up and down in a vertical plane by flexion and extension of the legs, in response to a snake predator. Snakes have been demonstrated to stop an approach to a lizard in response to this signal (Leal & Rodríguez-Robles, 1995). This earlier study also demonstrated that lizards that signal are attacked significantly less often than those that do not. These lines of evidence are certainly suggestive that the 'pushup' display acts, at least in part, as a signal to predatory snakes. Leal (1999) provided evidence that the intensity of the 'pushup' signal given in response to predators by this lizard species is correlated with individual physiological condition, as measured by endurance capacity; whereas the intensity of this signal given in intraspecific non-predatory contexts was not similarly correlated. Leal further argued that since struggles between a captured lizard and the snake can last several tens of minutes, and can result in the escape of the lizard, endurance capacity will correlate strongly with how easily an individual can be subdued. This suggests that the pushup signal may be an honest and reliable signal of escape ability. Although these studies do not definitively demonstrate that the pushup signal is costly, the authors argue persuasively that such signals are costly both energetically and in terms of increased conspicuousness attracting secondary predators.

Font et al. (2012) describe the anti-predator foot shake displays performed by wall lizards (*Podarcis muralis*) in the presence of their potential predators (e.g. snakes). Through simulated predatory attacks, it was found that these foot shake displays were consistent with expectations from pursuit deterrent theory: they were oriented toward the predator, they were performed more often when the predator was some distance from the lizard, and they gave way to flight behaviour when the predator closed in on the lizard.

Most putative elusiveness signalling in lizards consist of tail movements. Dial (1986) is often cited as empirical evidence supportive of prey signalling to potential predators. However, this paper's evidence amounts to demonstrating that of two lizard species

(*Cophosaurus texanus* and *Holbrookia propinqua*), the one living at lower population densities (*C. texanus*) produced a tail display more readily in response to predatory threats. We do not consider this conclusive proof that such signals are not aimed at conspecifics. Hasson et al. (1989) also suggested that tail-wagging in a lizard species (*Callisaurus draconoides*) was a pursuit deterrent signal. Their evidence is as follows. Wagging occurred more in individuals near to a refuge. Wagging increased with ground temperature, which was postulated to correlate with escape ability. Wagging was less common in a situation where the experimenters hypothesized that the lizard should consider itself to be in extreme danger. Lizards wagged more when moving than when stationary, which the experimenters suggest is because their risk of being detected by a predator is higher when moving. This seems contrary to the increased wagging near refuges and at higher temperatures, both of which were considered indicative of reduced predation risk. Despite this, Hasson et al.'s study certainly constitutes a stronger argument than that of Dial for tail-wagging in lizards being a predator-directed signal. Cooper (2000) raised further concerns about the interpretation of both the studies mentioned above, but he does consider predator deterrence to be a likely function of at least some tail-wagging behaviours in lizards. Very similar evidence to that presented by Hasson et al. is presented for another species (*Leiocephalus carinatus*) in a study by Cooper (2001), which was careful to consider and eliminate plausible alternative functions of the behaviour. Since then, Cooper and others have honed this approach and provided a jigsaw of very suggestive evidence for some tail-waving and arm-waving displays in a variety of lizards to have some function in communicating evasiveness to predators (Font et al., 2012; Cooper & Perez-Mellado, 2004; Cooper, 2007, 2010a,b, 2011; York & Baird, 2016; Kircher & Johnson, 2017), Hence, we agree with Cooper that elusiveness signalling is currently the most plausible explanation for some tail-wagging in lizards, but further work on eliminating other potential explanations is still required.

Telemeco et al. (2011), for example, investigated the tail-waving behaviour of hatchling Australian three-lined skinks (*Bassiana duperreyi*) and concluded that it appeared to function to deflect predatory attacks towards the expendable tail, rather than primarily functioning as a pursuit deterrent signal. They suggest that pursuit deterrence may be a less common and derived function of anti-predator tail displays in lizards, and that deflection is likely to have been the ancestral function. It may be that we really need to explore some way of observing predator behaviour in response to mechanical model lizards.

8.2.9 Predator inspection by fish

It is well known that a range of prey species actually respond to detecting a predator in their vicinity by approaching that predator. The important aspect to this predator inspection behaviour seems to be about gathering information about predatory threats. For example, predators may not always be motivated to attack (e.g. if they are currently sated), and prey can benefit from identifying such instances. However, the act of inspection has been demonstrated in staged laboratory encounters between freshwater fish to negatively influence the predator's likelihood of attacking the inspector (Godin & Davis, 1995a,b; Brown et al., 1999). That is, predator inspection can function in part as an elusiveness signal. These studies have not been substantially built on in the last two decades, perhaps because there are methodological challenges associated with designing experiments that allow the effect of such signalling to predators to be isolated from other mechanisms (Milinski & Boltshauser, 1995) and the ethical challenges associated with staged predation on vertebrate prey. However, this system should be worth revising to explore if new technologies can allow useful measurements to be taken of naturally occurring interactions.

8.2.10 Summary of empirical evidence

Due to the challenges of successfully observing predator–prey encounters involving fleeing prey, none of the studies above can be considered to demonstrate elusiveness signalling beyond any doubt, but in most of these cases (and particularly in the skylarks, vervet monkeys, and hares) the strength of empirical evidence is very suggestive of such signalling. It does seem that such signalling occurs, perhaps infrequently but widely spread, among the vertebrates. Perhaps the lack of evidence for its existence in invertebrates is related to the much greater use of post-capture defences among invertebrates. However, the absence of evidence may simply reflect disproportionate emphasis on vertebrate taxa in behavioural research, and later in this chapter we will mention a possible invertebrate system that may involve elusiveness signalling. Likewise, the paucity of aquatic taxa from this chapter most likely simply reflects the added practical challenges for humans to observe predator–prey encounters underwater.

That elusiveness signalling might have evolved several times may not be surprising. It is probably common for predators to encounter prey that vary in their ease of capture. Whenever attacks are sufficiently expensive for predators that predators should decline difficult opportunities and concentrate on attacks that are more likely to be successful, then prey should be motivated to inform predators that they are especially difficult to capture and predators would be motivated to respond to such signals. This issue of cost driving selectivity might hint at another reason why elusiveness signalling seems to occur especially in vertebrates. In all of the vertebrate systems considered above, prey are relatively similar in size to their predators, making the prey expensive to catch and subdue but offering a substantial reward when such investment is successful. For many systems involving vertebrate predators and invertebrate prey, the predators are much bigger than the prey, prey offer a relatively small meal size, and predators consequently invest little in capture. In such circumstances, elusiveness signalling will be less likely to be selected for.

We would only expect elusiveness signalling to evolve in situations where the signal is generally honest. The case of the hare looking directly at the approaching fox suggests that in some cases the honesty of the signal may be assured because it is unfakeable. However, this does not seem to hold true for all the signals we have considered in this section, so in the next section we will lay out current theoretical understanding of how reliable elusiveness signalling can be maintained in the face of prey's temptation to cheat.

8.3 Evolution

Let us turn from considering current prevalence of elusiveness signalling to evolutionary considerations, starting with theory on how such signalling might evolve and be maintained.

8.3.1 Theoretical stability of signalling that an approaching predator has been detected (perception advertisement)

Signals that inform predators that they have been detected have been suggested for several prey species (see section 8.2). The basic idea is that a predator has a reduced chance of successfully capturing an individual if that prey individual becomes aware of the approaching predator before the final attack is launched. Hence, the predator may benefit from being informed that it has been detected, since it may now be optimal for the predator to cancel its attack. This cancellation could be beneficial to the predator if the attack involves an investment in time or energy, or if there is some other potential cost to the predator, such as risk of injury or opportunity costs associated with revealing its presence to all potential prey in the vicinity. If this signal does cause the predator to cancel its attack, then the signaller also benefits in saving the costs (e.g. time, energy, risk of injury) of evading the attack and/or the risk (however small) of being captured in that attack. Since both parties appear to benefit, the evolution of such a signal seems uncontroversial. However, we might ask what stops prey individuals from cheating, and signalling that they have detected the predator when they have not. If signalling was energetically and otherwise inexpensive, this would seem an advantageous tactic, reducing the predation risk of cheats compared to non-cheats. As cheats prospered, though, so the signal would grow to have little value to the predator, which should be selected to ignore it. Hence, the signal could break down because of cheating by prey. Bergstrom & Lachmann (2001) developed a formal model, 'the watchful babbler game', to explore the conditions required for the signal to be maintained in the face of such a danger from cheating. We now describe this model in detail.

A game between predator and prey is played in a series of rounds. For each round, the predator is either in the vicinity of a single prey individual and hence able to attack (hereafter 'present') or not. The predator is present with fixed probability α. However, in a given round, the prey does not know with certainty whether the predator is present or not. Rather, it detects sensory stimuli (e.g. sounds, movements, etc.) that are caused both by the predator (if it is present) and by other components of its environment. These stimuli are characterized by a random variable $x \in [0,1]$. If x takes a high value (range B in Figure 8.4), then the prey is certain that the predator is present; if x takes a low value (range A), then it is certain that the predator is not present. There is, however, an intermediate range of values of the stimulus where the prey is uncertain. Specifically, x for a given round is drawn from different probability distributions according to whether the predator is present or not ($f(x)$ and $g(x)$ respectively). There is some overlap between the two distributions, so there is a range of x values for which the value of x does not unambiguously inform the prey whether the predator is present or not (the 'zone of uncertainty' in Figure 8.4).

The model assumes that signalling costs the prey. Specifically, if a prey individual signals in a given round then its probability of surviving that round is multiplied by a factor $(1-c)$ for some $0 \leq c \leq 1$. This cost has to be interpreted with care. The cost is paid regardless of any effect that the signal may have on the predator to which it is directed (hereafter the primary predator). Hence, this mortality cost is not

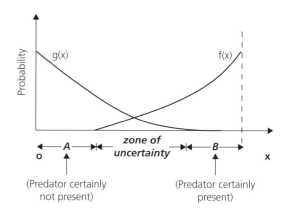

Figure 8.4 The situation modelled by Bergstrom & Lachmann (2001).

associated with the primary predator. Bergstrom & Lachmann suggest that signalling attracts a different type of predator (hereafter a secondary predator), attacks by which are responsible for the mortality cost of signalling. This interpretation justifies a situation where the cost is incurred whether or not the primary predator is present. If the primary predator had to be present for a cost to be imposed, then signalling would break down; prey could signal indiscriminately without heed to whether they thought a predator was present or not, without incurring prohibitive costs (since signalling would be cost-free when the predator was not present), and so the predator would be selected to stop heeding the signal. It is possible to imagine other costs of signalling also being expressed as a mortality cost of signalling regardless of the predator's response, for example this could be increased starvation risk if signalling and foraging are mutually exclusive.

All attacks by the primary predator require the predator to make an investment (d). However, attacks differ in their probability of success (t). Bergstrom & Lachmann argue that, for signalling to be stable in their model, the probability of success must be negatively correlated with the sensory stimulus x. This is a key result of the model, and formalizes the idea that the prey can gain information about whether a predator is likely to be present from the stimuli it detects (x) and reduce its risk from an attack (t) by being forewarned of the danger. Both x and t are generated from a bivariant distribution $f(x,t)$ for each round when the predator is present.

The rules just described translate into the payoff matrix shown in Table 8.3. The predator is unaware of the values of x (the strength of stimulus received by the prey, and hence the prey's knowledge of the

predator's presence) and t (the probability that an attack will succeed). The only information that the predator can use to decide whether to attack or not is whether the prey signals or not. The prey is aware of x but not t. Hence, the prey's decision to signal or not will be based on x and knowledge of the frequency with which predators occur (defined by fraction of rounds in which the predator is present (α)). A successful attack brings the predator a reward of one, not attacking brings no reward (but saves the investment of d).

The prey gains nothing from signalling if the predator is not present, but pays a price in raised mortality because of the signal attracting secondary predators (described by the value of c). Part of the prey's strategy should therefore be not to signal when it knows that the predator is not present; this occurs for the range A of x values in Figure 8.4. If it is worth the prey signalling for one value of x, then it is worth it signalling for all higher values of x (where it is more likely that the predator is present). Hence, the prey's strategy can be defined by a critical value x_c. If x in the current round is below x_c then the prey does not signal, otherwise it signals. We can see that costs of signalling (c) must be greater than zero for the signalling equilibrium to be stable. If there is no cost to signalling, then the prey could always signal, in which case the predator should ignore the signal as it contains no useful information about the prey's estimate of the likelihood of the predator being present. However, c should not be so high that prey never benefit from signalling.

The predator's strategy depends on the values of two related components: its probability of attacking when the prey signals and its probability of attacking when there is no signal. Since the predator has no

Table 8.3 The pay-off matrix for the model of Bergstrom & Lachmann (2001).

Payoffs per round						
	Predator present				Predator not present	
	Prey signals		Prey does not signal		Prey signals	Prey does not signal
	Attack	No attack	Attack	No attack		
Prey	$(1-c)(1-t)$	$1-c$	$1-t$	1	$1-c$	1
Predator	$1-d$	0	$1-d$	0	0	0

way to distinguish one non-signaller from another, or one signaller from another, then it should operate an all-or-nothing policy with respect to attacks on signallers and non-signallers (i.e. either attack all or no individuals of each type). If all values of probability of an attack in a specific round being successful (t) are greater than the investment made in each attack (d), then the predator should always attack regardless of any signalling. Similarly, if all t are less than d, then it should never attack. We are interested in cases where it should attack one group but not the other. Recall that a key requirement for stable signalling in the model is that t is negatively correlated with x. Now, high values of x induce the prey signal. Hence, the average value of t from rounds where the prey signal (call this average $t+$) is less than the average value from rounds where no signal is given ($t-$). The predator should attack when it is most beneficial to it, so if it attacks one group but not the other, then it should attack non-signallers.

Bergstrom & Lachmann demonstrate that there is a set of necessary conditions that are sufficient to produce a stable system in which 1) prey only signal when they are reasonably sure that they have detected a predator, and 2) the predator responds to that signal by not attacking. These conditions are that:

1) There is a cost to signalling.
2) This cost to signalling is not so high that signalling is never profitable.
3) Prey have some means (even if imperfect) of gauging the likelihood of a predator being present, and hence their risk of attack. That is, those prey that are most 'concerned' about predation are actually those that are most at risk.
4) Prey that strongly suspect the presence of the predator are more difficult to capture than those which have lower levels of suspicion.
5) The costs to the predator of attacking are not prohibitively high.

Two aspects of this model that may make it challenging to apply to real systems are: the description of time as a sequence of rounds; and the definition that t (the probability that an attack is successful) and d (the cost to the predator of attacking) are considered to be in equivalent units. These restrictions to the model's applicability were removed by Getty

(2002). As can be seen in Table 8.1, in the original model, if the predator is present but does not attack then it receives a payoff of zero. This can be considered as the opportunity cost of pursuit. If we move from a series of rounds of a game to a continuous time setting, then Getty demonstrates that the fundamental conclusions of the original paper are retained except that this opportunity cost now changes. The predator should pass up opportunities to attack signallers and resume searching for non-signallers whenever the opportunity cost of pursuing signallers exceeds the expected benefit from attacking a signaller. Getty's modifications therefore make the model more useful, in the sense that the opportunity cost of attacking a signaller is now costed as the expected value of the alternative activity (finding and attacking a non-signaller). The result is that parameters such as the expected time to find a non-signaller become important, with the further effect that population density of prey and search ability of the predator become explicit parameters of the model. Getty's work also provides a re-interpretation of the model in terms of commonly considered foraging concepts, like diet choice, and concepts from signal-detection theory, such as the receiver operating characteristic (ROC) curve.

This body of theory was further developed by Broom & Ruxton (2012). They showed that the essential arguments of the theory still hold when it is generalized, such that the signal is not simply binary but can vary in its intensity (with higher intensity signals being costlier to produce) and with the addition of a high-cost signal (such as staring directly at the predator) that can only be utilized when the prey is very confident about the existence of a nearby predator. These complexities still allow for stable signalling, but the prey do not use the range of signalling intensities to signal different degrees of confidence in the proximity of the predator. Rather, prey simply adopt a binary response of not signalling or always signalling at the same fixed intensity. This binary response occurs because the predator has only two responses open to it on detecting a signal (attack or not). They make the testable prediction, however, that the intensity of the signal will increase for prey that require greater confidence in the existence of the predator before signalling. Thus, they make the prediction that within an individual the intensity of the signal will not correlate with confidence in the proximity of a predator; however, across populations those populations whose signals are lower cost will be associated with higher rates of giving such signals mistakenly when no predator is present. The availability of the high-cost/high-certainty signal does prohibit the stability of signalling based on a lower-cost signal, but there are also circumstances where only the high-cost signal is used. Again, because of arguments based on the essentially binary options available to the predator, it is expected that where prey use a high-cost/high-certainty signal (such as directed staring) they will not also use a lower cost signal (e.g. tail-flagging) when their confidence in a predator's presence is reduced. Finally, they emphasize that any stable signalling equilibria co-exist with an equilibrium state where prey do not signal and predators do not respond to signals. We can assume that this non-signalling state is ancestral. However, Broom & Ruxton (2012) discuss how when the signalling solution is also stable, there are natural routes for natural selection to lead to the signalling state.

8.3.2 Theoretical stability of signalling that the prey individual is intrinsically difficult to catch (pursuit deterrent signals)

One benefit of signalling may arise simply by informing the predator that it has been detected. An alternative (yet not necessarily mutually exclusive) role of signalling may be to indicate to the would-be predator that the signaller is difficult to catch. A key theoretical work on honest signalling by prey is that of Vega-Redondo & Hasson (1993), and looks at the maintenance of signals of 'quality' rather than 'alertness'. Rather than communicating that it has detected a predator, the prey signals that it is a particularly high-quality individual that would be difficult to capture in an attack, and so the predator would be better served by passing up the chance of attacking it. Their model is structured as follows.

In each round of the game the predator confronts a group of N prey individuals drawn randomly from the entire population. The prey population is made up of two prey types, differentiated by their qualities q_1 and q_2. We assume that q_1 is greater than q_2, and that

this denotes that q_1 individuals are intrinsically more difficult for predators to catch than q_2 individuals. Prey can signal to the predators; this signal (s) must be selected from the continuous range $[0,s_{max}]$. In the example used by Vega-Redondo & Hasson, s represents the distance between predator and prey at which the prey flees. The constant s_{max} is then the maximum distance at which approaching predators can be detected by the prey. Small values of s are intrinsically more risky for the prey than higher values of s, and so are more likely to indicate a high-quality individual. It is on the basis of these signals that the predator chooses which prey to attack. The predator attacks only one individual from each group. We assume that if two individuals signal alike then (even if they are of intrinsically different quality) they are both equally likely to be the individual attacked. The payoffs from a round of the model are zero to any prey item in the group of N that is caught by the predator and one to all those that are not caught (either because they were not the one selected for attack or were selected but the attack failed). The predator gets a reward of one if it attacks successfully and zero if it is unsuccessful. The final part of the model is the function $h(s_j,q_i)$, which is the probability that a targeted individual of quality q_i that signals s_j and is targeted by the predator, escapes the attack.

Vega-Redondo & Hasson's key finding is that there is an evolutionarily stable signalling equilibrium where the two prey types issue different signals, and the predator responds differentially to those signals. The q_1 individuals issue signals s_1, and the q_2 individuals issue signals s_2 where $s_1 \neq s_2$. Indeed, in the case where s reflects the distance to which predators are allowed to approach: $s_1 < s_2$ (i.e. high-quality individuals allow the predator to approach more closely). At this signalling equilibrium, if the group contains individuals of both signal types, then only those of the s_2 type have a non-zero probability of being the one targeted. This probability is the same for all such individuals, and is the inverse of the number of s_2 signallers. Only if the group is entirely composed of s_1 signallers will one of them be targeted, each q_2 individual having a $1/N$ probability of being the one targeted.

The equilibrium exists and is evolutionarily stable providing $h(s,q)$ satisfies the following conditions:

1) An attack on a higher-quality individual is less likely to be successful than an attack on a poorer-quality individual if both signal identically (i.e. if both allow the predator to approach equally close). That is, the signal (s) is related to a quality of prey that affects their vulnerability to predators (q).
2) If s increases, then the relative decrease in the probability of an attack being successful is greater for the higher-quality individual than the lower-quality one. Thus, the cost of a more intensive signal (a smaller s) is greater for a low-quality individual than for a high-quality individual.
3) If s becomes small enough, then the predator's attack is always successful.
4) $h(s,q)$ converges to zero as s increases, and does so faster than its derivative.

These conditions would have to be modified (but in a very intuitive way) if the signal is one where increasing signal value is associated with increasing quality, e.g. stotting height or stotting frequency.

What keeps the signal honest is the risk that all individuals in a group will be vulnerable to attack, including a low-quality individual that pretends (by its signal) to be a high-quality individual. If, by chance, such an individual ends up in a group of high quality individuals then it has a $1/N$ chance of being the one selected for attack. If this occurs, then this individual pays the price of its deception through a high probability of predator success, because it has allowed the predator to approach so closely. This means that both a large group size N and a relatively low frequency of high-quality individuals strengthens the incentive for low-quality individuals to cheat. However, providing the four conditions above are met, the signalling equilibrium will remain stable, although the signals will be driven to higher s values by such circumstances. If both signals are greater than s_{max}, then the signalling equilibrium will break down and both types will do their best to flee as soon as they detect the predator. Hence, there is a requirement that s_{max} is sufficiently big to incorporate the two signals. Ecologically, we can see that if s_{max} is very small, then signalling breaks down because the optimal strategy for all prey types is to flee as soon as the predator is detected. However, there may also be a restriction that s_{max} cannot be too big; if the prey are too good at

spotting the predator, then the analysis becomes more problematic and is not explored in the original paper.

Notice that, at equilibrium, the low-quality individuals do not necessarily flee as soon as they detect the predator. A mutant that did this would always be the one targeted by the predator in preference to the other low-quality individuals that allow the predator to close a little more before bolting.

There is much that can be done to develop this model. It seems likely that predators can only discriminate signals if they are sufficiently far apart, and this minimum separation distance is likely to change with signal intensities. Prey individuals need not associate at random. It may be possible for cheating to take root more effectively if cheats band together or preferentially associate with low-quality individuals, to avoid or reduce the risk of being the targeted individual. There may be costs to both predator and prey individuals involved in an unsuccessful attack. Prey populations are likely to be composed of more than two quality types, and predators may meet prey singly rather than in groups. All of these elaborations are worth exploring, but even the original model warrants further investigation. The original paper demonstrates that a signalling equilibrium can exist; it does not explore the nature of that equilibrium or how it is affected by plausible parameter values and specific functional forms.

8.3.3 Summary of theoretical work

There is a quantitative theoretical basis for considering that both signals that predators have been detected and signals that individual prey are hard to catch or subdue are at least theoretically possible. Notice, for all the modelling discussed in the preceding sections, that we would only expect signals to be honest 'on average'. Johnstone & Grafen (1993) argue that we should always expect some low level of cheating. Hence, evidence that cheating sometimes occurs is not evidence that the signal carries no useful information to the predator 'on average'. Indeed, the fact that palatable (Batesian) mimics can persist strongly suggests that they do not completely undermine the effectiveness of the signal generated by the unpalatable model (see Chapter 9).

8.3.4 Empirical explorations of evolutionary trajectories

Stang & McRae (2009) present an insightful comparative analysis of the evolution of white under-tail plumage associated with tail-flicking behaviours in response to predators in a group of ground-dwelling birds (the rails), considered in section 8.2.6. They found that white under-tail plumage generally evolves after adaptation to open wetland habitat but before evolution of gregariousness. This relative timing is important, since the most plausible alternative explanation for selection of flicking behaviour and associated coloration is that it is an interspecific signal. However, the evolution of this coloration prior to adopting a gregarious lifestyle argues against this. This paper thus adds to the strong body of recent evidence that tail-flicking in rails is directed towards predators. Further, the study found no support for the white tails being sexually selected, since they were not associated with sexual dimorphism, with the use of the tail in courtship displays, or with their mating system.

Key to the evolutionary stability of the types of signalling described in this chapter is that those signals are sufficiently costly for the signaller to produce. However, such signals may serve multiple functions, and these may reduce the potential costs of such signals for the mechanism under consideration here. For example, a signal may act both to inform the predator about the prey individual, but it may also act to inform kin or other conspecifics about the presence of the predator. There may be a kin-selected benefit to alerting relatives; alerting other conspecifics to the predator's presence may benefit the focal individual if the simultaneous fleeing of several individuals offers protection to the individuals concerned, through predator confusion. As a further complication for the study of such signals, identification of the means by which costs are expressed (never mind quantifying them) may be challenging. Consider the examples below.

Stankowich (2008) provides a thoughtful critique on the extensive literature on the possibility that tail-flagging by fleeing deer and bright rump patches may be a signal to predators. They make the interesting prediction that only relatively short tails can be lifted during flight to reveal a bright rump patch

and act as a signal, and animals with a short tail cannot act to disperse biting insects the way those with a longer tail can. Having a tail short enough to be useful in signalling may come at a cost of increased irritation from flies. Stankowich & Coss (2008) report on the characteristic walking gait adopted by Columbian black-tailed deer (*Odocoileus hemionus columbianus*) in response to an approaching predator. A biomechanical analysis of gait leads them to speculate that it has traits that would not be possible for a lame or arthritic animal, and so the walk may be an honest signal of fleeing ability. The take-home message from this is that researchers must be alert to the number of different ways in which costs of signalling to predators may be expressed.

8.4 Ecology

If (as we expect) elusiveness signals are expensive, then we would expect their use to be dependent on ecology, since ecology will influence the magnitude of costs and benefits to both parties. Indeed, in some situations prey may be signalling not simply intrinsic traits about the prey individual (such as how quickly it can run), but ecological aspects of a particular predator–prey interaction (e.g. how close the prey individual is to protective cover). The prey's motivation to signal is that it may deter the predator from attacking, saving the prey the costs of any attack. These costs will be influenced by the ecology of the situation. Setting aside, for now, the effects of ecology on the likelihoods of the prey detecting the predator and of the predator responding to any signal, the costs of an attack for the prey will be influenced by aspects of the ecology. For example, the probability that attack will lead to capture of the prey might be dependent on: the nature of the terrain, the distance of the prey from protective cover, and/or environmental temperature for ectotherms. Even if the attack is unsuccessful, the prey may pay a cost in energy spent and/or foraging time lost. These costs will be dependent on the energetic state of the predator and on the availability of its food. The cost might be expressed as increased apparency of the prey to other predators, which will depend on the local predator community. Aspects of ecology also have the potential to affect the distance at which the predator is first detected by the prey (and so the maximum distance between prey and predator at which the signal can be given). Obvious aspects of the prevailing ecology in this regard include the physical complexity of the terrain and light levels. Further, considering the costs to predators of failed attacks, aspects of ecology will also influence the willingness of the predator to heed such a signal. Risk of injury during an attack might be influenced by terrain and microhabitat, and the risk of a predator attracting its own predators will be dependent on the local community of such predators. The size of a cost to the predator in revealing itself to other prey that have yet to detect it will be dependent on the local prey community, as will the ability of the predator to pay energy costs of failed attacks. The value of a successful attack will be dependent on the local prey community (which will influence the availability and value of other possibilities for prey capture) and the local competitor community (which may steal captured prey before it is fully consumed). Thus, we should expect that for those prey that have elusiveness signals, those signals are not used indiscriminately, but their use (and perhaps their nature) will be influenced by variation in the ecology of different predator–prey interactions.

Because many aspects of the interaction between prey and predator will be influenced by ecology, making clear predictions about the effects of a given ecological parameter on signalling can be difficult. For example, consider the effects of prey density. If the prey density is higher, then this may reduce the risk of attacks being successful, because prey can mount collective defences (say, through confusion effects). However, a higher local density of prey may require the predator to approach a given prey more closely before it is clear to the prey individual that it has been singled out for attack, and predator–prey distance is likely to influence attack success rate, and so the value of signalling. The effect of attack success rate on prey's propensity to signal is likely often complex. We agree with Cooper (2011) that signalling may be most attractive at intermediate success rates: at low success rates the cost of signalling is not worth paying; and at very high success rates prey should not delay their fleeing by signalling first, but should flee immediately on detection of the predator. A high local prey density might increase the costs of failed attacks, since all those

prey in the vicinity will be alerted to the presence of the predator. Further, a higher local prey density may also make the predator more willing to forgo one particular attack, because there are likely to be further (sometimes better) opportunities to make a kill soon. The situation is too complex and non-linear for verbal reasoning to furnish us with clear, general predictions about how this ecological factor (prey density) should influence signalling. Rather, predictions will be dependent on the final detail of the interaction, and qualitative modelling will be required to yield predictions for specific situations. However, the value of such theory will be dependent on finding a suitable empirical system, in which key quantities and functional relationships can be estimated empirically and used to parameterize the theoretical model.

Several recent publications on signalling by lizards are beginning to gather valuable empirical information on how signalling can be used flexibly in different ecological circumstances (Cooper, 2010a, 2011). The zebra-tailed lizard *Callisaurus draconoides* sometimes waves its tail in response to approaching predatory threats. Cooper observed that individuals that signalled prior to fleeing allowed him to approach closer before fleeing than non-signalling individuals. He interprets the signal as one of elusiveness, and argues that signalling will increase the probability that real predators will break off attacks, and thus lowers the risk associated with a given approach. He assumed that this reduced risk translates into lizards reducing their flight initiation distance. He also observed that signalling was most frequent when prey were at intermediate distances from a refuge from predation, which he interprets as fitting his conjecture that signalling is most appropriate at intermediate levels of the risk of an attack being successful. He also observes signalling to be more common at higher substrate temperatures, which is consistent with higher ambient temperatures leading to higher athleticism in an ectothermic lizard. Lastly, he observed that signalling was more likely when on soil than when the lizard was on rock; this he interpreted as indicative that rock provides few refuges and is difficult to run quickly on—and so is a high-risk microhabitat, where immediate fleeing is a better option than signalling.

Murphy (2006) argued that his study species (turquoise-browed motmots, *Eumomota superciliosa*) might have a number of life-history characteristics that make signalling of detection to potential ambush predators (perception advertisement) particularly beneficial. It is a large, slow-flying bird that might be particularly vulnerable to surprise attacks. Also, its nesting burrows are near the ground, exposing it to many mammalian ambush predators. Further, Murphy suggests that a motmot commonly repeatedly hunts (terrestrial insects and small vertebrate prey) from the same perch, making its positioning predictable to its predators. More generally, it may be possible to make cross-species predictions about which types of ecologies are suitable for signalling, but we suspect (because of the multiplicity of different mechanisms involved, as discussed in the section 8.3) within-species comparisons may be more effective except in cases where the species being compared have very narrow ecological niches. However, comparing signalling behaviour between different species can reveal similarities and differences in the evolution of form and function of anti-predator signalling, and how different ecologies may influence their evolution (Clark et al., 2016).

8.5 Co-evolutionary considerations

The signals discussed in this chapter require prey to benefit, on average, for giving the signals and predators to benefit, on average, from responding to them. Thus, the evolutionary considerations of section 8.3 have extensively considered the co-evolutionary requirements for such signalling to be maintained. However, this does not preclude the possibility that in some circumstances prey can benefit from using a signal dishonestly. As discussed in section 8.4, Murphy (2006) suggested that turquoise-browed motmots use a tail signal to inform potential ambush predators that they have been detected. In a follow-up paper, Murphy (2007) presents data that suggests that the same signal can be given in another context, even when a predator is not present: when an adult returns to its nesting tunnel to feed chicks. They argue that this signal is unlikely to be directed to conspecifics in this instance, and suggest that it may be used in this context as a dishonest perception-advertisement signalling: in effect, pretending that a potential ambushing

predator has been detected when it has not. However, responses of any predators to this signal have yet to be reported, and in the absence of this, it is difficult to say anything definitive about whether predators are truly being fooled in this way. It seems relatively challenging to explain how predators have sufficient familiarity with this prey species to respond to the signal when it is used honestly, but can still be fooled when it is used dishonestly; especially since a visual cue to differentiate between the two cases will be available: when the adult returns to the nest to feed the chick it will be carrying food in its beak. In contrast, Barbour & Clark (2012) observed that over 90 per cent of the tail-flagging displays seen in their California ground squirrels were given when no rattlesnake predator was nearby. However, they do not interpret this as widespread dishonesty, but rather as evidence that the signal honestly indicates a state of heightened vigilance in a situation where a predator might be present. Further, Putman & Clark (2014) argue that tail-flagging displays in the absence of predators can honestly signal vigilance, or the readiness to avoid an attack, even if predators are undetected, as, in their study, squirrels signalled often when interacting with a snake, but also in areas of recent snake encounters where they would then respond faster to simulated strikes (see Figure 8.5). They suggest that anti-predator displays may be widespread across species, due to the benefits of advertising alertness even when uncertain of the location of predators.

The Californian ground squirrel presents a fascinating example of signals being tuned to the sensory systems of the predators concerned. In the study of Rundus et al. (2007) the squirrels were faced with rattlesnakes and gopher snakes; only the first predator type is sensitive to infrared. The squirrel tail-flags in response to detection of both types of snake, but only with the rattlesnake does it add an IR component, by piloerection and enhanced blood flow to the surface in the tail. Experiments with a robotic squirrel demonstrated that this IR component enhanced the predator deterring effect of tail-flagging towards rattlesnakes.

Co-evolution leading to the maintenance of elusiveness signalling will be facilitated when the predator is relatively specialized on a given type of prey and when the signal is distinctive and

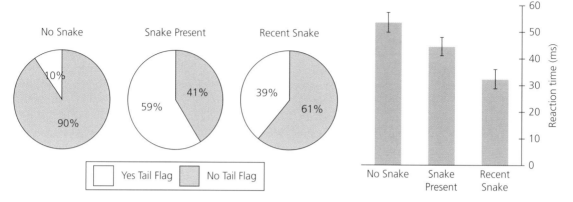

Figure 8.5 Adapted by Bree Putman from results presented in Putman & Clark (2014): shows the proportion of squirrels that exhibited tail-flagging at a bait station when no snake was present, when a snake was present, and when they had recently encountered a snake that was no longer present, and it shows the reaction times to simulated snake strikes under these three treatments.

unambiguous. However, displays used by parental birds to distract predators away from their offspring can be seen as dishonest signalling of evasiveness. This is likely possible only because the predators concerned are generalists and because the behaviour of the prey is not a ritualized and distinctive signal, but rather a suite of behaviours that the predator does not interpret as a signal but as the normal behaviour of an unevasive prey type. We consider such behaviour in the next section.

8.6 Unresolved issues and future challenges

As we saw in section 8.2, almost all the empirical systems where signals of evasiveness have been explored involve large vertebrate prey and predators. Laboratory experimentation of such systems is near impossible. However, there have been a number of reports that different types of fly display to salticid spiders that provide a predatory threat to them, and that the spiders can break off attacks in response to such displays: e.g. Figure 8.6 and Aguilar-Argüello et al. (2016) and references therein.

The flies appear to offer no defence against spiders that make aposematism a likely explanation. Although it has been repeatedly suggested that the signalling may be perception advertising, it would be useful to evaluate startle effects as an alternative explanation. One would expect that if the signal acts to startle the spider, then habituation

to the signal should be more likely than if it is perception advertisement. If startling mechanisms can be excluded, then this system may hold real possibilities for experimental manipulations, control of interaction circumstances, and substantial replication; all of which are difficult or impossible to achieve in current systems. The idea that some butterflies signal evasiveness to avian predators might also offer potential for developing a laboratory system (Pinheiro et al., 2016).

Recording data on naturally occurring interactions involving vertebrate predators and prey is very challenging. One expects the occurrence of such interactions to be difficult to predict, both in space and in time, since any predictability by predators can be exploited by prey to reduce encounter rates. Coursing vertebrate predators can often chase their prey over several kilometres, and ambushing predators are easily disturbed by human observers. It may be that technological advances may solve some of the current challenges associated with such data. We saw earlier in this chapter (section 8.2.4) an example where researchers benefited from the ability to radiotrack ambush-hunting snakes. The sophistication of data-recording equipment that can be carried on animals is increasing rapidly and, for example, drone technology may allow effective remote monitoring of interactions with reduced risk of disturbance. It is likely that, to make substantial further progress in our understanding of elusiveness signalling, we may need to move to larger-scale, more expensive studies

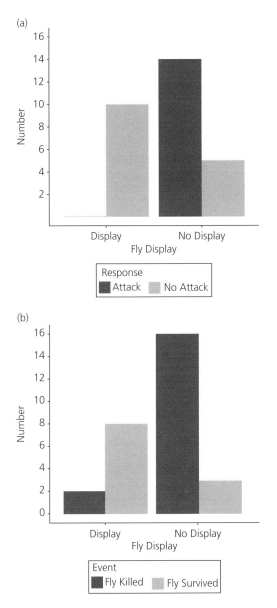

(a)

(b)

Figure 8.6 Developed by Dinesh Rao from data presented in Aguilar-Argüello et al. (2016), these data suggest that flies that display are less likely to be attacked by spiders and more likely to survive an encounter with a spider.

order to truly understand such signalling, we need to quantify the costs and benefits to the prey of signalling and those to the predator of responding to the signal. As we have discussed earlier in this chapter, costs in particular will often be multivariate, and we acknowledge that their measurement will not be easy. However, we feel that there are examples elsewhere in behavioural ecology where detailed measurements of suitable model systems have yielded wide-reaching improvements in understanding of general principles. Such a case might be the study of foraging by the shorebird the Eurasian oystercatcher (*Haematopus ostralegus*), which has been a very successful model system for enhancing our understanding of the economics of foraging (Goss-Custard, 1996).

Enhancing our understanding of elusiveness signalling cannot be achieved by empirical measurement alone: we must use such measurements of specific systems to parameterize and test underlying general theory. We now have a relatively solid theoretical framework for understanding both pursuit deterrent and perception advertisement signals. However, as Broom & Ruxton (2012) discuss, a number of other functions may influence selection on signals triggered by detection of predators (such as signalling to conspecifics). Further, as discussed in various sections above, such signals may also be influenced by sexual selection. Hence, there is a need to build on existing theory and explore the influence of multiple selective pressures on the existence and forms of signals between prey and predators. Further, as discussed in section 8.5, we must expand the aims we have for theory in this area. Until now, investigation of theory has been focussed on identifying the conditions required for the maintenance of signalling. We have argued that we should expect that prey will use such signalling flexibly, in some encounters with predators but not in others. It is also possible that the nature of the signal (rather than whether it is given at all) will vary between encounters to encode encounter-specific information. We need to expand our theory to offer predictions about how prey signalling will change in different ecological scenarios. These predictions should allow the integration of theory with empirical observation and, in doing so, strengthen our understanding of a relatively uncommon but evolutionarily fascinating tactic for avoiding attack.

than have historically been the case (the recent study of Isbell & Bidner, 2016 is an example of what can be achieved). This change will not just be driven by the availability of useful technologies, but also by a need to develop a more detailed understanding of some model systems. Elusiveness signalling will only be selected if it brings net benefits to both parties. In

CHAPTER 9

Batesian mimicry and masquerade

9.1 Introduction and overview of the chapter

In this chapter we consider deceptive mimicry as a means of gaining protection from predation. We start the chapter by briefly reviewing the alternative forms of mimicry, showing that they can be challenging to classify. Subsequently we present examples of deceptive protective mimicry, and examine multiple lines of evidence that these resemblances serve to protect prey in the manner ascribed.

The possibility that individuals of one species may gain selective advantage by mimicking members of another species was raised by Henry Walter Bates just three years after the publication of *On the Origin of Species* (Darwin, 1859).[1] In his classic paper on species resemblances among Amazonian butterflies, Bates (1862) described several different types of mimicry, but the form that he is now most closely associated with ('Batesian mimicry') occurs when members of a palatable species (the 'mimic') gain a degree of protection from predators by resembling an unpalatable or otherwise defended species (the 'model'). Charles Darwin was clearly much impressed by Bates' article, writing in a letter to Bates on November 20 1862: 'You have most clearly stated and solved a most wonderful problem. Your paper is too good to be largely appreciated by the

mob of naturalists without souls; but rely on it that it will have lasting value' (Darwin, 1887). Indeed, it has.

As Bates had anticipated, mimicry is thought to evolve for several different reasons, including gaining protection from predators ('protective' mimicry), and gaining access to prey, hosts, or other resources ('aggressive' mimicry). Examples of aggressive mimicry include deception to gain access to hosts, nuptial gifts, pollinators, parental care, and even burial as a seed (Midgley et al., 2015). However, in keeping with our chosen topic ('avoiding attack') we will restrict much of our discussion of mimicry in this book to protective mimicry, i.e. a consideration of selection on potential prey species to resemble other species or objects, thereby gaining protection from would-be predators. Examples of protective mimicry include: (i) the resemblance of the species to an object of no inherent interest to a potential predator such as leaves, thorns, sticks, stones, or bird droppings ('masquerade'; Skelhorn, 2015); (ii) the resemblance of species to other unpalatable or otherwise defended species that employ warning signals to indicate their unprofitability ('Batesian mimicry'; Bates, 1862); and (iii) the evolution of a shared warning signal among defended prey types ('Müllerian mimicry'; Müller, 1879). Here we consider protective masquerade and Batesian mimicry, which we collectively refer to as 'protective deceptive mimicry' since we discuss 'Müllerian' mimicry (which we might consider as 'protective informative mimicry') at length in Chapter 7.

We must first make it clear that there are numerous ways of classifying protective deceptive mimicry and it is not always obvious to which class of defence

---

[1] Examples of mimicry had been recognized long before Bates (1862), although they were not understood to have arisen as a consequence of natural selection. For example, Kirby & Spence (1817) noted the striking similarity of adults of *Volucella* hoverflies to bees, considering it (p. 223) *'a beautiful instance of the wisdom of Providence'* allowing them to enter *'nests and deposit their eggs undiscovered'*. The natural theology of mimicry before Bates is reviewed by Blaisdel (1982).

Avoiding Attack: The Evolutionary Ecology of Crypsis, Aposematism, and Mimicry. Second Edition. Graeme D. Ruxton, William L. Allen, Thomas N. Sherratt, & Michael P. Speed, Oxford University Press (2018). © Graeme D. Ruxton, William L. Allen, Thomas N. Sherratt, & Michael P. Speed 2018. DOI: 10.1093/oso/9780199688678.001.0001

a given example belongs. In particular, as the value of masquerade rests on not being recognized as a prey item (or as a potential predator) yet also involves resembling items or objects in the local environment such as bird droppings and pebbles, then masquerade can be thought of either as a form of camouflage or as a form of mimicry. However most recent definitions stress the latter (see Skelhorn et al., 2010a; Skelhorn, 2015).

A graphic illustration of the difference between masquerade and background matching was proposed by Allen & Cooper (1985), who drew inspiration from a much-loved fictional bear. In the first chapter of Milne (1926) we are introduced to Winnie-the-Pooh who was attempting to reach honey that was high in the trees and guarded by bees. Pooh first thought of floating up on a blue balloon that would be cryptic against the blue sky, hence hard to detect. However, he then figured that the bees would identify him hanging below. As a solution, Pooh rolled himself in mud so that he would resemble a small black cloud, which would be easy for the bees to detect but hard to recognize. It is the difference between general background matching and adaptive resemblance to a specific object in the background that is now used to distinguish crypsis[2] from masquerade. Potential examples of 'aggressive masquerade' initiated by Pooh are occasionally reported, such as that seen in the Amazon fish *Monocirrhus polyacanthus* which resembles a dead leaf, most likely to launch surprise attacks on fish and invertebrate prey (Catarino & Zuanon, 2010).

While masquerade is now seen as a form of mimicry, it generally involves resemblance of traits that have not evolved to convey information (i.e. cues) rather than warning signals (Jamie, 2017), so it is separable from both Batesian and Müllerian mimicry. Just as crypsis and masquerade are hard to disentangle, however, it is clear that appearances alone are insufficient to infer the specific reason why

a given form of mimicry has evolved. For example, in 1892 Bateson and Poulton engaged in a caustic debate in the journal *Nature* on whether *Volucella* hoverflies are aggressive or Batesian mimics of bumblebees (Poulton, 1890; Bateson, 1892). Naturally, in principle the two tactics are not mutually exclusive, and the same species may benefit both from avoiding predators and gaining access to hosts (Nelson & Jackson, 2009). For example, the fang-blenny mimic *Plagiotremus laudandus* gains from its resemblance to the toxic poison fangblenny *Meiacanthus atrodorsalis* through its ability to parasitize local coral reef inhabitants (aggressive mimicry) but also from its reduced predation risk (protective mimicry) (Cheney, 2010). Even more remarkably, the dusky dottyback (*Pseudochromis fuscus*) can alter its body coloration to mimic differently coloured reef fishes and in so doing gains access to prey as well as protection from predators (Cortesi et al., 2015). Some species are thought to employ two different forms of mimicry at different stages in their life cycle. For example, although cuckoos are widely known as brood parasites that often engage in aggressive mimicry of eggs and chicks to avoid being expelled from the nest, they have also long been suspected as Batesian mimics of raptors, especially *Accipiter* hawks (Wallace, 1889; Gluckman & Mundy, 2013; Trnka et al., 2015).

To further complicate matters, Batesian and Müllerian mimicry are generally treated as distinct forms of mimicry, but it is important to note at the outset that it is sometimes difficult to distinguish between these two forms of mimicry. For instance, the viceroy butterfly (*Limenitis archippus*) has traditionally been regarded as a palatable Batesian mimic of the unpalatable monarch (*Danaus plexippus*) yet Ritland & Brower (1991) presented red-winged blackbirds with abdomens of these two species and found that the viceroy and monarch were approximately equally unpalatable, a finding more consistent with Müllerian than Batesian mimicry. In other instances it may be quite fruitless to make the distinction. For example, some predator species may find a given species profitable to attack, while other predator species find it unprofitable. Even the motivation of the same individual predator to attack a given type of prey may vary according to its state of hunger or

---

[2] Likewise, disruptive coloration and crypsis have often been distinguished as two separate camouflage strategies, the former based on hindering detection or recognition of an object's outline or characteristic shape, and the latter based on minimizing visual difference between object and background (see Chapters 1 and 2).

toxin burden (see section 7.5.6.1), so the profitability of a given mimic species to a given predator may be highly variable.

Bearing in mind that no simple categorization of mimicry will neatly classify all examples, we begin by briefly reviewing some well-known examples of protective deceptive mimicry. In some cases, while the mimic is assumed to be palatable, and the model is assumed to be unpalatable or otherwise unprofitable, this is not always known for sure. Moreover, the major predators of mimics are not always known, and it is generally implicitly assumed that the apparent visual (or other sensory) resemblance of the mimic and model(s) are also perceived by the major predators in question. Despite these reservations, many of these examples have all the hallmarks of selection to deceive predators, including the adoption of behavioural traits to help further the illusion. We then compare and contrast the various theories that have been proposed to understand the phenomenon. Next, we examine the evidence for the phenomenon and its predicted properties, and finally we address several important questions and controversies relating to the evolution and ecology of mimicry, many of which remain only partly resolved.

9.2 Examples of protective deceptive mimicry

9.2.1 Masquerade

Many organisms appear to mimic inedible objects such as leaves, thorns, sticks, and bird-droppings; that is, objects of no inherent interest to the potential predator. Remarkable examples include the sea dragon (*Phyllopteryx eques*), a seahorse found off the coast of Australia that has numerous outgrowths that together create the impression of seaweed. Likewise, the northern potoo bird (*Nyctibius jamaicensis*) from Central America rests motionless on trees in the day, and closely resembles a broken branch (Edmunds, 1974). The potter wasp *Minixi suffusum* disguises its nest—effectively its extended phenotype (Dawkins, 1982)—in the form of a bird dropping (Auko et al., 2015). Naturalists have long been impressed with the way in which certain leaf mimics not only have a leaf shape, but also exhibit

patterns resembling the leaf's veins and even their blemishes. One characteristic that appears to help predators identify objects of interest is their bilateral symmetry (Cuthill et al., 2006a,b). Intriguingly therefore, many moths that resemble leaves also hold their wings in an asymmetric posture (Preston-Mafham & Preston-Mafham, 1993). Moreover, many species adopt characteristic postures or behaviour that enhance their resemblance to inedible objects, including twig-mimicking caterpillars that hold themselves at angles similar to those of the twigs of their host plant and those that select microhabitats where they are most likely to fool would-be predators (Skelhorn & Ruxton, 2013). Plants themselves appear to have evolved masquerade to avoid being attacked by herbivores. For example, plants from the genus *Lithops* look remarkably like stones (see Lev-Yadun, 2014a for an excellent review).

One particularly striking case of masquerade, which remains to be more formally confirmed, involves a drepanid moth from Malaysia, *Macrocilix maia*, with a wing patterning at rest that appears to mimic an entire scene: a pair of flies feeding on bird droppings (Piel & Monteiro, 2011)—see Figure 9.1. A rose by any other name smells just as sweet, but what makes this an example of masquerade as opposed to Batesian mimicry? In this case the flies have not evolved a signal to indicate their unprofitability, so the moth is copying a cue rather than a warning signal (Jamie, 2017).

9.2.2 Batesian mimicry

Perhaps the classic examples of Batesian mimicry occur in adult butterflies and hoverflies, and we will deal in more depth with several specific cases of these in later sections. Batesian mimicry appears to be a very widespread taxonomic phenomenon and there is no shortage of examples in other groups outside the Lepidoptera. Recent recorded examples include the potential Batesian mimicry of a predacious ground beetle (*Craspedophorus sublaevis*) by a cockroach (an undescribed Blattinae), although the possibility that both species are unpalatable (hence Müllerian mimics) cannot entirely be ruled out (Schmied et al., 2013). The Tomentose burying beetle *Nicrophorus tomentosus* (Silphidae) engages in a

Figure 9.1 The drepanid moth *Macrocilix maia* with wing patterns that appear to resemble a pair of flies feeding on a bird dropping. Photo courtesy of William H. Piel and Antónia Monteiro (see Plate 15).

twist and flip manoeuvre that exposes the yellow underside during flight, thereby enhancing its resemblance to the flight behaviour and appearance of a bumblebee (Heinrich, 2012). The phenomenon of behavioural mimicry is reviewed in section 9.6.3.

Although widely associated with insects, mimicry also occurs in vertebrates. Nestlings of the Amazonian bird *Laniocera hypopyrra* have modified feathers comprising elongated orange barbs with white tips when newly hatched, resembling local aposematic caterpillars (notably a caterpillar from the Megalopygidae). These nestlings even move their head slowly in caterpillar-like fashion, further supporting the contention that they are Batesian mimics (Londoño et al., 2015)—see Figure 9.2. Numerous nonvenomous colubrid snakes of the Neotropics and Nearctic (e.g. kingsnakes and milk snakes) have evolved the characteristic tri-colour ringed patterns of venomous elapid coral snakes, which has led many to suggest that these snakes are Batesian mimics (see sections 9.3 and 9.4 for more extensive discussion). Several convincing examples of Batesian mimicry are also known in fish (see Randall, 2005 for review). For instance, the plesiopid fish *Calloplesiops altivelis* exhibits a colour pattern which make the caudal portion of the fish resemble the head of a toxic moray eel, *Gymnothorax meleagris* (McCosker, 1977). When approached by a predator the fish flees to a crevice and leaves its tail exposed, thereby resembling the moray eel in several ways, including an eyespot. Conversely, the harmless ringed snake eel *Myrichthys colubrinus* appears to have evolved a resemblance to the venomous sea snake *Laticauda colubrina* and other dark-ringed sea snakes (McCosker & Rosenblatt, 1993).

Finally, note that while Batesian mimics sometimes fit the classical dyadic model–mimic scenario, they frequently form part of more complex mimetic communities, often involving mimicry rings (Mallet & Gilbert, 1995; Gilbert, 2005). For example, edible longicorn beetles in Arizona appear to participate as Batesian mimics, not of any specific species but of a more general Müllerian complex consisting of lycid beetles and arctiid moths (Linsley et al., 1961). Similarly, Neotropical butterfly communities often consist of mimicry rings that support both Müllerian and Batesian co-mimics (DeVries et al., 1999). Pekár et al. (2017) described a particularly impressive multi-order mimicry complex that includes at least 140 different putative mimics drawn from four arthropod orders including ants, wasps, bugs, tree hoppers, and spiders.

Figure 9.2 A putative example of Batesian mimicry. Nestlings of the Amazonian bird *Laniocera hypopyrra* (left) look and behave like a toxic caterpillar (Megalopygidae) found in the same area (right). Photo credits: left, Santiago David Rivera; right, Wendy Valencia.

9.3 The origin of protective mimicry: selection or shared ancestry?

Though it is tempting to assume that all instances of shared appearance can be interpreted as mimicry evolved via natural selection, we must remain open minded, and be prepared to accept the possibility that some instances of apparent mimicry have arisen simply as a consequence of common ancestry. To quote Fisher (1930), page 156, '*Obviously the more closely allied are the organisms which show resemblance, the more frequently are homologous parts utilized in its elaboration, and the more care is needed to demonstrate that a superficial resemblance has been imposed upon or has prevented an initial divergence in appearance*'.

Clearly, examples of species evolving to resemble taxonomically distinct species such as birds resembling caterpillars, caterpillars resembling snakes, and spiders resembling ants are highly unlikely to have arisen due to shared ancestry. In cases where a species masquerades as an inanimate object such as stone or bird dropping, then naturally there is no shared ancestry to consider. Additional comparative evidence also supports the contention that mimicry has arisen as a specific adaptation (presumably to avoid attack), rather than as a general response to a shared environment. As Cott (1940) emphasized, mimetic species often differ markedly from closely related non-mimetic species. For instance, those Hymenoptera that resemble the

orange and black of unpalatable lycid beetles have coloured wings, rather than the typical colourless wings of the majority of Hymenoptera. Of course, it is possible that a common environment generates similar selection pressures on appearance. Yet it is worth noting that while models and mimics often share the same geographical and ecological environment, only specific palatable species resemble the model. As Nicholson (1927) remarked, '*Actually, we find that in many cases closely related mimics resemble a series of unrelated models which differ from one another greatly in appearance and which have only one factor in common, that they are all found in the same environment*'.

Finally, it is important to note that mimetic resemblances are often only 'skin deep'. Thus, as Cott (1940) noted, similar appearances in putative mimics and models are often produced by widely different mechanisms. For instance, the transparent wings in mimetic Neotropical Lepidoptera (consisting of both Müllerian and Batesian mimics) have been achieved by an array of different structural and optical mechanisms in different species (Cott, 1940). Similarly, the red spots near the base of the wing in the great Mormon butterfly *Papilio memnon* mimic the body spots of their models (Joron, 2003). Nevertheless, it will come of no surprise to find that even distantly related mimics can arrive at a phenotype similar to their model using similar mechanisms. For example, the scarlet kingsnake *Lampropeltis elapsoides* produces its red, yellow, and black rings through the same pigments and incorporates these pigments in the tissues in the same way as its model, the eastern coral snake *Micrurus fulvius* (Kikuchi & Pfennig, 2012). Manceau et al. (2010) provide an excellent overview of phenotypic convergence in relation to crypsis, emphasizing that different genes can be responsible for convergent phenotypes even among closely related populations of the same species, while the same mutation can create similar phenotypes in distantly related species. It is clear from the above that the similarity of individuals to other species rather than their background is likewise achieved in a range of different ways.

9.4 The evolution of protective deceptive mimicry

There is considerable evidence that mimicry protects prey from predators. Here we briefly review some basic evidence, before moving on to review whether the model–mimic relationship has the sort of properties one would expect. In many instances the studies we describe were multifaceted, but we discuss only aspects of studies that are pertinent to the point we are making, rather than giving a comprehensive overview of the entire paper. The vast majority of work to date has been on Batesian mimicry rather than masquerade, but where possible we have highlighted relevant examples based on masquerade.

9.4.1 Evidence that masquerading prey dupe predators

Until recently scientific research on masquerade has been somewhat limited, perhaps in part because this adaptation appears relatively straightforward to understand. de Ruiter (1952) found that captive jays (*Garrulus glandarius*) could not discriminate twig-like caterpillars from the twigs on which the caterpillars normally fed, but could distinguish these caterpillars from the twigs of other trees, confirming that masquerade can dupe would-be predators but also suggesting that it can involve close model–mimic relationships. As we discuss in more detail in section 9.4.8, the American peppered moth caterpillar *Biston betularia cognataria* shows polyphenetic (i.e. environmentally induced) masquerade, with caterpillars found on birch trees looking like birch twigs and those on willow trees looking like willow twigs (Skelhorn & Ruxton, 2011).

There is an important methodological challenge to studying masquerade, related to how it is defined. Say we observe a potential herbivore overlooking a plant that mimics a stone. Is this background matching (i.e. failure to detect the plant as an entity) or masquerade (i.e. detecting it as an entity but misclassifying it as a stone)? As Getty (1987) noted 'we do not have good operational definitions that allow us to recognize and count encounters and rejections unambiguously'. A study by Skelhorn et al. (2010b) met this challenge head-on. Caterpillars of the brimstone moth (*Opisthograptis luteolata*) and the early thorn moth (*Selenia dentaria*) resemble twigs, but do predators see them and fail to recognize them? To test this idea, Skelhorn et al. manipulated the prior experiences of naïve domestic chicks (*Gallus gallus domesticus*) by presenting groups of domestic chicks with a hawthorn branch, a hawthorn branch bound in cotton thread to change its visual appearance, or no branch at all (an empty arena). Each of these three groups of chicks were then further subdivided into birds that were presented with either a brimstone caterpillar, an early thorn caterpillar, or a single hawthorn twig, all in clear view. Birds with prior experience of twigs took longer to attack the caterpillars of both species and handled them more cautiously, indicating that the caterpillars were seen but not as readily recognized.

9.4.2 Evidence that Batesian mimics dupe predators

Laboratory studies

Perhaps the most direct evidence that predators can be duped into avoiding palatable Batesian mimics after experience with similar-looking but more noxious prey comes from laboratory studies. Classic research of this kind includes the work of Mostler (1935)[3]. Mostler showed that, in addition to their propensity to sting, the abdomens of wasps and honeybees (but not bumblebees) were unpalatable to a wide variety of species of bird. When young birds were offered wasp-like flies (hoverflies) or

---

[3] We are grateful to Francis Gilbert, University of Nottingham, UK for providing an English translation of this work.

other Diptera for the first time they were readily eaten. However, mimetic hoverflies tended to be attacked less often when birds had been given experience with their noxious hymenopteran models (to which they bore either a particularly close resemblance, or only an approximate similarity). Better known work includes the studies of J. V. Brower (Brower, 1958a, b, c; Platt et al., 1971) who found that captive blue jays learned to avoid attacking the pipevine swallowtail (*Battus philenor*), monarch (*Danaus plexippus*) and queen (*Danaus gilippus berenice*) butterflies after they had tasted them, and subsequently tended to reject species (which were either moderately unpalatable or palatable) that resembled them. In a more recent study, Kuchta et al. (2008) presented scrub jays (*Aphelocoma californica*) with the warningly coloured and toxic Pacific newt (genus *Taricha*) followed by morphs of the polymorphic salamander *Ensatina eschscholtzii*. The birds were much more hesitant to handle the presumed Batesian mimic morph with yellow eyes than the cryptic morph.

While birds have been widely used as predators in these types of study, it is clear that predators other than birds can similarly learn to avoid distasteful prey and thereby be deceived into rejecting palatable prey. For example, Brower et al. (1960) showed that toads (*Buffo terrestris*) learn to avoid attacking bumblebees and that these toads subsequently had a greater tendency to reject edible asilid fly mimics. In an experiment that any fly-fisherman would enjoy, Boyden (1976) cast palatable *Anartia fatima* and unpalatable *Heliconius* spp. butterflies to free-ranging *Ameiva ameiva* lizards as well as their re-constructed butterfly mimics (wings glued on bodies of different species). The degree of protection afforded to mimics was greater in those lizards known to have had prior experience with the unpalatable model.

Far fewer laboratory studies have been conducted with invertebrate predators (such as dragonflies, spiders, bugs, and wasps), despite the fact that many species are abundant and visual hunters. Bates (1862, page 510) himself observed that '*I never saw the flocks of slow-flying Heliconiidae in the woods persecuted by birds or Dragon-flies, to which they would have been easy prey; nor, when at rest on leaves, did they appear to be molested by Lizards or the predacious Flies*

of the family Asilidae, which were very often seen pouncing on Butterflies of other families', indicating that he had observed that invertebrate predators refrain from attacking certain types of prey, and had appreciated its potential significance. It has been shown, for example, that praying mantids learn to avoid food associated with electric shocks (Gelperin, 1968) and distastefulness (e.g. Bowdish & Bultman, 1993). In an early experiment, Berenbaum & Miliczky (1984) found that the Chinese mantid *Tenodera aridifolia sinensis* rapidly learned to avoid unpalatable milkweed bugs (*Oncopeltus fasciatus*) that had been reared on milkweed seeds. Furthermore, of six mantids that had learned to reject milkweed-fed bugs, none of these individuals attacked palatable sunflower-fed bugs (automimics, see Chapter 5) when they were offered. Recent experiments by Morris & Reader (2016) suggest that prior exposure of the predatory crab spider *Synema globosum* to the paper wasp *Polistes dominula* resulted in a decline in attack rates on mimetic hoverflies that was most marked in species that most closely resembled the wasp model (to human eyes).

Many species of spider appear to have evolved a visual (and sometimes behavioural) resemblance to ants ('myrmecomorphy'). Huang et al. (2011) presented three large jumping spider species with a range of smaller spiders as potential prey and found that the rates of attack on ant-mimicking jumping spiders were significantly lower than those of non-ant-mimicking jumping spiders, indicating that the resemblance of spiders to ants could provide protection against salticid predation. Nevertheless, the deception may not have been perfect since the spider predators tended to treat ants themselves rather differently.

Field studies

Laboratory experiments provide direct evidence that predators can be duped into avoiding otherwise palatable mimics, but an even more appropriate way to test whether mimicry enhances the survival of edible prey is to see if artificial mimics benefit from their resemblance to real models in the field. In an ingenious series of experiments, L.P. Brower and colleagues (Brower et al., 1964, 1967a; Cook et al., 1969) released palatable day-flying male moths *Callosamia promethea* (a species native to eastern

North America) that had been painted in different ways, into areas of Trinidad. A major advantage of using this species was that the moths could be recaptured using traps baited with pheromone-releasing females. Experimental moths painted to mimic the unpalatable *Heliconius erato*, and the unpalatable *Parides neophilus*, were not re-caught at significantly higher rates than control-painted (similar amounts of paint but no change in colour or pattern) alternatives (Brower et al., 1964). Follow-up experiments with different unpalatable models also generated mixed results, leaving Cook et al. (1969, p. 344) to conclude that *'under wild conditions no clear selective differential can be demonstrated with the promethean moth mimicry system'*. Of course, failure to reject the null hypothesis does not prove the null hypothesis, and to obtain sufficient data the densities of moths may have been unrealistically high. Indeed, it is possible that birds became familiar with feeding on the moths (of all colours) where they were released, in a fashion similar to the 'feeding stations' that may have incidentally been formed in Kettlewell's classic experiments on melanism in *Biston betularia* (Grant & Clarke, 2000).

Half a decade later, Waldbauer & Sternburg (1975) proposed an alternative explanation for the results of Brower and his colleagues, noting that the black non-mimetic controls of *C. promethea* may have gained protection through their resemblance to unpalatable *Battus* spp. which occur in Trinidad. Thus the survivorship of artificial mimics may have been compared against co-mimics, rather than non-mimics (Waldbauer, 1988). Impressed with the elegance of earlier work and motivated by this re-interpretation, Sternburg et al. (1977) and Jeffords et al. (1979) used painted promethea males to evaluate the advantage of mimicry in the pipevine swallowtail complex in Illinois. This time, some of the promethea moths were painted with yellow bars to resemble the non-mimetic forms of the palatable tiger swallowtail butterfly—others (in some experiments) were painted orange to resemble the unpalatable monarch, while other male moths of this species were again control-painted (considering them natural mimics of the pipevine swallowtail). In these experiments, the yellow-painted moths were significantly less likely to be recaptured than orange and black-painted moths (and surviving yellow-painted

moths also showed more evidence of wing injury), providing perhaps *'the first field demonstration of the efficacy of Batesian mimicry'* (Jeffords et al., 1979). By contrast, in similar work conducted in Michigan (Waldbauer & Sternburg, 1987) male *C. promethea* moths painted to resemble the palatable tiger swallowtail, the monarch, and the pipevine swallowtail all had similar recapture rates to one another. One explanation may be that the pipevine swallowtail does not occur locally, while the local monarchs feed on plant species that are most likely not sufficient to provide an emetic dose of cardenolides. Taken together, these studies suggest that the selective advantage of Batesian mimicry may be dependent on the availability of suitable noxious models. We now consider this property in more detail.

9.4.3 Evidence that the success of the mimic generally requires the presence of the model

One might expect that Batesian mimicry can only evolve and be maintained in the presence of a noxious model. Bates (1862) drew much from the fact that models and mimics co-occurred. Likewise, Wallace (1867) p. 8 argued: *'The first law is, that in an overwhelming majority of cases of mimicry, the animals (or the groups) which resemble each other inhabit the same country, the same district, and in most cases are to be found together on the very same spot'*.

This view has been supported by several experiments that have demonstrated a greater advantage to mimics when they are sympatric with models. For example in a field study, Pfennig et al. (2001) put out hundreds of plasticine models of snakes in triplicates. These triads included a replica of the putative mimetic snake in the area, a distinct striped version with the same proportions of colours, and a plain brown version. The replica mimics were either based on colour patterns of the scarlet kingsnake (*Lampropeltis elapsoides*) or the Sonoran mountain kingsnake (*Lampropeltis pyromelana*) that resemble the venomous eastern (*Micrurus fulvius*) and western (Sonoran) coral snake (*Micruroides euryxanthus*) respectively. For both model–mimic pairs, the proportions of total attacks by carnivores on the plasticine mimics were significantly higher in locations where the model was absent compared to where it was present (see also Pfennig et al., 2007). In an even more comprehensive analysis, Davis Rabosky et al. (2016) combined geographic, phylogenetic, ecological, and coloration data to elucidate the spatiotemporal association between New World coral snakes and their Batesian mimics. They found that the shifts to mimetic coloration in nonvenomous snakes were highly correlated with the colour patterns of coral snakes in both space and time, although they noted that mimicry appears to have been gained and lost in this group, suggesting that it was not necessarily a stable 'end-point'.

Perhaps most enlightening evidence of the importance of the presence of the model is the forms (sometimes sub-species) a putative mimic species tends to adopt in areas where the model is not present. For example, the harmless snake *Lampropeltis doliata* appears to mimic the coral snake *Erythrolamprus aesculapii* in the southern United States, but further north where no venomous models are present the snake takes a more blotchy appearance, more consistent with disruptive coloration (Hecht & Marien, 1956; Edmunds, 1974). In butterflies, the mimetic black female form of the eastern tiger swallowtail *Papilio glaucus* tends to occur where the pipevine swallowtail *Battus philenor* is abundant (Platt & Brower, 1968; Edmunds, 1974). However, in southern Canada where *Battus philenor* is rare, the black female form of *Papilio glaucus* is rarely encountered and the genetically distinct yellow non-mimetic female form predominates (Layberry et al., 1998). Likewise, another of the pipevine swallowtail's mimics, the red-spotted purple (*Limenitis arthemis astyanax*) is common in the United States, but it is the non-mimetic (and possibly disruptively patterned) subspecies of white-banded admiral (*L. a. arthemis*) that predominates in Canada (Platt & Brower, 1968; Layberry et al., 1998). Analysing 29 years of observational data from the 4th of July butterfly counts, Ries & Mullen (2008) confirmed that there is indeed a sharp phenotypic transition zone between mimetic (red-spotted purple) and non-mimetic (white-banded admiral) subspecies distributions, which corresponds to the limit of the model's range, although it is notable that mimicry appears to be selected for even when the swallowtail model is rare (Figure 9.3). Has mimicry been gained in the red-spotted purple, or lost in the white-banded admiral? Although an initial analysis

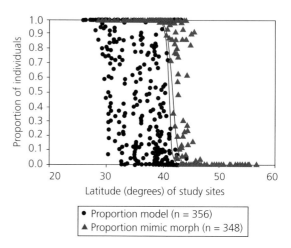

Figure 9.3 Changes in the proportions of individuals of the mimetic morph (red-spotted purples), in relation to the total number of mimetic and non-mimetic (white admirals) individuals of the *Limenitis arthemis* species complex, with latitude (red triangles). Also shown are changes in proportion of the model (pipevines) relative to the total numbers of the model and its mimic recorded at that latitude (blue circles). Where the model is present at lower latitudes, the mimetic morph comprises 100% of the *L. arthemis* population. Where the model is absent at higher latitudes, the non-mimetic morph of *L. arthemis* dominates. Data drawn from the 4th of July butterfly counts. Figure from Ries & Mullen (2008), with permissions (see Plate 16).

suggested that mimicry was lost (Prudic & Oliver, 2008), a more detailed phylogenetic analysis of the different morphs conducted by Savage & Mullen (2009) suggests a single origin of mimicry and retention of the ancestral non-mimetic form in the north.

In a recent analysis of the frequency of different forms of the polymorphic hoverfly *Volucella bombylans*, Edmunds & Reader (2014) found that the frequency of the common black and yellow hoverfly (*plumata*) morph was positively related to the relative frequency of black and yellow bumblebees across 52 sites in central England and northern Wales. Likewise, they found that the frequency of the rare red-tailed (*bombylans*) *Volucella* morph was significantly positively related to the frequency of red-tailed bumblebees, although this relationship disappeared when spatial autocorrelation (morph frequencies at nearby locations are not strictly independent data points) was accounted for. Collectively their results indicate geographical variation in selection by predators as a consequence of variation in the frequencies of different bumblebee models.

While the co-occurrence of models and mimics in the same location generally appears important for the maintenance of mimicry, the co-occurrence of models and mimics at the same time may be far less important. Indeed, it may sometimes be advantageous for a mimic to delay its appearance until after the appearance of the model. Moreover, by appearing first, the model is 'uncontaminated' by palatable mimics. The buff ermine moth (*Spilosoma lutea*) is considered a mimic of the distasteful white ermine moth (*S. lubricipeda*), although it may be a Müllerian rather than Batesian mimic because it is unpalatable to some predators (Rothschild, 1963). As might be expected in this situation, the white ermine moth reaches peak abundance before the buff ermine moth (Rothschild, 1963). Similarly, Brodie (1981) noted that the palatable salamander *Desmognathus ochrophaeus* is active later in the season than its unpalatable salamander model *Plethodon cinereus*. Strangely, many good hoverfly mimics of stinging Hymenoptera in North America reach a peak in spring before their models are most abundant (Waldbauer & Sheldon, 1971; Waldbauer et al., 1977; Waldbauer & Laberge, 1985). One explanation for this puzzling phenomenon (see Waldbauer, 1988) is that while adult birds may avoid mimics based on their previous seasons' experience with models, early emergence avoids the time when fledgling birds (which have not yet learned to avoid models) begin to capture their own insect prey. This is an intriguing explanation, although the timings of emergence of hoverflies in relation to their models are different in other parts of the world. Working in the north-west of England, Howarth & Edmunds (2000) could find little evidence that hoverfly mimics were rare compared to models when fledgling birds are abundant, and generally found that hoverfly mimics were synchronous in their flight season with their hymenopteran models. Indeed, after accounting for site, season, microclimatic responses, and overall abundance, Howarth et al. (2004) concluded that the residual numbers of mimics were significantly correlated positively with their models nine times out of 17, with 16 of 17 estimated relationships positive.

Finally, while mimetic protection has long been assumed to disappear when the model is permanently absent, it must be borne in mind that mimetic advantage in the absence of the model may *in theory* be possible—for example when predators migrate from areas that contain the model (e.g. Poulton, 1909, page 217; Carpenter & Ford, 1933, page 43; Waldbauer, 1988). Anecdotal observations suggest that some vertebrate predators may remember unpleasant experiences for periods over a year (Mostler, 1935; Rothschild, 1964). While there is clearly potential for the evolution of Batesian mimicry in the absence of the model, we do not know of any study that has formally demonstrated mimetic advantage to a given prey species on the basis of learnt (or innate) aversion to models in another geographical area. The maintenance of eyespots in caterpillars in areas such as Canada where venomous tree snakes are absent is one possibility (Hossie & Sherratt, 2014) although the eyespots may have evolved to resemble a general predator (Janzen et al., 2010) and/or may have additional adaptive benefits (Stevens & Ruxton, 2014).

Other puzzling examples of mimics persisting in the absence of a model have occasionally been observed, but they are exceptions rather than the rule. For instance, Clarke & Sheppard (1975) reported that females of the nymphalid butterfly *Hypolimnas bolina*, which gave 'no convincing evidence' of unpalatability to birds (although it may contain cardioactive substances obtained from certain food plants) occurred entirely in their mimetic form in Madagascar even though its putative model (*Euploea* butterflies) was absent (a 'rule-breaking' mimic, Clarke et al., 1989). Rather than being maintained by the presence of migrating birds, or regular immigration of the butterfly from India, the authors argued that the resemblance was likely to persist either because of a lack of non-mimetic variants (especially likely if the species had recently immigrated to the area), or due to some selection unassociated with mimicry (it is also possible that the model once existed but is now extinct in Madagascar; Clarke et al., 1989).

9.4.4 The relative (and absolute) abundance of the model and mimic affect the rate of predation on these species

While the presence or absence of an aversive model is important, one might wonder whether deception shows a more continuous form of frequency-dependent advantage, such that it is more beneficial to the mimics/masqueraders when they are rare compared to their models. For example, one might expect that as the frequency of prey that masqueraded as twigs increased with respect to twigs, then it would pay potential predators to pay more attention to twig-like objects. This is precisely what Staddon & Gendron (1983) predicted in an early application of signal detection theory. Decades later, Skelhorn et al. (2011) presented domestic chicks with different mixtures of twigs and twig-mimicking caterpillars (the early thorn moth, *Selenia dentaria*) and showed that the benefit of masquerade (measured in terms of latency to attack) increased as the ratio of twigs to masquerading caterpillars increased. This effect appeared to arise through two separate mechanisms. Increasing the ratio of twigs to caterpillars reduced the predators' motivation to search for prey as predicted, but increasing the density of twigs also produced a more complex environment, making it harder for the chicks to find the caterpillars.

Of course, many masquerading prey items have evolved to resemble extremely common objects in their environment, such as bird-droppings, twigs, and leaves, so masqueraders may always be rare in the field compared to the objects they model, whatever their population size. Moreover, if the model is a small part of an organism (such as thorns or twigs) then masquerading species that resemble them are unlikely to have an effect on selection for model appearance (but see Abbott, 2010 for some potential counter-examples in which plants are selected to reveal herbivores to their predators). Indeed, it is the lack of deleterious effects on the model that is sometimes used to distinguish masquerade from Batesian mimics (Endler, 1981; Skelhorn et al., 2010a).

A fascinating recent study investigated the maternal nest-brooding cichlid *Neolamprologus furcifer* which inhabits shaded areas of large rocks in Lake Tanganyika (Satoh et al., 2017). Unlike the brooding mothers, the young have clear white stripes on the dark brown body which appears to enhance their resemblance to a local white-striped and inedible snail *Reymondia horei*. Indeed, the guarded young had black/white coloration only in populations where the model snails were present. In manipulative experiments, when the putative model snail was removed from the female territories, the mothers

had to 'up their game' and work at higher rates to repel would-be predators from attacking their young. Moreover, the mimic/model ratio appears such an important factor in deterring predators that the mothers actively sought to remove non-model snails from their territories!

It has been widely predicted that the attack rates of predators on encounter with Batesian mimics and their models should depend not just on the presence of the model, but on the relative abundances of mimics and models. Bates (1862, page 514) himself noted: '*It may be remarked that a mimetic species need not always be a rare one, although this is very generally the case*'. Note, however, this does not mean that mimics should never outnumber their models. Indeed, when the model is particularly unprofitable then it may pay predators to continue to avoid similar-looking prey even when there is a high 'mimetic load'. The mimics of venomous coral snakes, for instance, often vastly outnumber their models (Harper & Pfennig, 2007; Davis Rabosky et al., 2016).

Several studies have examined the nature of the quantitative relationship between predator attack rates and the relative proportion of models and mimics. In a classic study, Brower (1960) showed that the rate of attack of caged European starlings (*Sturnus vulgaris*) on palatable ('mimetic') and unpalatable mealworms (the model, dipped in quinine dihydrochloride) tended to increase as the percentage of models offered in a sequential mixture decreased. As might be expected, this relationship was non-linear. Thus, even when models comprised only 40 per cent of the models/mimics, then mimics often received similar protection compared to trials where there were a greater percentage of models. This is consistent with a non-linear 'zone of protection' (see for example Turner, 1984; Sherratt, 2002a,b).

In more recent fieldwork, Jones et al. (2013) presented wild garden birds with dishes containing 16 mealworm larvae that comprised a mixture of palatable mimics and unpalatable models (rendered bitter-tasting using Bitrex), and varied the proportion of the mimics from 0 to 1 by increments of 0.25. The authors found strong frequency-dependent selection with the proportion of mealworms attacked after 4 hours of foraging again increasing non-linearly with the frequency of mimics. Several experiments have shown similar frequency-dependent effects when the mimics were not perfect replicas. For instance, Pilecki & O'Donald (1971) showed that neither poor mimics (artificial palatable pastry mealworms, paler in colour than models) nor their models experienced high attack rates by sparrows when the poor mimics were relatively rare (their Experiment III). By contrast, significantly more mimics than models were taken when mimics and models were presented in a 1:1 ratio (Experiment IV). In this case the predators became more discriminatory when mimics were more abundant, although even here the poor mimics suffered less predation than expected compared to the palatable controls. Likewise, Lindström et al. (1997) sequentially presented captive great tits (*Parus major*) with a combination of model and mimic mealworm larvae (6 in total). Models were dipped in (and injected with) chloroquine and 3 small light blue 'nonpareils' (small round balls of brightly coloured sugar used in food decorations) were placed close to their heads. Imperfect mimics were simply mealworm larvae with the nonpareils in the middle of their bodies. As might be expected, they found that imperfect Batesian mimics were attacked least frequently by the birds on a *per capita* basis when the models were more common.

In sum, a number of experiments have demonstrated that the relative abundances of models and mimics influence the mean rates of predation on these types—as the ratio of models to mimics increases then the attack rates on both models and mimics on encounter tends to decrease, typically in a non-linear fashion.

9.4.5 The distastefulness of the model affects the rate of predation on the model and mimic

Common sense suggests that, other things being equal, the protection an edible mimic gains from mimicry will vary with the intensity of the model's defences. This might arise for two complementary reasons. First, predators may more readily learn to avoid models that are particularly unprofitable, and mimics might incidentally benefit from this accelerated learning.

For example, Skelhorn & Rowe (2006) presented coloured crumbs that had been soaked in different concentrations of quinine sulfate and found that chicks more rapidly learned to avoid those crumbs containing higher concentrations of aversive chemicals, and were less prepared to reverse their aversions when all crumbs presented were palatable. Second, predators may choose to avoid prey types that look remotely like highly unprofitable prey for fear of misclassification, even after learning is complete (see Sherratt, 2002b).

The above common-sense prediction has been supported on a number of occasions. For instance, Alcock (1970) first trained captive white-throated sparrows (*Zonotrichia albicollis*) to overturn seed shells (painted with dots) under which was hidden half a mealworm or no mealworm at all. The birds were then presented with a binary choice of a model and mimic. The mealworm models were either weakly unpalatable (salted) or highly unpalatable (smeared with tartar emetic) and hidden under shells decorated with specific markings. The mealworm mimics were consistently palatable and were either good or poor mimics of the models (manipulated by changing the dots on the shells). When the mimic was poor and the model weakly aversive, then the mimic was attacked at a relatively high rate (and significantly more frequently than the model). However, when the model was emetic then neither the model nor the mimic were frequently attacked, even when the degree of mimicry was poor. The Lindström et al. (1997) study described in the previous section also found that the concentration of chloroquine in the mealworm model affected the rate of attack of great tit predators on the model, but in this case the attack rates on imperfect mimics did not change significantly—possibly because birds could distinguish them.

9.4.6 The model can be simply difficult to catch rather than noxious on capture

Can a prey species that is easy for its predators to catch gain protection by mimicking a species that is sufficiently difficult to catch that predators decline the opportunity to pursue them? We will call this 'evasive Batesian mimicry'; although it has also been called locomotor mimicry (e.g. Brower, 1995), this term has also been used to mean mimicry of the movement behaviour of unpalatable prey (Srygley,

1999). Naturally, it is also possible that collections of species, all of which are difficult to catch, evolve similar appearances to warn predators more effectively of their unprofitability. This form of mimicry could be considered evasive Müllerian mimicry (see Chapter 7).

van Someren & Jackson (1959) were amongst the first to suggest that a species could gain selective advantage simply by resembling another species that was difficult to catch. Rettenmeyer (1970) similarly included an effective escape mechanism in his list of features worth mimicking. Evasive mimicry has been widely suspected among coleoptera. In one of the first specific examples, Lindroth (1971) remarked on the visual similarity between flea-beetles (Alticinae, Chrysomelidae) and sympatric ground beetles (Lebia, Carabidae). He proposed that neither beetle species was unpalatable to their predators and hypothesized that the ground beetles had evolved to mimic the flea-beetles because they can jump as an effective form of escape from predators. Holm & Kirsten (1979) argued that the remarkable similarity of a complex of fast-moving scarab beetles in the Namib desert, all with orange-brown elytra, was best explained by a combination of Batesian and Müllerian evasive mimicry. Likewise, Hespenheide (1973) described a diverse collection of Neotropical beetle species, not normally considered distasteful,

that appear to have evolved to resemble flies. Vanin & Guerra (2012) detail one such example of a weevil, *Timorus sarcophagoides* that appears to have evolved to resemble flesh flies from the family Sarcophagidae. The weevil even appears to behave like their model with stereotypical leg-scrubbing behaviour widely associated with flesh flies. Why evolve this resemblance? Both sets of authors suggest it is a case of evasive mimicry because many flies are hard to catch, and while it remains to be tested, the extent of mimicry is compelling (Figure 9.4).

More recently, Pinheiro & Freitas (2014) reviewed some purported cases of evasive mimicry in Neotropical butterflies and suggested several potential mimicry rings based on escape behaviours including a 'bright blue bands' ring and a 'creamy bands' ring. The blue-banded one-spotted prepona (*Archaeoprepona demophon*), for example, flies at high speed and moves erratically, and has been observed to escape 73 of 83 (88 per cent) attacks by rufous-tailed jacamars (Chai, 1990) and 8 of 18 attacks (44 per cent) by tropical kingbirds (Pinheiro, 1996). Many other blue-banded species also appear to have the same ability to evade predators, and the common warning patterns may therefore represent a case of Müllerian evasive mimicry. Naturally, we should be aware that associations in appearance and flight characteristics can arise simply through shared phylogeny rather

Figure 9.4 (A) The flesh-fly mimicking weevil *Timorus sarcophagoides* (Coleoptera. Curculionidae) from Southeastern Brazil. Not only does the weevil morphologically resemble flesh flies, but it also behaves like them when moving and engages in stereotypical leg-scrubbing behaviour. (B) A potential flesh fly (Sarcophagidae) generic model. The adaptive significance of the mimicry is unclear, but flesh flies are hard to catch and it seems likely that the weevil has evolved this resemblance because it provides a degree of protection from predators (thus Batesian evasive mimicry, although it could be Müllerian evasive mimicry if the weevils are also hard to capture). Photographs courtesy of Tadeu Guerra (see Plate 17).

than selection, especially for closely related species (Brower, 1995). However, Pinheiro & Freitas (2014) argue that the candidate species have strongly diverged in the underwing coloration pattern, but not on the upperwing, suggesting active selection for monomorphism in appearance.

Experimental evidence that mimicry of a more evasive species can protect a slower-moving mimic from predation was provided by Gibson who found that Australian star finches (*Neochmia ruficauda*; Gibson, 1974) and robins (*Erithacus rubecula*; Gibson, 1980) continued to avoid attacking prey types (seeds and mealworms) of particular colours that had previously been difficult to catch (by means of a hinged platform). Interestingly, the more conspicuous evasive model appeared to provide higher and longer levels of protection to the mimic. In a similar experimental design, Hancox & Allen (1991) presented red and yellow pastry baits on a specially constructed bird table, which allowed one colour of bait to drop out of reach when attacked. In two sets of 20-day sessions (yellow evasive, red non-evasive, followed by yellow non-evasive, red evasive) it was clear that birds rapidly learned to avoid attacking those pastry baits with the 'evasive' colour. The implications of the above results for mimicry were contested by Brower (1995), who stressed that the conditioned aversion to escaping prey is temporary, whereas aversion to chemically defended prey is much longer-term. However, it should be noted that, in contrast to prey with a chemical defence, predators gain no nutritional benefit from chasing after an evasive model. Moreover, predators do not need to attack mobile prey to be reminded of their escape ability since direct observation of their manoeuvrability (and/or a short, failed chase) will suffice.

One might intuitively expect that evasive Batesian mimicry would be a rare phenomenon if the cost of ascertaining whether a prey individual is easily caught (for example, by attempting to catch it) were low: in these instances, both models and mimics would be pursued and mimicry would not evolve. Indeed, Ruxton et al. (2004b) developed a simple optimal foraging model which showed that while evasive Batesian mimicry was plausible, it was much more likely to evolve when it is costly (in time or energy) for the predator species to pursue evasive prey, when mimics are encountered less frequently

than evasive models, and where there are abundant alternative prey. Conversely, Ruxton et al. (2004b) argued that evasive Müllerian mimicry would be most likely to arise when evasive prey species differ in abundance, predators are slow to learn to avoid evasive prey, and evading capture is costly to the prey. So, while evasive mimicry remains a real possibility in theory, and there are plenty of plausible examples, it is clear that more empirical work is needed to ascertain whether putative examples of the phenomenon are indeed a result of selection to signal difficulty of capture.

9.4.7 The success of mimicry is dependent on the availability of alternative prey

Carpenter & Ford (1933, page 28) noted that, since the motivation of predators to distinguish model from mimic depends on how hungry they are then *'the success of Mimicry depends on the abundance of other palatable food'*. These conclusions were later supported in a wide range of quantitative models (Holling, 1965; Emlen, 1968; Dill, 1975; Luedeman et al., 1981; Getty, 1985; Kokko et al., 2003; Sherratt, 2003; Sherratt et al., 2004a) all of which showed that predators maximizing their rate of energy gain per unit time (or simply acting to avoid starvation) would be expected to more readily attack models and mimics when alternative palatable prey were in short supply. Indeed, Holling (1965, page 36) proposed that '…*the marked effects produced by removal of alternate prey or changes in their palatability are great enough that alternate prey should be included as an essential feature of mimicry theory'*.

Several authors have attributed low predation pressure on mimics at least in part to the presence of alternative prey. For example, Nonacs (1985) reported that the attack rates of predators (long-eared chipmunks, *Neotamias quadrimaculatus*) on mimics and models (palatable and unpalatable dough balls) decreased when alternative palatable prey were offered in abundance. Lindström et al. (2004) likewise presented trained wild-caught great tits with palatable and unpalatable (soaked in chloroquinine solution) almonds covered by paper with different patterns. The unpalatable models were green while the palatable prey were either blue (considered imperfect mimics) or distinct alternative prey with crosses, resembling the background—the green and blue colour treatments were reversed in other experiments. The birds were allowed to attack 50 prey items in a given trial. As one might anticipate, the unpalatable models were consistently under-represented in the diet compared to what one would expect if the birds exhibited no selectivity. When alternative prey were scarce (50 models, 50 mimics, 50 alternatives) then the imperfect palatable mimics were actually over-predated compared to what one would expect by chance, presumably due to their imperfect resemblance to unpalatable models and their higher conspicuousness compared to alternative prey. By contrast, when the density of alternative prey was raised (50 models, 50 mimics, 100 alternatives) then the distribution of attacks on mimics decreased. It would be of interest to evaluate the success of mimicry when alternative prey replaced mimics or models thereby maintaining a constant initial density of prey. Nevertheless, the results confirm that the density of alternative prey available affects the level of protection afforded to mimics.

Many defended prey contain toxins but they also contain nutrients, and such prey may be profitable to attack when alternative prey are in short supply. Indeed, Speed (1993) noted that '*in periods of actual (or anticipated) hunger, and of scarcity of alternative prey, eating a species that contains some toxin and surviving is a better strategy than eating none at all and starving*'. Over the past decade, considerable work has been conducted to elucidate the effect of a predator's nutritional state on its propensity to attack chemically defended prey (and presumably mimics that look like them). Barnett et al. (2007), for instance, showed that European starlings increase their attack rates on distinctly coloured chemically defended (quinine sulfate injected) mealworm larvae (*Tenebrio molitor*) when their body masses and fat stores were experimentally reduced (and subsequently restored in cycles). Nevertheless, the birds were still able to discriminate between differently coloured chemically defended and undefended (water injected) larvae when presented in simultaneous choice tests, indicating that this increase in consumption was not due to a reduced ability to discriminate. In an extension of this work, Barnett et al. (2012) presented three types of mealworm to captive starlings—undefended mealworms, mildly

defended mealworms injected with 1 per cent quinine solution, and moderately defended mealworms injected with 3 per cent quinine solution, all distinguishable by colour. After a period of training the birds ate more undefended than defended prey and more mildly than moderately defended prey. However, following a period of food restriction, the birds increased the number of defended prey that they consumed, although as one might expect, their relative preferences for the different prey types were maintained.

In another experiment aimed at elucidating precisely why the availability of alternative prey might reduce the attack rates on chemically defended prey, Halpin et al. (2013) found that wild-caught starlings ate fewer unpalatable mealworms when the alternative palatable mealworms were large compared with when they were small. Intriguingly, however, the size of the unpalatable prey had no effect on the numbers eaten, implying that decisions to attack chemically defended prey are not based on energy considerations alone. Likewise, Halpin et al. (2014) found that starlings increase consumption of quinine-injected prey mealworms when the nutritional content of the prey was artificially increased. In another experiment, Chatelain et al. (2013) found that starlings increased their consumption of unpalatable mealworms when the ambient temperature was reduced to 6°C (14°C below their thermoneutral zone), indicating that thermal stress can induce similar effects as dietary restriction. Energy is not the only state variable that influences predatory decisions and, in a complementary experiment, Skelhorn & Rowe (2007b) revealed that the dose of quinine sulfate administered to European starlings affected the number of chemically defended prey that they were subsequently prepared to consume, with birds receiving high toxin doses less inclined to eat chemically defended prey.

9.4.8 Frequency-dependent selection on Batesian and masquerader mimics can lead to mimetic polymorphism

In general, if Batesian and masquerading mimics increase in frequency then predators will learn of their presence and tend to increase their attacks, since their chances of attacking a model concomitantly decrease. Since rarity compared to the model maximizes the survival of a Batesian mimic, then one might expect selection to promote polymorphisms in which some conspecifics resemble one noxious model and other conspecifics resemble another noxious model. Alternatively, if Batesian mimicry carries some form of cost (such as enhanced conspicuousness) then a polymorphism between mimetic and cryptic forms is possible (Holen & Johnstone, 2004). For example, the highly polymorphic ground snake (*Sonora semiannulata*), a Batesian mimic of coral snakes, has both cryptic and conspicuous morphs (Cox & Davis Rabosky, 2013). Likewise, the trimorphic salamander *Plethodon cinereus* is considered to have forms that mimic the salamander *Notophthalmus viridescens*, but it also has a non-mimetic morph (Kraemer & Adams, 2014). Theoretical considerations suggest that the equilibrium frequency of different mimetic forms, if influenced solely by predation, should be such that each morph gains as much protection as other forms.

Mimetic polymorphism occurs in a number of Batesian mimicry complexes, but it is by no means a diagnostic characteristic (Joron & Mallet, 1998). Of 62 species of Holarctic hoverfly species considered to resemble bumblebees, Gilbert (2005) estimated that 25 were polymorphic, with variation sometimes confined to one sex. The palatable salamander *Desmognathus ochrophaeus* is considered to have colour morphs that resemble several species of unpalatable salamander (Brodie & Howard, 1973; Brodie, 1980, 1981). Other potential examples of polymorphic Batesian mimics include the non-aggressive social wasp *Mischocyttarus mastigophorus* that exhibits dark and pale colour morphs, thought to resemble the aggressive swarm-founding wasps *Agelaia xanthopus* and *A. yepocapa* (O'Donnell, 1999). In these and other cases it should be borne in mind that morphs often co-occur locally but often the models only partially overlap in distribution. Hence, polymorphism is likely to be maintained by a complex interplay of local balancing selection and the mixing of morphs from different areas.

Polymorphisms (genetic and environmentally determined) have also been reported in masquerading species. For example, the appearance of the American peppered moth caterpillar *Biston betularia cognataria* is known to be influenced by the colour of light in the feeding environment (Noor et al., 2008). Likewise, *Nemoria arizonaria* caterpillars resemble oak twigs when fed on oak leaves, and oak catkins when fed on oak catkins (Greene, 1989). Skelhorn & Ruxton (2011) showed that this environmentally induced variation ('polyphenism') may be adaptive, in that chicks misclassified birch-fed larvae as birch twigs and willow-fed larvae as willow twigs based on their similar visual appearance.

Perhaps the best-known example of mimetic polymorphism occurs in swallowtail butterflies (Figure 9.5). Many of these species have a single male-like non-mimetic female form and one or more mimetic female forms that mimic unpalatable models with different appearances (e.g. *Papilio polytes* and *P. glaucus*). In other mimetic species (e.g. *Hypolimnas misippus*, and several mainland populations of *Papilio dardanus*), all female forms resemble different unpalatable models (Kunte, 2009a, b). Intriguingly, the males in female-limited mimetic species are generally believed to represent the ancestral wing colour pattern of the species, the mimetic female forms having diverged from their ancestral, male-like wing patterns (Kunte, 2008). Why mimicry should be restricted to one sex is a separate question (see section 9.4.9) from why one or both sexes are polymorphic, but the answers are related. For instance, in south and central America, the kite swallowtail butterflies of the genus *Eurytides* consist of 15 recognized species, 14 of which appear to be Batesian mimics, largely of papilionid *Parides* species (West, 1994). Of these Batesian mimics, six species are poly- or dimorphic without sex limitation and the remainder have sex-limited polymorphism (West, 1994).

Formal experimental evidence for the prediction that Batesian mimicry is capable of generating and maintaining polymorphisms in systems with two or more models is understandably rare. However, as noted in section 9.4.4, there is considerable evidence that attack rates on mimics increase as the ratio of mimics to models increases. Intuitively polymorphism would be more likely to be selected for when there is a rapid increase in the attack rate of a model/ mimic mixture as the mimetic load increases—such cases will arise when the model is only mildly aversive. O'Donald & Pilecki (1970) presented palatable and unpalatable (dipped in 1 per cent or 3 per cent

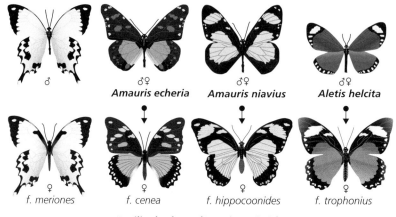

Figure 9.5 Polymorphic Batesian mimicry in the African mocker swallowtail, *Papilio dardanus*. Dubbed "the most interesting butterfly in the world" (Poulton, 1924), this species (bottom row) exhibits female-limited polymorphism, with the different female forms resembling different unpalatable models (top row). The subspecies of *P. dardanus* from the Madagascar region (illustrated in first column) has male-like non-mimetic females. Image courtesy of Krushnamegh Kunte.

quinine hydrochloride) pastry 'mealworms' to house sparrows (*Passer domesticus*). There were two forms of distasteful model (blue and green) and three forms of palatable prey (blue mimic, green mimic, and non-mimetic yellow). They found from comparing separate experiments that if the models were only slightly distasteful then the rare mimic was at an advantage over the common one, confirming the plausibility of active selection for mimetic polymorphism. However, when the models were highly distasteful then a rare mimic had no advantage over a more common one. There has been some question of the appropriate interpretation of this experiment (Edmunds & Edmunds, 1974; O'Donald & Pilecki, 1974) due primarily to the differences in the consumption rate of the two unpalatable models (blue and green) used, and statistical questions relating to the appropriate population from which the observations were drawn. O'Donald & Pilecki (1970) does not therefore provide the unequivocal evidence of frequency-dependent predation risk in mimics that it purports to and further work is required.

Naturally, when there are two or more separate potential models then they can be exploited in ways other than via polymorphisms. Perhaps the most impressive example of a species varying the model it resembles (although it is clearly not a genetic polymorphism) is the Indo-Malayan octopus *Thaumoctopus*

mimicus, in which an individual may impersonate a range of venomous animals that co-occur in its habitat (Norman et al., 2001). There is some evidence that this mimicry can be employed facultatively, with the octopus adopting a form best suited to the perceived threat at any given time, although this proposal needs further testing (Hanlon et al., 2008). An alternative solution is to evolve an intermediate phenotype, most especially if predators generalize widely over both models (Sherratt, 2002b). Indeed, stimulus generalization by predators (which is expected to be wide in the case of highly toxic prey) may be one reason why Batesian polymorphisms are not more frequent. Darst & Cummings (2006) describe two Ecuadorian poison frogs, *Epipedobates bilinguis* and *E. parvulus*, which act as models for their phylogenetically distant relative, *Allobates zaparo*. Intriguingly, *A. zaparo* is geographically dimorphic, matching each model's warning signals where only a single model occurs. However, in areas where the two models co-occur, then the mimic has evolved to resemble only one model. Somewhat unexpectedly, the model in these cases was the less toxic and less abundant *E. bilinguis*. Note that a vague similarity to an extremely noxious species is usually sufficient to guarantee significant protection, whereas a much closer resemblance to a mildly noxious model species is necessary to afford a similar level of benefit.

Therefore if the generalization curves for these models overlap, then a mimic might maximize its overall protection by adopting an intermediate phenotype that more closely resembled the less defended model (Sherratt, 2002b).

9.4.9 Sex-limited polymorphic mimicry

Some of the most fascinating and complex examples of polymorphic mimicry include cases where the model is sexually dimorphic and the two sexes of mimic resemble each of the sexes of the model. Typically the male and female mimic their respective sexes. However, sometimes the male resembles the female and vice versa (Carpenter & Ford, 1933, p. 91). In other cases both males and females are mimetic, but they resemble different model species. For example, the Central American Erostratus swallowtail (*Papilio erostratus*) females are thought to mimic pink-spotted cattleheart (*Parides photinus*), while the males are considered mimics of the gold rim swallowtail (*Battus polydamas*). Conversely, there are examples of multiple mimetic forms of the same species, but these mimetic forms are shared between the sexes. Examples of such polymorphic mimics are most common (albeit relatively unusual) in Müllerian mimics, but they have also been documented in Batesian mimics, for example *Papilio clytia* (Kunte, 2009a), which has two mimetic forms per sex.

Intriguingly, in other species mimicry is limited to one sex. In butterfly species where sex-limited Batesian mimicry occurs, it is invariably the female that exhibits mimicry with both mimetic and male-like (non-mimetic) female forms (Turner, 1978). Male-limited Batesian mimicry has however been reported in other taxonomic groups including the buprestid beetle *Chrysobothris humilis* whose males resemble members of the chrysomelid subfamily Clytrinae (Hespenheide, 1975). Hespenheide (1975) explained this unusual case of male-limited mimicry as a consequence of sexual differences in behaviour: males often sit on legume twigs waiting for ovipositing females, and thereby occur more frequently where the model occurs and are more frequently exposed to predators. Another potential example is the promethea moth *Callosamia promethea,* where the males may gain a selective advantage by resembling the pipevine swallowtail butterfly

(*Battus philenor*) yet the females appear non-mimetic. Male moths are almost certainly exposed more to visual predators, primarily through their need to fly towards pheromone-releasing females during the day—by contrast, females lay their eggs at night—(Waldbauer, 1996, page 127). Many hoverfly mimics of bumblebees in the Holarctic also exhibit sex-limited Batesian mimicry: 17 cases of male-limited and 11 cases of female-limited mimicry have now been recognized (Gilbert, 2005). While the reason for sex limitation in these species is unclear, it is notable that the majority of polymorphic species are bee mimics. Bees are widely considered less aversive models than wasps, so it is possible that they face selection to diversify because they are attacked more frequently as the mimetic load increases.

If Batesian mimicry is advantageous, why don't both sexes always adopt it? So far two main theories have been put forward to explain the occurrence of sex-limited mimicry—one based on differences in the ecology of the two sexes, and the other based on sexual selection. In his classic paper, on Malayan swallowtails, Wallace (1865) explained the widespread occurrence of female mimetic forms: '*with their slower flight, when laden with eggs, and their exposure to attack while in the act of depositing on leaves, render it especially advantageous for them to have some additional protection*'. Later Belt (1874, pp. 384–385), added the suggestion (not necessarily mutually exclusive) that sexual selection may help explain non-mimetic forms in males: mimicry does not arise in this sex because it would render them at a disadvantage during courtship. Turner (1978) took up the challenge of modelling this latter phenomenon, noting that sexual selection may resist colour changes more strongly in males than females. Using a population genetic framework, he showed how female-limited mimicry could readily arise when modifiers that suppress the expression of mimicry in males get selected for. The predicted outcomes include female-limited mimicry, or better mimicry in females than males, and males retaining some of their colours that play a role in mate choice.

Experimental evidence that males and females of the same species may experience different selective pressures with respect to predation is widespread. Bowers et al. (1985) capitalized on an unusual finding in 1980 of large numbers of detached wings (matched for 309 individuals) of the checkerspot butterfly *Euphydryas chalcedona* in California, almost certainly the remains from being attacked and eaten by birds (coldness may have contributed to making the species more vulnerable that year). By comparing the coloration and sexual composition of this sample with a representative sample that had not been killed, Bowers et al. (1985) argued that birds had attacked significantly more females than males, and fewer of those males with high amounts of red on their forewing. The authors proposed that females may have been actively selected by predators (they are larger and probably represent a more nutritious meal) while redder males may have benefited from warning coloration. More direct evidence that sexual differences in potential rates of predation can select for mimicry in just one sex was presented by Ohsaki (1995) who examined the frequency of beak marks on the wings of palatable Papilionidae and Pieridae and unpalatable Danaidae in Malaysia. The analysis of beak marks is fraught with problems, not least because it also indicates successful escape from a predator[4]. However, females of each palatable family had a greater rate of wing damage than males. The numbers of recorded observations of *Papilio polytes* were relatively low, but non-mimetic females of this species had a higher per capita incidence of beak marks than mimetic females or males. Ohsaki (1995) points out that these observations indicate that '*females benefit greatly when they become mimetic, whereas males will benefit much less even if they become mimetic*', but this rather leaves open the question why all females are not mimetic (Ohsaki, 1995 suggests this may possibly be a consequence of local disequilibrium, or some other cost to mimicry).

Looking at the bigger picture, Kunte (2008) analysed a partial phylogeny of *Papilio* (swallowtail) butterflies, a group well known for sexual dimorphism, based predominantly on multiple instances of female-limited mimicry (as noted earlier in this section). He argued that where sexual dimorphism

was observed, it was often females that were selected to be different (mimetic), with males tending to conserve their ancestral colour pattern. Kunte (2008) interpreted the observation of active selection on female appearance as evidence for Wallace's theory, but we still need to ask why males do not engage in mimicry and it remains possible that sexual selection on male colour pattern serves to conserve the ancestral form.

Evidence that sexual selection (either through female mate choice or male–male interactions) can maintain non-mimetic males, comes from several quarters. Krebs & West (1988) blackened the wings of male *Papilio glaucus* butterflies to look like the mimetic female form and examined their mating success compared to yellow-painted controls. No significant difference was found in the proportion of presentations to both mimetic and non-mimetic females, which resulted in mating. However, female solicitation of males that themselves had failed to court occurred at a significantly higher rate for yellow-painted than black-painted males.

In general, there is relatively little evidence that papilionid females employ colour in male mate choice outside the ultraviolet, with most of the close-range mate choices being based on olfactory rather than visual signals (see Kunte, 2009b and references therein). There is slightly more evidence for non-mimetic male forms having some advantage in male–male competition. For instance, Lederhouse & Scriber (1996) investigated the benefits of male coloration in the eastern black swallowtail butterfly *Papilio polyxenes asterius*. The dorsal sides of males are less convincing mimics of the pipevine swallowtail *Battus philenor* (to human eyes at least) than the dorsal side of females, and the species is therefore generally considered as one exhibiting female-limited mimicry. Males of the species seek out and defend hilltop territories ('hilltopping') where the vast majority of courtship takes place. Lederhouse & Scriber (1996) found that males that had been altered using marker pens to have more female-like (and therefore mimetic) coloration were less likely on release to be able to establish (and maintain) a local hilltop territory (at least close to the release site) than mock-treated controls. One reason for this may be that altered males were involved in significantly longer male–male encounters, through being

---

[4] The analysis of statistician Abraham Wald in World War II comes to mind, who successfully argued that the military should (somewhat paradoxically) reinforce areas of the fuselage where returning bombers were unscathed, as opposed to where they showed damage—because the bombers were clearly capable of surviving damage in areas where it was observed.

misidentified as females. By contrast, the experimental creation of female-like males did not appear to affect female choice per se: while altered males did not as readily obtain territories, the mean durations of courtship flights that progressed to copulations were not significantly different in the territories of males that were treated and mock-treated.

In summary, it is perhaps fair to say that while there are several good candidate explanations for why Batesian mimicry might sometimes be limited to one sex, we still do not know which explanation is the most general and important one (if indeed there is such an explanation). As Kunte (2009b) noted, more experiments are needed. One might also ask if sexual selection acts to conserve male appearance and thereby prevent mimicry, why does it not tend to do so in Müllerian mimics? One explanation may lie in the differing nature of the frequency dependence (Turner, 1978; Joron & Mallet, 1998). By its nature, Batesian mimicry may involve selection against common morphs. In contrast to Batesian mimicry, the mutualistic nature of Müllerian mimicry ensures that the more individuals that adopt a common form of advertising, the higher the fitness of all.

9.4.10 The genetics of polymorphic Batesian mimicry

Butterflies have become the central model for understanding the genetics of Batesian mimicry, not only because they provide some of the most celebrated examples of protective resemblance, but also because many species can be crossed under laboratory conditions. As we have seen, many Batesian mimics are polymorphic—with the polymorphism often limited to one sex—so one of the primary motivational questions is how several discrete colour patterns can be generated.

In an early study in Sri Lanka, Fryer (1914) conducted breeding experiments on the swallowtail *Papilio polytes* whose females are polymorphic (one form resembles the male and other forms resemble different unpalatable models). Somewhat surprisingly, no intermediate forms resulted from the crosses, and it was also deduced that males (all of which are non-mimetic) could transmit genes relating to mimicry to their female offspring. Perhaps understandably, observations such as this fuelled

Punnett's (1915) belief that mimetic resemblance was more likely to have arisen by a sudden 'sport or mutation', rather than a gradual accumulation of mutations providing closer resemblances. Gabritchevsky (1924) crossed three forms of the polymorphic hoverfly *Volucella bombylans*, namely *bombylans*, *haemorrhoidalis*, and *plumata*, and found that the frequencies of resultant offspring matched simple Mendelian ratios (see also Keeler, 1925). We now know that the underlying genetics of polymorphic mimicry are far more complex than at first believed. Most of our early knowledge of the genetics underlying Batesian (and Müllerian) mimicry in Lepidoptera comes from the dedicated and detailed studies of E.B. Ford, C.A. Clarke, P.M. Sheppard, J.R.G. Turner, and H.F. Nijhout, among others. However, with the advent of modern genomics, there have been a number of breakthroughs in elucidating the genetic mechanisms involved (see Deshmukh et al., 2017 for a recent review).

In the previous section we discussed the case of polymorphic mimicry in the African mocker swallowtail butterfly *Papilio dardanus*, in which females have evolved colour patterns that mimic several different species of unpalatable danaid butterfly. Detailed heritability studies of Ford (1936) and Clarke & Sheppard (1959, 1960) revealed that within a single population, inheritance of wing pattern generally behaved as if it were controlled at a single locus with alleles of large phenotypic effect and a characteristic Mendelian dominance hierarchy. However, when the researchers crossed individuals from different subspecies and from populations with differing morph compositions, these crosses produced different forms with a dominance hierarchy that was less clear cut, indicating that the nature of the expression of this Mendelian factor depends on other co-evolved genetic 'modifiers' specific to that population (Nijhout, 1991). As Turner (2000) notes, mimicry is not simply a consequence of variation at one locus, but the result of fine tuning of the genome by further modifiers. Intriguingly, many of these modifiers have no obvious phenotypic effect in the absence of mimicry. For instance, modifier genes in a population of *Papilio dardanus* in Ethiopia shorten the tails of mimetic forms (so that they resemble their tailless models) but do not shorten the tails of non-mimetic forms (Turner, 2000).

The fact that the main pattern 'gene' has such a complex and diverse set of simultaneous effects on the phenotype (colour and pattern) has led to the suggestion that the gene is a complex locus (a 'supergene') consisting of several closely linked genes each with a specific effect on the pattern (Clarke & Sheppard, 1960). Given the tight linkage of genes with developmentally different characteristics, the 'supergene' hypothesis can explain why only the advantageous mimetic forms are seen, with few recombinants that combine properties in disadvantageous ways. Supergenes have long been discussed in the context of mimicry (Nijhout, 1991), but until recently formal experimental evidence of supergenes in *Papilio* species has remained relatively tentative. In the last decade, however, our understanding of the genetics underlying mimicry has improved dramatically. For example, as discussed in our chapter on Müllerian mimicry (Chapter 7), supergenes in *Heliconius numata* have been identified, with their integrity maintained by chromosomal inversions that lock multiple adjacent genes into single, non-recombining clusters (Joron et al., 2011).

The common Mormon (*Papilio polytes*) is a swallowtail butterfly distributed across Asia that exhibits female-limited polymorphic mimicry. The males have a single non-mimetic wing pattern, while females arise in a variety of discrete forms, exhibiting the non-mimetic male-like wing pattern but also wing patterns that resemble different toxic species of swallowtail in the genus *Pachliopta* (Figure 9.6). In a breakthrough paper, Kunte et al. (2014) found that variation in phenotypes was associated with genetic variation around the *doublesex* locus, a member of the *Dmrt* family of genes which control aspects of sexual differentiation and development. In retrospect, this was a good candidate to screen: the hunt was on for such a gene (Nishikawa et al., 2013; see Mallet, 2015), especially since these genes have long been known to play a role in generating discrete polymorphisms based on multiple traits, namely males and females. Rather than a tightly linked cluster of genes collectively controlling wing pattern, *doublesex* appears to have been co-opted as a master regulator to control distinct networks that pattern various aspects of the wing, and hence the entire phenotype from a single locus. Precisely how one gene controls so many distinct networks remains

unclear. Kunte et al. (2014) found that the *doublesex* transcripts spliced differently in different sexes, but they find no evidence for differential splicing between mimetic forms in females. Instead, non-coding, regulatory DNA that controls when and where *doublesex* is expressed may play a role in mimic

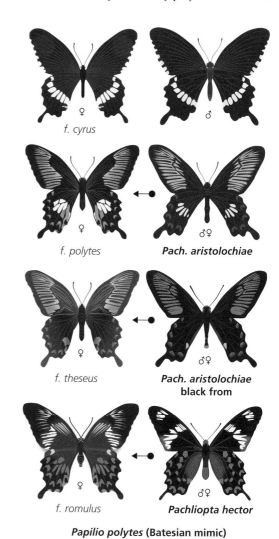

f. cyrus

f. polytes   **Pach. aristolochiae**

f. theseus   **Pach. aristolochiae black from**

f. romulus   **Pachliopta hector**

Papilio polytes (Batesian mimic)

Figure 9.6 Polymorphic, sex-limited mimicry in *Papilio polytes* (left column). The non-mimetic (form *cyrus*) females of *P. polytes* look like males, whereas mimetic female morphs (forms *polytes*, *theseus*, and *romulus*) mimic distantly related, toxic *Pachliopta* swallowtails (right column). In some populations (including those studied by Kunte et al., 2014), the non-mimetic *cyrus* form lack hindwing tails, yet the presence/absence of tails segregated perfectly with female wing pattern in their crosses. Image courtesy of Krushnamegh Kunte (see Plate 18).

differentiation. Do other *Papilio* species with sex-limited polymorphism employ the same genetic systems? Intriguingly, female mimetic polymorphism in *Papilio dardanus* has been associated with variation in the genomic region containing the genes *engrailed* and *invected* (Clark et al., 2008) which are not linked to *doublesex*, so natural selection may have found rather different solutions to the same underlying problem in different (yet closely related) species.

Overall, it is clear that polymorphic Batesian mimic species may require a special form of genetic architecture if its members are to simultaneously reach optima on different adaptive peaks. Researchers do not currently know enough about the genetics of mimicry in monomorphic mimetic species to discuss the various ways in which mimicry is achieved, but intuitively in monomorphic species with a single adaptive peak it may not require supergenes or master regulators. We are still at an early stage of combining our knowledge of molecular and developmental biology and how it relates to mimicry. This may add a whole new layer of complexity. For instance, while the characteristics of the different eyespots are developmentally coupled in non-mimetic *Bicyclus anynana*, Beldale et al. (2002) found that artificial selection for increased size of one individual eyespot tended to proceed in a manner largely independent from selection imposed on another eyespot. The authors argued that this flexibility was probably related to the compartmentalization of the wing, possibly mediated compartment-specific genetic components that regulate the expression of the eyespot-forming genes. Clearly, there is more to learn and very little work has been done to understand the genetics of mimicry in non-lepidopteran groups such as snakes and hoverflies. Nevertheless, the time is coming where we can close the circle and understand the evolution of mimicry not just from a phenotypic perspective, but also from a genotypic one.

9.5 Ecological and phylogenetic considerations

If mimicry is so beneficial, then why don't all species adopt it? Carpenter & Ford's (1933, p. 28), response to this familiar 'ridiculous question' was a *reductio ad absurdum*: the success of mimicry depends upon abundance of alternative palatable food. In effect, if all species became mimics, then there would be no advantage to mimicry. This, however, is not a real evolutionary argument—it does not in itself explain why only a subset of species become mimetic, and which species these are.

Phylogenetically controlled analyses are beginning to shed light on the selection pressures leading to mimicry, and why only some species become mimetic. For example, Higginson et al. (2012) compared the life history and coloration of British species of Geometridae and Drepanidae (Lepidoptera) and found that the tendency of species to employ masquerade was associated with the utilization of a greater diversity of host plants. The reason(s) for this association are not entirely clear, but crypsis (a common alternative strategy) may involve much more host-specific adaptation than masquerade. Indeed, it is possible that looking like a twig (or bird dropping—although only five species in the sample did this) confers protection to masqueraders over a broader array of different host plants.

In one of the most thorough comparative analyses of the evolution of masquerade to date, Suzuki et al. (2014) attempted to elucidate the how leaf mimicry patterns varied among species of the nymphalid genus *Kallima* and 45 related species. Comparative morphological analyses of the species wing patterns suggest that they share the same basic 'ground plan', which allowed the authors to reconstruct the likely evolutionary pathways as new species are formed. Their character state-change analysis showed that the leaf pattern has evolved over time from a non-mimetic ancestor through accumulation of changes in multiple pattern elements to enhance mimetic resemblance.

In a broader scale analysis, Merilaita & Tullberg (2005) compared the coloration of seven phylogenetically matched pairs of diurnal and nocturnal adult Lepidoptera species and found that Batesian mimicry and aposematism were much more likely to evolve in the day fliers than in the nocturnal species. A likely explanation for this simple pattern is that crypsis is harder to achieve in active, day-flying species, but readily realized in night-active species that rest during the day. Given the preponderance of night-flying moths and the enhanced detectability of diurnally active species, why was daytime activity ever selected for and maintained? Clearly a range of physiological and ecological factors may play a role, including relative freedom from competition. Indeed, in some species there may be high opportunity costs of crypsis; by hiding from predators, prey may lose opportunities to forage, to thermoregulate, and to find mates. Hence, Batesian mimicry may evolve in species that have much to gain from conspicuousness, but for whom the costs of secondary defences are too high. Intriguingly, Yack & Fullard (2000) note that ultrasound hearing is absent in butterflies but prevalent in moths, leading them to suggest that diurnal activity may have arisen through selection to avoid bat predation. So certain Lepidoptera may have jumped from the frying pan into the fire as they exploited the daytime, switching their predators from bats to birds.

Of course, breaking free of crypsis can be realized either by aposematism or by Batesian mimicry. Some Batesian mimic species we see today may have originally been aposematic, but because of the costliness of secondary defences, became cheats. Alternatively, for some species the costs of acquiring secondary defences in the first place may simply be too high so that Batesian mimicry is the next best method of removing the opportunity costs that follow from crypsis (Turner, 1984). In some intriguing comparative work, Prudic et al. (2007) noted that most swallowtail (*Papilio*) caterpillar species masquerade as bird droppings in their early instars. However, as they enter their fourth or fifth instar they adopt alternative phenotypes including an eye-spotted form that putatively resembles a snake (Janzen et al., 2010) and a conspicuous form (probably aposematic, since it is likely that many *Papilio* caterpillar species are unpalatable). Intriguingly, the tendency to become aposematic was not related to the species' extent of diet specialization, but their signal environment—those that develop on narrow-leaved plants were more likely to evolve aposematism since exposure of caterpillars on such plants would tend to render caterpillars conspicuous.

In another comparative study, Hossie et al. (2015) examined the distribution of eyespots across final instar hawkmoth caterpillars (Macroglossinae). Although the evolutionary significance of eyespots is still being debated, it is likely that the eyespots improve the resemblance of the caterpillars to verte-

brate predators and hence have an intimidatory effect. After controlling for phylogeny, they found that eyespots were significantly more likely to occur in larger caterpillars. Why would this be so? Complementary laboratory work by the authors confirmed that naïve chicks showed a greater latency to attack large (pastry) caterpillars with eyespots, than those without eyespots. By contrast, chicks more rapidly attacked small caterpillars with eyespots than those without. Eyespots on small prey may not be especially intimidating, and yet serve to enhance the prey's conspicuousness. However, since they may pose a legitimate threat, large prey items are generally more intimidating to small birds, and the addition of eyespots further enhances the effect. This cross-species pattern is entirely consistent with the observation that eyespots tend to be substantially reduced or absent altogether during early instars of hawkmoths, only becoming prominent in late instars when the caterpillars are much larger.

Taken together, for morphological protective mimicry to be selected for, we require the prey species to benefit from exposure to visually hunting predators, no means to protect themselves directly and a suitable noxious model to resemble. Many hoverfly species, for instance, may be mimetic because they are regularly exposed on flowering heads and because these sites are also visited by wasps and bees. It almost goes without saying that some species (such as nocturnal-flying and day-resting moths—see above) would not benefit from adopting the visual appearance of an unpalatable species, since they would not be clearly seen by predators. There may also be anatomical restrictions (size and shape) that may drastically restrict the range of traits that would confer any initial mimetic advantage at all, and at least have some influence on choice of model. Phylogenetic approaches are increasingly used in the study of mimicry, and we are already seeing how detailed spatial and phylogenetic analyses have helped reveal that mimicry is not necessarily a stable endpoint (Davis Rabosky et al., 2016). Future work may help elucidate the nature and frequency of co-evolutionary chases of model and mimic.

It has recently been argued that selection for mimicry may be more intense in some geographical regions than others. Echoing the earlier work of

Nicholson (1927), Herberstein et al. (2014) noted a high reported prevalence of aggressive mimicry in cuckoos, crab spiders, and orchids in Australian biota. The authors discussed several explanations for the apparent over-representation of mimicry in these taxonomic groups, ruling out the possibility that the taxa are simply more diverse in Australia than elsewhere. Instead, they suggest it is likely the indirect result of arid climate and isolation (through a series of tentative but plausible mechanisms). If the authors are correct, then Australia may have provided a 'perfect storm' for the evolution of deception, at least in terms of aggressive mimicry.

9.6 Associated phenomena in the evolution of Batesian mimicry and masquerade

9.6.1 Tastes and toxins: the role of prey rejection

It is well known that aposematic insects, including butterflies, are frequently released unharmed after being seized by birds (Boyden, 1976; Fink & Brower, 1981; Järvi et al., 1981; Wiklund & Järvi, 1982; Pinheiro & de Campos, 2013). Many defended prey species, including butterflies and wasps, have tough and flexible bodies (Fisher, 1930; Rothschild, 1971; Chai, 1986) so the release response of birds may in part be on the basis of tactile rather than gustatory stimuli (Kassarov, 1999). Whatever the mechanism, predators are able to distinguish defended and undefended prey post attack, while defended prey can survive this encounter.

There is also considerable laboratory evidence that birds can sample and selectively reject unpalatable prey. Gamberale-Stille & Guilford (2004), for example, found that domestic chicks tended to sample and reject a high proportion of crumbs rendered unpalatable through quinine, and yet consume most palatable crumbs. Skelhorn & Rowe (2006b) conducted an analogous experiment using dried crumbs, this time with two concentrations of quinine, and found that the chicks could discriminate between crumbs with different concentrations of quinine, rejecting a higher proportion of crumbs with higher concentrations of quinine following attack. In these experiments, the rejection rates of

palatable automimics decreased with increasing automimic frequency, a phenomenon that might arise at least in part because the underlying chances of misclassifying palatable prey as unpalatable prey are lower the higher the frequency of mimics.

Clearly this sample-rejection behaviour has important implications for the evolution of Batesian mimicry. One might think that if predators could discriminate between unpalatable models and palatable mimics *after* attacking them, then they could avoid the costs of consuming the models and Batesian mimicry would no longer provide any protection. It is rather like a trader maintaining honesty of a given currency simply by biting into any coins they receive and rejecting forgeries. Of course even if a degree of taste-rejection were possible, then there will still be a cost to sampling any prey item that is ultimately rejected, both in terms of time and risk of injury or ingestion of noxious substances (cf the conditions favouring Batesian evasive mimicry, section 9.4.6). Therefore, cautious sampling or 'go-slow' behaviour (Guilford, 1994) would be more likely to arise when predators not only have the ability to successfully taste-reject harmful prey, but also the motivation to take the risk and handle such prey. Indeed, Holen (2013) presented a model based on the classical optimal diet model which predicts precisely this—predators should avoid the mimicry complex altogether if the mimicry complex is unprofitable and alternative prey are profitable and/or common. Conversely, when the mimicry complex is profitable to attack by a taste-rejecting predator, then the predator should attack these prey, but reject prey (largely the models) according to their perceived distastefulness. Naturally, when predators decide to attack a model–mimic complex, then we would expect the taste-rejection strategy to be frequency-dependent, in that (following signal detection theory) the acceptance/rejection thresholds should change with mimic/model ratio.

Although prey palatability, taste, and profitability are often treated synonymously, researchers are increasingly starting to highlight the differences between these phenomena, so we should be more careful how we label them (Ruxton & Kennedy, 2006). Palatability and taste are largely characteristics of the prey item itself (although both can depend to an extent on the dietary history of the predator), while the profitability of a prey item is as much a characteristic of the predator's state (its hunger or toxin levels, for instance) and environment (the abundance and profitability of alternative prey) as it is the nutrients and defences the prey item exhibits.

Indeed, sometimes even that a prey item tastes bad (but does not contain toxins) is sufficient to make a prey unprofitable to attack, especially when there are alternative palatable prey available. For example, Skelhorn & Rowe (2009) found that coating mealworms with Bitrex could protect them from predators, but most particularly when they are presented alongside alternative, distinct palatable mealworms which predators prefer.

Intuitively, one might expect that predators would evolve to sense the unprofitability of a prey item through some sensory modality such as its taste. Like warning signals, such cues would be mutually beneficial to both predator and prey. However, again like warning signals, there is scope for exploitation, especially since taste receptors cannot be specific to every type of toxin. Skelhorn & Rowe (2010) manipulated palatability of mealworms (through coating them with non-toxic Bitrex) and their toxicity (by injecting quinine sulfate through the mouthparts), and presented them to European starlings along with entirely palatable mealworms. The palatable mealworms were visually discriminable, but the defended mealworms were not. When taste was a reliable predictor of toxicity (mildly toxic prey had fewer drops of Bitrex than moderately toxic prey), then the birds used the taste of Bitrex to preferentially ingest more of the mildly toxic prey while taste-rejecting moderately toxic prey—maintaining a relatively high nutrient intake while managing their toxin burden. By contrast, when toxicity and taste were independent, the birds did not taste-reject prey and instead ate approximately equal numbers of mildly and moderately distasteful prey. This result is important because it shows that birds do not use toxicity directly to discriminate between visually identical mimics and models, but learn about toxicity through associating it with taste. Conversely, here the bad taste of Bitrex itself was not sufficient to motivate predators to taste-reject—it had to be associated with quinine toxins before taste-rejection ensued.

It seems likely that predators would adopt a 'go-slow' approach if there were outward signs that the prey could be costly to attack. For example, Sillén-Tullberg (1985) found that great tits tended to taste-reject both red and grey forms of an unpalatable bug (*Lygaeus equestris*) after attacking them, but that the red form was more likely to survive this sampling. It was suggested that the difference in survival was due to birds handling the red morph more cautiously at the outset, perhaps forewarned by the conspicuous signals. Halpin & Rowe (2017) recently conducted similar experiments, presenting domestic chicks with conspicuous or cryptic (green or purple, that either contrasted with their background or not) live waxmoth larvae (*Galleria mellonella*) with certain of these caterpillars rendered unpalatable by the application of Bitrex.

While the total number of presentations was relatively low, there was no strong evidence for differential rejection rates of conspicuous and cryptic prey and (in contrast to Sillén-Tullberg, 1985) no differences in the survivorship of these prey types following rejection.

While Batesian mimicry has been discussed for over a century, the phenomena of taste and toxicity have only begun to be separated. Indeed, what we may be calling 'taste-rejection' may more aptly be referred to 'toxin rejection on the basis of taste'. The model of Holen (2013) confirms that the two phenomena may inter-relate in some subtle ways. For example, if a given toxin enhances the ability of a predator to distinguish the model from its palatable mimics, then the profitability of the model–mimic complex may actually increase as the toxin concentration of the models increase, so that mimics and models may be better protected if the model was moderately rather than highly toxic.

More work needs to be done to elucidate the roles of taste and toxicity in predator decision making, especially if their relationship is capable of exhibiting such non-intuitive properties. There are a number of fundamental questions. How often is the toxicity of a prey type determined directly from its taste? Are there any taste mimics in the natural world that share the taste but not toxicity with their models? Some fruiting plants produce hypersweet peptides rather than more expensive sugars, possibly to fool frugivores into consuming them

(Ruxton & Kennedy, 2006), but we clearly need more examples. When predators choose to sample prey rather than attack it outright, does this cautious ('go-slow') strategy involve any trade-offs to the predator, such as a higher probability of prey escape? If so, what are the implications of this trade-off for the survival of mimics and models? What prey species have evolved post-attack defences, such as a thickened cuticle, and have any mimics also evolved these defences? More generally, if rejection of models/mimics before attack can allow the mimics to gain a foothold, while taste-rejection following attack can prevent Batesian mimics and automimics from dominating, then is taste-rejection the key reason why Batesian mimicry so rarely undermines the protection afforded to the model at high frequencies of mimics? Clearly there are more questions than answers in this emerging field.

9.6.2 Imperfect mimicry and the limits of natural selection

One widely held belief is that there should always be strong selection pressure on mimics to resemble their models as closely as possible. Indeed, there are some spectacular cases of close mimicry in a range of sensory modalities. For example, the parasitoid wasp *Gelis agilis* (Ichneumonidae) closely resembles the sympatric black garden ant (*Lasius niger*) in body size, morphology, and movement and even emits an ant alarm pheromone when threatened (Malcicka et al., 2015). Yet mimicry only rarely involves the convergence of such a full range of traits and there are many cases in nature in which potential Batesian mimics (and masqueraders) do not appear to resemble their models particularly closely. For example, many species of hoverfly are generally regarded as Batesian mimics of wasps and bees, yet to the human eye at least, they do not resemble their models particularly well (e.g. Dittrich et al., 1993; Edmunds, 2000; Penney et al., 2012; Wilson et al., 2013). Likewise, as we have seen (section 9.2.2), many harmless Neotropical snakes resemble highly venomous coral snakes. Although some of these Batesian mimics are considered high fidelity (Greene & McDiarmid, 2005) many of these snake species have only a coarse resemblance (Savage & Slowinski, 1992).

The widespread occurrence of apparently imperfect Batesian mimics has so far been explained in a variety of different ways (Edmunds, 2000). Kikuchi & Pfennig (2013) list and classify eleven separate (but not mutually exclusive) explanations (see Table 9.1). These hypotheses include the possibility that imperfect mimicry arises simply as an artifact of human perception (Cuthill & Bennett, 1993) or as a consequence of constraints that prevent selection for more perfect mimicry (Holen & Johnstone, 2004). Imperfect mimicry could also arise as a consequence of some trade-off, such as the selection to simultaneously resemble a number of different models (Edmunds, 2000), selection for mimicry and thermoregulation (Taylor et al., 2016), or selection to deal with multiple predators with different sensory ecologies (Pekár et al., 2011) with imperfect mimics gaining the greatest overall protection. In other cases imperfect mimicry may provide just as much protection as perfect mimicry, so there is simply no further selection to improve the degree of mimetic fidelity (Bates, 1862; Duncan & Sheppard, 1965). A recent paper stressed that predators do not have complete information about the physical properties of all the prey types they encounter and argued that optimal sampling behaviour can lead to situations under which there is little selection to improve mimetic fidelity (Sherratt & Peet-Pare, 2017). For example, predators might quickly learn to adopt the rule 'avoid prey with black and yellow stripes' but evaluating the profitability of more complex conditional rules (of which there will be many) such as 'avoid black and yellow stripes but only when the prey has long antennae' may be perceived as too costly to warrant the extensive sampling required (Sherratt & Peet-Pare, 2017).

The hypothesis that there is no selection on imperfect mimics to improve their fidelity beyond a certain point has had a long history. Indeed, Bates (1862) himself invoked this 'relaxed selection' argument to explain variation in the degree of mimetic fidelity: '*In many cases of mimetic resemblance, the mimicry is not so exact as in the Leptalides…It would show that the persecutor is not keen or rigid in its selection; a moderate degree of resemblance suffices to deceive it, and therefore the process halts at that point.*' In support, the relationship between the attack rates on mimics and their extent of mimetic fidelity often tends to appear

Table 9.1 The eleven specific hypotheses for the evolution of imperfect mimicry tabulated by Kikuchi & Pfennig (2013) along with a simple classification of the nature of selection. See their Tables 1 and 2 for further details.

Specific Hypothesis	Brief description	Nature of selection
Eye-of-the beholder	Imperfect mimicry is an anthropocentric projection	Adaptive peak at perfect mimicry to relevant perceiver
Developmental and genetic constraints	Perfect mimicry cannot be achieved (or is only achieved at a cost)	No opportunity to improve imperfect mimicry (or adaptive peak at imperfect mimicry)
Chase-away	Models evolve away from their mimics leaving a 'phenotypic gap'	Non-equilibrium dynamics
Relaxed selection	The same level of protection afforded whether the mimic is perfect or imperfect	Selection neutral beyond certain point (a plateau)
Mimetic breakdown	Mimicry is no longer beneficial due to paucity of models or absence of predators	No selection, or selection away from mimicry
Perceptual exploitation	The imperfect mimic has more of an effect on receiver than a perfect mimic	Adaptive peak at imperfect mimicry
Satyric mimicry	A chimera with mixed signals confuses the receiver even more than a perfect mimic, at least when rare	Adaptive peak at imperfect mimicry
Multiple models	Mimics evolve a 'jack of all trades' appearance somewhere between several models to maximize their protection	Adaptive peak at imperfect mimicry
Multiple predators	Imperfect mimics face selection to dupe generalist predators while avoiding specialist predators	Adaptive peak at imperfect mimicry
Kin selection	While individual selection favours perfect mimicry, imperfect mimicry can be selected if it better protects others with the same trait	Adaptive peak at imperfect mimicry
Character displacement	Competition displaces phenotypic optimum away from perfect mimicry	Adaptive peak at imperfect mimicry

sigmoidal—so while mimicry pays, there is little difference in the success of perfect and imperfect mimics (Dittrich et al., 1993; Caley & Schluter, 2003). One reason for this sigmoidal relationship is that mimics and models vary continuously in appearance and the way they are perceived can also introduce additional uncertainty. For example, receivers may view mimics and models at unusual angles, at a distance (Skelhorn & Ruxton, 2014) or with insufficient or deceptive illumination (Kemp et al., 2015). Speed–accuracy trade-offs may also actively promote only a cursory inspection of objects that are encountered (Chittka & Osorio, 2007). Since discrimination is not always clear-cut, then predators cannot rely on identifying a single phenotype but must employ thresholds. Perhaps most importantly, it is not what a predator believes that is important to selection (for example, it may be reasonably convinced that the prey it has encountered is a mimic) but how it responds to that prey in terms of its decision to attack—despite its beliefs, it may decide to leave partial mimics alone because the consequences

of attacking a prey should it turn out to be a model are too costly.

Intuitively, one would expect relaxed selection to be particularly likely when the model is extremely numerous or highly unprofitable (and conversely when the mimic is relatively rare and low in profitability)—(Schmidt, 1958; Duncan & Sheppard, 1965; Sherratt, 2002b). Likewise, imperfect mimicry might be expected to be maintained when alternative prey are so abundant that it is not worth taking the risk and attacking anything vaguely model-like (Dill, 1975; Sherratt, 2003). Although the above relationships arise for a number of reasons, the fidelity of scarlet kingsnakes to their coral snake models tends to increase as the coral snake becomes rarer (Harper & Pfennig, 2007). Likewise, the mimetic fidelity of hoverfly species tends to increase with body size (Penney et al., 2012; but see Taylor et al., 2016), a result readily understood if one assumes that larger mimics are more profitable to predators and/or more visible, and so intrinsically more likely to be attacked. Using a model of avian colour perception,

Su et al. (2015) quantified the extent of mimetic resemblance among butterfly models and mimics within seven mimicry rings (comprising Papilionidae, Pieridae, and Nymphalidae) in the Western Ghats, India. They found that females of sexually mono-morphic mimetic species were better Batesian mimics than males of the same species, a result that may arise because females are more exposed to predators, and/or because males are under stronger sexual selection (see section 9.4.9).

What specific traits should be mimicked to confer a high degree of protection (and conversely, which ones might be less important)? Imperfect mimicry might be selected and maintained if predators learn to associate prey defences with certain salient dis-criminative features and effectively overlook others that are equally informative but functionally redun-dant. The phenomenon of reacting (for example, a dog sitting down) more strongly to one salient con-ditioned stimulus (such as its owner waving a treat) than another, less salient, stimulus (such as its owner saying 'sit'), when both have been presented together, is known as 'overshadowing' (Mackintosh, 1976; Shettleworth, 2010). Building on earlier studies on the degree of mimetic protection gained by new par-tial mimics (Schmidt, 1958, 1960; Ford, 1971; Terhune, 1977) there has recently been a surge of interest in understanding how objects are categorized and what traits are attended to when predators make their discriminative decisions (Bain et al., 2007; Aronsson & Gamberale-Stille, 2008; Balogh et al., 2010; Aronsson & Gamberale-Stille, 2012). Bain et al. (2007), for example, analysed the same training and test images of non-mimetic flies, wasps, and hoverflies that had been presented to pigeons in Dittrich et al.'s (1993) experiment using a feed-forward neural network classifier. They were able to closely replicate the pigeon response data only if they assumed that pigeons placed weight on some discriminative features (such as antennae length) but effectively overlook others (such as wing trans-parency). Likewise, Kikuchi & Pfennig (2010) were able to explain why non-venomous scarlet kingsnake mimics with concentric yellow/black/red/black rings can persist despite the fact that their models have concentric yellow/red/yellow/black rings in a different order. Their field experiments involving clay models revealed that predators are averse to ringed mimics with the right proportions of red and black, but the order of the coloured rings was effect-ively immaterial. For the same reason that humans employ mnemonics ('Red and black, friend of Jack, red and yellow, kills a fellow'), conditional rules (avoid red but only if next to yellow) are harder to remember than simpler unconditional rules. Indeed, after presenting human subjects with computer-gen-erated mimics and models that varied continuously in their colour (proportion of yellow and blue pixels) and/or size, Kikuchi et al. (2015) found that subjects could successfully use either trait when it was the sole distinguishing characteristic, adopting a thresh-old as classical signal detection theory would pre-dict. However, when mimics and models differed in both size and colour (such that, for instance, subjects were expected to avoid attacking prey with a high proportion of yellow, but only when large), then subjects did not follow the predictions of multidi-mensional SDT and instead placed more emphasis on one or the other characteristic (typically colour over size). Once again, simpler avoidance rules are easier to learn and adopt than more complex ones, even by humans.

One of the most ambitious studies to date on this topic was conducted by Kazemi et al. (2014) who trained wild-caught blue tits, *Cyanistes caeruleus*, to discriminate between rewarding prey (symbol-bear-ing laminated cards over wells containing pieces of mealworm) and non-rewarding prey (symbol-bear-ing cards over empty wells). The symbols over rewarding and non-rewarding wells differed from one another in their colour, pattern, and shape, and the blue tits learned to feed almost exclusively on the rewarding wells over four separate trials. Naturally, this high level of discrimination could have been achieved by using any given combination of colour, pattern, and/or shape in the symbols. To evaluate what traits were being used in the discrim-ination process, the authors then presented the same birds with an identical collection of rewarding prey, along with non-rewarding (Müllerian) mimetic prey that only had one trait (colour, pattern, or shape) in common with the earlier models that the birds had learned to avoid (the former models were also used as a control). The results of this test showed that rather than learning to avoid only those prey with all three attributes, the birds generalized their avoidance to non-rewarding prey that had colour in common with the former models, to such an extent that they were not attacked any more frequently than the perfect controls. Collectively, this work suggests that high-salience discriminatory traits (in this case colour) can overshadow other informative traits, allowing incipient mimics a significant survival advantage even if they share only one trait with the model. Sherratt et al. (2015) extended their design to include two-trait mimics and no-trait novel prey, this time using humans as predators, with similar results. However, the fact that novel prey were frequently avoided in these experiments indicates that predators do not simply learn to associate bad outcomes with certain traits, but that they also learn to associate good outcomes with certain traits.

There are now such a plethora of explanations for imperfect mimicry, one might wonder why perfect mimicry arises at all. Naturally, when simple, safe discriminative rules are indefinitely favoured, then imperfect mimicry can be maintained. However, what is the outcome if predators are exposed repeatedly to the mimics? Skelhorn et al. (2016b) performed just this sort of experiment when they presented naïve domestic chicks with pastry caterpillars over 8 days. The caterpillars were either with or without eyes, and with or without a swollen head, with the presence of both traits conferring the closest resemblance to a snake model. The chicks rapidly attacked the caterpillars that had neither mimetic trait, but initially exhibited similar levels of wariness towards mimics with just one or both deceptive traits. Over repeated exposures however, chicks discovered that the single-trait mimics were profitable more quickly than they discovered the double-trait mimics were profitable. So, predators' initial reactions to novel or deceptive prey will not necessarily be the same as their long-term reactions. Indeed, in systems where the incentive or opportunity to gain more information is high then one can anticipate strong selection for perfect mimicry.

The study of mimicry has allowed considerable insight into not just the power of adaptation, but also its limits. Like many well-known puzzles in evolutionary biology, ranging from cooperation to sexual reproduction, imperfect mimicry is rapidly becoming a problem with many potential solutions. The various hypotheses that have been proposed to

explain imperfect mimicry are not mutually exclusive and while we are not in a position to evaluate their relative importance, imperfect mimicry is far from paradoxical—indeed, it is the perfect mimics we now need to explain!

9.6.3 What selective factors influence behavioural mimicry?

So far we have dealt almost exclusively with morphological mimicry, yet some masquerading species and some Batesian mimics also behave like their models. Although one might anticipate that resembling a leaf, thorn, or stone would not involve any behavioural adaptations, this is not always the case. For example, twig-mimicking caterpillars of the early thorn moth *Selenia dentaria* are able to select microhabitats with size-matched twigs (Skelhorn & Ruxton, 2013). Likewise, females of the phytophagous stick insect (*Extatosoma tiaratum*) typically hang inverted among the foliage with curled abdomen and have been frequently observed rocking from side to side as if swaying in the wind. In a recent experiment, Bian et al. (2016) investigated this behaviour, finding that wind would initiate (but not indefinitely maintain) a swaying response in the phasmid, and that its movement had quantitatively similar properties to that of wind-blown plants. Since a motionless object on a moving plant is likely to attract attention while a moving object on a motionless plant will do the same, it seems highly likely that the behaviour has evolved to enhance its disguise. Finally, the orb-web spider *Cyclosa ginnaga* not only has a silver body but also adds a white discoid-shaped silk decoration to its web to enhance its resemblance and this combination was found to reduce its probability of wasp predation by increasing its resemblance to bird droppings (Liu et al., 2014).

Many Batesian mimics also engage in behavioural mimicry. For example, we have already described chicks that use behaviour to enhance their resemblance to an aposematic caterpillar (Londoño et al., 2015). The Indian rat snake *Ptyas mucosus* may be an acoustic Batesian mimic of the king cobra *Ophiophagus hannah* (Young et al., 1999). The ant-like jumping spider *Myrmarachne melanotarsa* not only resembles ants from the genus *Crematogaster*, but it also forms aggregations that appear to enhance the overall protective effect of the mimicry (Nelson & Jackson, 2009). Similarly, certain spiders that mimic ants also hold their front legs before them to resemble their antennae (McIver & Stonedahl, 1993). Srygley & Chai (1990) report that members of the butterfly subfamily Dismorphiinae, which mimic highly unpalatable Ithomiinae butterflies, have slow regular flights like their models. In a recent paper, Kitamura & Imafuku (2015) quantified the movement behaviour of mimetic and non-mimetic females of the polymorphic swallowtail butterfly *Papilio polytes*. They found that the flight paths of mimetic and non-mimetic females were different from each other, but the locomotory behaviour of the mimetic females was not significantly different from that of the model (*Pachliopta aristolochiae*) they morphologically resemble. Behavioural mimicry is not limited to adults either—certain eye-spotted caterpillars adopt a defensive posture when approached that increases their resemblance to snakes (Hossie & Sherratt, 2014).

In each of the above cases it seems clear that behaviour has augmented the effectiveness of their disguises. Behaviour—from the way an organism moves, to how it behaves under threat—may often serve to discriminate profitable and unprofitable prey. So if a given behaviour is sufficiently characteristic of unprofitable prey then there may be selection on mimics to adopt these diagnostic behavioural traits.

Indeed, sometimes behaviour may be one of the most important traits that need to be mimicked. Pekár et al. (2011) suggested that because spiders that mimic ants are also selected to run quickly like ants, then there may be little or no selective advantage to them more closely resembling the ants from a morphological perspective. Other behaviours, such as leg waving to provide the illusion of antennae, may serve to obfuscate key morphological differences between mimics and models that would otherwise be used for discrimination.

Despite its potential advantages, it is clear that not all species of a given group engage in behavioural mimicry, so one might wonder why some species have evolved behavioural mimicry and others have not. For example, Penney et al. (2014) surveyed the incidence of behavioural mimicry (mock stinging, wing wagging, leg waving) in 57 field-caught species of hoverfly, and found that these highly associated traits occurred in just two genera (*Spilomyia* and *Temnostoma*)—see Figure 9.7. Intriguingly, the behavioural mimics were all good morphological mimics of wasps (to human eyes at least), but not all good mimics were behavioural mimics. This led the authors to conclude that while the behavioural mimicry may have evolved to augment good morphological mimicry, it does not advantage all good mimics. When behavioural mimicry arises then it may often reflect overall stronger selection for

Figure 9.7 A hover fly (*Spilomyia longicornis*) and its wasp model (*Vespula* sp.). Note how the hover fly, left, is waving its darkened forelegs in front of its head to resemble the longer antennae of its hymenopteran model. Photographs courtesy of Dr Henri Goulet.

mimetic fidelity on all levels, rather than compensation for poor morphological similarity.

9.7 Overview

Batesian mimicry is now a well-studied phenomenon. While field demonstrations that Batesian mimicry provides a degree of protection from predators remain rare, there is ample evidence that predators given prior experience with noxious models subsequently tend to avoid the palatable mimics. It has also been repeatedly argued, and shown, that the efficacy of Batesian mimicry is dependent on the abundances of the models and mimics, the noxiousness of the model, and the abundance of alternative prey.

Since the last edition, more fascinating examples of Batesian mimicry have come to light and there has been much more work to test the basic properties of masquerade. Significant progress has been made in understanding the genetics of mimicry notably in polymorphic Batesian mimics. There has also been increasing use of phylogenetic techniques to understand the evolution of Batesian mimicry and how certain traits are gained and lost. There have also been significant advances in understanding the cognition of predators and how they arrive at the decisions that they do. These insights have helped to explain why imperfect mimics can persist, but also why mimics face selection to resemble their models in multiple traits.

As always, there is work to do. We still do not understand the genetics of mimicry in most systems, and the application of phylogenetic methods to mimicry is still in its infancy. The role of taste-rejection behaviour in the evolution of warning signals and mimicry is still somewhat uncertain, as is the prevalence of evasive mimicry. Indeed, while the possibility of mimicry based on evasiveness has long been discussed, and is plausible in theory, there are no field studies demonstrating its effectiveness in deterring predators. Most work has focussed on avian predators, but invertebrate predators are extremely common and voracious consumers and deserve further attention. We are still only beginning to learn what discriminative features predators pay attention to when deciding whether to attack a prey or not. For example, we still need to fully understand the conditions under which multi-trait mimicry is favoured compared to just a single trait. Since sophisticated discriminative rules (such as those involving interactions of traits—avoid yellow, but only when it has long antennae) require more information to evaluate: are there limits on what rules predators would ever benefit from learning? While most attention has focussed on selection for traits directly involved in deceiving predators, do other correlated traits such as nutritional value of the mimic, whether it can survive handling, and even its rate of senescence evolve as a consequence of this selection? Placing Batesian mimicry in a more general evolutionary or co-evolutionary life-history framework could well be enlightening. Finally, work on Batesian mimicry is often based on pairs of species, but since many Batesian mimics are parts of rings then we should be taking a whole community approach (see Pekár et al., 2017 for an excellent example), viewing mimicry in its appropriate context.

CHAPTER 10

Startling predators

10.1 What do we mean by startle?

Startling signals are secondary defences that occur after the focal prey individual has been singled out for attack. They are also often called 'deimatic signals'—coined by Maldonado (1970)—from the Greek for 'I frighten'; and less commonly 'bluff displays'. Startling signals involve stimulation of the predator's senses that cause it to delay or break off an attack. The assumption is that even a delay in attack can confer a survival advantage to the prey. This might occur because delay gives the prey an added opportunity to flee, or added opportunity for some other event to occur (perhaps the arrival of a predator of the predator) that causes the predator to break off the attack permanently. The mechanism by which sensory stimulation might cause the predator to break off or delay the attack is not specified, and indeed is a current area of controversy. It might be: overloading of sensory systems; simple confusion; a caution associated with novel stimuli; or it might be the prey stimulating the predator in a way that triggers the predator's own anti-predatory responses. There is no reason why the underlying mechanism of startle responses should be the same in all cases.

In general, startling stimuli involve a change in behaviour on the part of the prey. However, demonstrating that a given behaviour is triggered by imminent or actual attack by a predator and is associated with reduced likelihood of attack, delayed attack, or less effective attack is not, of itself, sufficient to conclude that the behaviour has a startling effect. First, it must be demonstrated that the behaviour influences the predator's behaviour primarily through sensory and/or cognitive manipulation. This differentiates it from behaviours whose primary effect is physical—such as fleeing or struggling. Secondly, argument that a given prey behaviour has a startling effect should involve consideration of why other mechanisms (such as aposematism and deflection) can be discounted. Behaviours that influence the predator's behaviour can include the revealing of aposematic or deflective signals that are not constantly displayed. Some prey may be selected not to display such signals permanently if they are effective only against a subset of predators or in a subset of ecological situations, and their permanent display would involve a cost (e.g. increased conspicuousness to predators against which the display is ineffective in deterring or deflecting attacks). Thus, startling signals must be separate from other mechanisms that may also influence predators' behaviour so as to curtail, delay, or diminish an attack. In theory, a startling effect could be produced by the action of the predator rather than the prey. For example, a bird rooting through leaf litter might be startled when it suddenly reveals a brightly coloured insect. However, we know of no example where a startle signal is not associated with a behavioural change by the prey, itself triggered by close approach by the predator.

Separating startle signals given by unpalatable prey types from aposematism is conceptually challenging, and not completely resolved in the literature. For Skelhorn et al. (2016c), startle displays should be defined by their ability to evoke fear responses, as they are suggested to result from a predator misclassifying prey as a potential threat to its immediate safety. Here, Skelhorn et al. also

Avoiding Attack: The Evolutionary Ecology of Crypsis, Aposematism, and Mimicry. Second Edition. Graeme D. Ruxton, William L. Allen, Thomas N. Sherratt, & Michael P. Speed, Oxford University Press (2018). © Graeme D. Ruxton, William L. Allen, Thomas N. Sherratt, & Michael P. Speed 2018. DOI: 10.1093/oso/9780199688678.001.0001

differentiate startle displays from aposematism, as they argue that aposematic displays do not provoke predators to fear for their immediate safety, but have evolved in response to the way in which predators learn and decide about what prey to consume. However, Umbers & Mappes (2016) argue that aposematic and deimatic displays are not necessarily mutually exclusive strategies, and additionally emphasize the difficulty of measuring the physiological states of fear responses in predators, especially in field studies. Umbers et al. (2017) then further distinguished aposematism from startling signals, in emphasizing that—unlike aposematism—startle signals do not require a predator to have any learned or innate aversion.

Another complication is to differentiate between startling signals and Batesian mimicry. Batesian mimicry relies on the predator generalizing between the sensory stimulation produced by the focal unprotected prey (the mimic) and the stimulation produced by a defended prey type (the model). In startle such a model cannot be identified and the protection afforded to undefended prey does not stem from them being misidentified as a defended prey type. If (as some have claimed) 'eyespot' patterning of the wings of some butterflies causes a potential predator of the butterfly to mistakenly interpret these eyespots as the eyes of one of their own predators, we would consider this a case of a startle signal, because here the mimicry is not of something that would be unattractive as food but of something that possesses a direct and immediate lethal threat.

Having considered in general terms what we consider to be a startling signal, we will now consider the empirical evidence for the existence of such signals.

10.2 Empirical evidence for the defence

10.2.1 Sound production in insects

Bura et al. (2011) demonstrated that caterpillars of the North American walnut sphinx (*Amorpha juglandis*) can be induced to produce a whistling sound by simulated avian attack, using physical contact with blunt forceps. This was accompanied by thrashing movements but not any obvious chemical defence,

such as regurgitation. Contact with an avian predator (the yellow warbler *Dendroica petechia*) in laboratory trials also caused this behaviour in the caterpillars, and the whistling sound itself then caused the birds to hesitate, jump back, or dive away. Bura et al interpret the whistle as a startle signal. This seems plausible, although further experiments where the noise was experimentally broadcast when birds approached an otherwise undefended food item, or where caterpillars were experimentally muted, would be interesting. Bura et al. also note that caterpillars showed no escape response after birds broke off attacks, but such escape behaviours may not be necessary to select for startle signals (as discussed in section 10.1).

Recently, Dookie et al. (2017) also investigated the whistles of walnut sphinx caterpillars, in a study where variations of these sounds were presented to wild-caught predator red-winged blackbirds (*Agelaius phoeniceus*), following activation of a sensor while they fed on mealworms (*Tenebrio molitor*). They found that birds exposed to whistles exhibited significantly more startle behaviours—such as flying away, flinching, and hopping—and took longer to return to the area than during control conditions where no sounds were played. Interestingly, during the hour-long sessions birds habituated to the whistles, but after two days the startling effects were restored (Figure 10.1). Dookie et al. suggest that the re-sensitization underscores the importance of novelty for effective startle displays, and that the infrequency of encounters with these cryptic caterpillars—that occur solitarily on their host plants—would likely result in a restoration of the startle response after a few days in the wild as well as in captive situations.

Many insects produce a clicking or buzzing sound, through stridulation, immediately prior to or after contact by a predator. In some circumstances where the insects have a clear defensive capability this is most likely to be an aposematic signal. Even when the individual prey types do not have obvious defence, the sound may function as acoustic mimicry of the aposematic signals of defended prey. However, it is certainly clear that this noise can be associated with less effective handling of the prey by predators, and startle may be one of the mechanisms leading to this. The most convincing

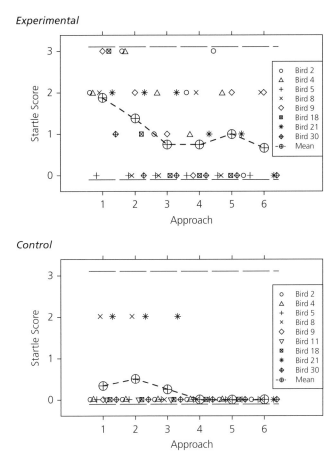

Figure 10.1 Red-winged blackbirds (*Agelaius phoeniceus*) exhibit startle responses (by flying away, flinching, and hopping) when exposed to playbacks of defensive whistles recorded from the walnut sphinx caterpillar (*Amorpha juglandis*). Sounds were presented to birds that activated a sensor while feeding on mealworms (*Tenebrio molitor*). Birds exposed to whistles (top panel) exhibited significantly higher startle scores and took longer to return to the feeding dish than during control conditions (bottom panel) where no sounds were played. Birds habituated to sounds during a one-hour session, but after two days the startling effects were restored. Figure provided by Jayne Yack.

case of startle in this context was reported by Lewis & Cane (1990). They used a beetle, *Thanasimus dubius*, as predator and another beetle, *Ips calligraphus*, as prey in staged laboratory encounters. These species were chosen for a number of reasons: 1.) they naturally have predatory encounters; 2.) the prey has been extensively investigated biochemically and seems to have no chemical, or other, defences; 3.) only female prey are capable of producing sound, but otherwise there are no obvious differences in morphology or behaviour between males and females that are likely to impact on predators; and 4.) the predator is an oligophagous predator of *Ips* and *Dendroctonus* bark beetles and, thus, is not expected

to encounter aposematically stridulating prey. Lewis and Crane found that females always reacted to handling by the predator by stridulating and, thus, predators were more likely to drop females than males whilst handling them. The authors argue that such drops in natural circumstances are likely to lead to the loss of the prey by the predator. In the lab, dropped prey were recovered by the predators and all prey were readily consumed.

When approached by predators, several species of butterfly, including peacocks (*Aglais io*), respond by a display of opening and closing their wings revealing a bright, spotted pattern and generate a high-pitched noise. Mohl & Miller (1976) demonstrated that peacock butterflies produced intense ultrasonic clicks that startled three captive vespertilionid bats. Bats' responses to displaying peacocks included jumps, audible screams, retreats, head jerks, and pinna movements. The bats responded in this way even when out of visual contact with the butterflies. Olofsson et al. (2012a) showed that the sound of displaying peacocks also startled wild-caught mice of the species *Apodemus flavicollis* and *A. sylvaticus* in staged encounters. Mice were more likely to flee from sound-producing butterflies than those that had had their sound production disabled. These butterflies are perfectly edible to the mice, which are likely to be significant predators in the wild (Wiklund et al., 2008;, Olofsson et al., 2012a). One limitation of this experiment, however, is that the authors did not play back control sounds, so we cannot tell if mice are deterred by a sudden noise generally or by the specific type generated by the butterflies.

10.2.2 Clicking sounds of aquatic organisms

Vester et al. (2004) reported that the close approach of harp seals (*Pagophilus groenlandicus*), hooded seals (*Cystophora cristata*), or human divers to captive cod (*Gadus morhua*) caused the cod to generate conspicuous clicking sounds. They report anecdotally that harp seals, but not hooded seals, often aborted approaches towards cod coincident with the onset of clicking, and had a lower success rate in capturing cod than salmon (*Salmo salar*) that were never heard to emit clicks. These possible startle signals seem worthy of further exploration.

Bouwma & Herrnkind (2009) demonstrated that Caribbean spiny lobsters (*Panulirus argus*) produce sound by stridulation during tailflip escape manoeuvres, in response to being grasped by potential predators. In laboratory interactions, they found that stridulating lobsters were more able to escape from the grasp of Caribbean reef octopuses (*Octopus briareus*) than experimentally muted individuals. Further, sound production in the lobster seemed to only be elicited by restraint, and experimentally muted lobsters exhibited the same escape behaviours as the unmanipulated controls. Startle might be expected to be effective in this case, since once free of an octopus's grasp, the lobster should be able to easily outdistance it in the open ocean. However,

the authors argue against such a startle effect. Sound production generally occurred just before the octopus made initial contact, but the authors found no change in octopus behaviour coincident with the onset of sound production, or at any stage prior to contact being made. Rather, they suggest that sound may be a non-functional by-product, and the key advantage of stridulating is that it sets up vibrations in the lobster's carapace that make it more difficult for the octopus to maintain a good grip. This is plausible, but not definitively demonstrated, and it is possible that startle may provide an additional benefit. This warrants further exploration, especially when considering that Bouwma & Herrnkind cite references for 35 fish species and 39 spiny lobsters that produce sound in the context of heightened predation risk.

10.2.3 Posture, appearance, and inking in cuttlefish and squid

Staudinger et al. (2011) reported staged encounters between longfin squid, *Loligo pealeii*, and either a cruising predator (bluefish *Pomatomus saltatrix*) or an ambush predator (summer flounder *Paralichthys dentatus*). The discharge of ink was a common response to both predators, causing the bluefish to abandon attacks and the flounder to misdirect attacks towards the ink plume. The second of these might be a case of distraction (see Chapter 11), but the first may be an example of startle, at least in part. However, cephalopod ink is known to contain chemicals that disrupt some predator sensory systems and can be unpalatable, or otherwise aversive. Nonetheless, since some fish were not deterred by ink, we agree with Straudinger et al. that the ink is more likely to function as a signal rather than as a defence in itself. Another possibility might be that the ink acts as a cloaking device that disguises the direction of flight of the squid. This seems unlikely in this case, as inking was often paired not with fast escape behaviour, but with posture change which did not seem well designed to reduce ease of visual localization of the squid. Hence, this paper provides very suggestive evidence of startle signalling in this squid.

Straudinger et al. also report that some postural displays were effective in causing bluefish to break off an attack. The authors speculate that such postures may increase the apparent size of the prey and, thus, be effective in dissuading gape-limited predators. This is an interesting aspect to startle signals, and staged encounters with model squid in a variety of sizes and different postures (and against predators of different gape) would be illuminating.

10.2.4 Adult lepidopteran wing patterns

The most active current field of investigation of possible startle effects in predators relates to some of the patterning on the wings of adult Lepidoptera. Many species sit with their wings closed such that the undersides only are visible, and these undersides are often assumed to be relatively cryptic. However, on close approach of a predator, the insect will flick its wings open (sometimes repeatedly) to expose conspicuous patterning. This pattern can take a diversity of forms, but often includes approximately round elements called eyespots because of their visual similarity to vertebrate eyes. There is now a considerable body of evidence that these conspicuous patterns induce a startle response in potential predators (Vallin et al., 2005, 2007; Olofsson et al., 2012b).

10.2.5 Posture and stridulation in mantids

Robinson (1969), Edmunds (1972), and Edmunds (1974) all discuss observations that mantids (*Mantidae* spp.) and stick insects (*Phasmatodea* spp.) can commonly respond to approaching predatory threat by adopting an erect posture, often with wings spread and with noise production by stridulation. In some cases, these insects have potent toxic defences, but not in all cases, and it is likely that some of these behaviours are best interpreted as startle rather than aposematism. Edmunds (1974) summarizes the results of a number of essentially observational studies that suggest that some, but not all, predators are intimidated by mantid displays. Grandcolas & Desutter-Grandcolas (1998) provided the most detailed account. They reported on field observations of interactions between a generalist bird—the yellow-vented bulbul (*Pycnonotus barbatus*)—and a mantid of the species *Polyspilota aeruginosa*. The mantid adopted a posture with all four wings opened and slightly moving and forelegs outstretched. Colourful stripes on the wings and ocelli on the forelegs are prominently displayed during this. The bird approached and retreated several times before eventually flying away. Further work on this system would be worthwhile, especially since it might offer a valuable model system of multi-component and even multimodal startle signalling, as different body parts (legs and wings) are generally evolved and appear to have coloration that might be selected, at least in part, for a startle function, and because stridulation is often induced alongside the visual display.

10.2.6 Summary of empirical evidence

The examples given here are very unlikely to be an exhaustive list of startle signals. For example, it has repeatedly been suggested that marine organisms can use bioluminescence triggered by the close approach of predators as a startle defence (Haddock et al., 2010). However, we often simply lack systematic study of encounters with potential predators.

It is difficult to identify commonality among the eclectic group of taxa that appear to show startle defences, or to identify common traits in the form of these defences. This variation perhaps should not be surprising, given that startle defences are essentially a bluff, and variation in the form of such defences experienced by a given predator might help to avoid selection pressure for predators to ignore such displays. Edmunds (1972) reported on apparently startling responses of 25 species of praying mantid from Ghana. He noted that the patterning and colours used in display were highly variable between species, which he suggested might be a counter-selection against predator habituation to a particular signal, although no evidence for this was presented, and other (potentially complementary) explanations—such as species-identification signals—are possible.

10.3 The evolution of startle defence

10.3.1 The mechanism underlying startle

To many researchers, the explanation of startle is self-evident and follows from well-established concepts of signal detection theory, speed-accuracy trade-offs, and the life–dinner principle. When encountering a stimulus that might be generated by prey, sometimes that stimulus can be ambiguous, in that it can possibly be generated by prey and possibly by a predator of our focal predator. This is the signal-detection problem, where two types of error are possible—mistakenly approaching a predator or mistakenly giving up on a prey item. Predators are likely to have evolved cognitive processes that trade off these two types of error. However, the cost of mistakenly approaching a predator (or simply not fleeing in time) may be very high compared to the cost of missing out on a meal (this is the life–dinner principle) and so we might expect predators to have evolved cognitive processes that leads to many false alarms where the predator flees from sudden conspicuous signals. Startle defences in prey are exploiting these processes to the benefit of defenceless prey. We find this line of reasoning to be very plausible; however, where we disagree with some authors is in the extent to which the startle signal causes *misidentification*. This issue has been clearly discussed in terms of eyespots but holds more generally.

To many, eyespots work because they cause the predator to misidentify these markings as the eyes of a potential predator (e.g. Janzen et al., 2010). To such authors, the eyespots have been adapted to appear particularly eye-like to potential predators of the eyespot-bearing organism; the more eye-like the signal, the more likely the predator is to misidentify these markings as eyes of potential predators and flee. An alternative explanation is presented particularly clearly by Stevens (2005). This suggests that effective startle signals need not resemble particular features of predators sufficiently closely to trigger misidentification, but must simply be conspicuous and salient enough to trigger a neophobic reaction, confusion, or sensory overload. Essentially, the question is whether a blue tit flees from a displaying peacock butterfly because the display of the peacock in some way shares appearance features with a predator, or just because it was unexpected and 'surprising'. Both mechanisms are plausible, and a signal could be both unexpected and reminiscent of a predator. Hence, the selective importance of the two mechanisms for a particular case needs to be resolved empirically on a case-by-case basis, and it would be unwise to assume one mechanism is operating without careful consideration of the other. There has been an explosion in empirical research on eye mimicry associated with startle signals: Stevens & Ruxton (2014) offer a valuable review.

Since this review was published there have been a number of further interesting studies, e.g. De Bona et al. (2015), Mukherjee & Kodandaramaiah (2015).

Currently, the most suggestive evidence that startle displays cause predators to misidentify the stimulus as a predator comes from exposure of adult domestic fowl (*Gallus gallus domesticus*) to the eyespot display of the peacock butterfly (Olofsson et al., 2013). In this study, some peacocks had their eyespots obliterated with black marker pen, while the controls had a similar amount of black added to the wings but in positions that leave the eyespots intact. Birds were much more likely to emit a particular type of call in response to wing flicking by eyespot-intact peacocks than eyespot-covered ones. Further, the authors argue that this particular type of call is otherwise very closely associated with detection of a ground-based predator. This seems a very promising test system. However, experimental confirmation that such calls cannot be produced by the sudden appearance of appropriate control stimuli that look nothing like a potential predator (but are no less conspicuous and novel) would be useful, in order to increase our confidence that this call is truly functionally referential and indicates that the caller believes that it has detected a ground-based predator. If this is so, then domestic fowl might be a model system for exploring what features of wing patterning are required to produce this type of call. This will allow investigation of how well human perception of similarity to the appearance of eyes matches with at least one non-human animal. However, we acknowledge that design of a range of appropriate control stimuli of the type described above is a non-trivial challenge.

10.3.2 Satyric mimicry

A separate, but related, putative mechanism underlying startle effects is that of satyric mimicry—introduced by Howse & Allen (1994) and most recently articulated in Howse (2013). This is based on the reasonable assumption that misidentification of some stimuli as prey may be costly to a predator, perhaps because this causes it to expose itself to its own predators or to attack a well-defended species. This theory rests on the further assumption that accurate identification of prey by predators takes time, and that there is a speed–accuracy trade-off, such that predators are less likely to make mistakes if they take longer in the identification phase prior to deciding whether to attack or not. This too is relatively uncontroversial. The next step in Howse's reasoning is that 'images of any part of a prey animal's immediate environment may be found on its body and it is hypothesized that the ambiguity that is presented can interfere with the process of visual perception in potential predators, resulting in withdrawal or increased latency to attack.' So, if part of the body of the prey has the appearance of something else—which could be something innocuous and need not be a predator—this causes confusion in the predator that makes the signal-detection challenge for the predator greater, and so more time-consuming and/or error prone. This hypothesis is plausible, and cannot be rejected on logical grounds. The key issue is one of empirical evidence, and here we do not fully agree with Howse.

Firstly, he claims that '[i]t has now been found that eyespots and other designs on the wings of many insect species are often coupled with other wing patterns and designs. These composite images often closely resemble heads and bodies of vertebrates (including birds and reptiles) and of various invertebrates.' Although Howse offers a great many examples of such putative cases, we often find these less convincing than he does, and it must be borne in mind that the cognitive and sensory systems of ecologically relevant predators can be quite different from our own. As an example, he claims that the white spots on a black background that make up part of the wing pattern of the monarch butterfly *Danaus plexippus* represents potential satyric mimicry of the slimy salamander *Plethodon americanus*, whose sticky skin secretions would be aversive to many potential predators of monarchs. In order to build a more convincing case for this, or any other example of satyric mimicry, we need experiments. Firstly, this theory would predict that birds with recent experience of the salamander will be less willing to attack the butterfly than control birds exposed to a similarly defended amphibian with different visual appearance. Secondly, if we modify the appearance of the spotting pattern of the butterflies, then this should have more of an effect on the behaviour of birds with previous experience of the slimy salamander than the control birds. Without

such experimental exploration, many will see the current putative examples of satyric mimicry as no more than imaginative over-interpretation similar to the way humans can often imagine that they see patterns in random data (a phenomemon called apophenia).

10.3.3 Evolutionary history, genetic control, and environmental plasticity of startle signals

Butterflies are becoming a model system for understanding the factors controlling expression of visual signals. As discussed in Chapter 11, the wingspots of the intensively studied species *Bicyclus anynana* can be experimentally influenced by variation in rearing temperature, to produce either the wet-season form with prominent eyespots or the dry-season form with much-reduced eyespots. Although the selective benefit of these spots has yet to be identified, this study demonstrates that the type of markings relevant to startle signals might be very easily evolved.

It has been shown that the evolution of eyespot patterning in butterflies of the tribe Junoniini (Nymphalidae) has not been subject to consistent strong directional selection either with respect to eyespot number or size (Kodandaramaiah, 2009). This is not surprising given that, firstly, eyespots can function in a number of different ways in butterflies including species recognition, mate choice, and deflection, as well as startle; secondly, that there are possible costs to eyespots in terms of reduced crypsis (Stevens et al., 2008a); and, thirdly, that the costs and anti-predatory benefits of eyespots are likely to change differentially with changes in predator community structure. Add in the mimicry that is commonplace in butterflies, and they become a very problematic group in which to search for evidence of consistent evolutionary pressure for startle signals. This difficulty would be greatly alleviated if we could identify the particular appearance features that are likely to be associated with a startling effect.

10.3.4 Selective pressures

The frillneck lizard (*Chlamydosaurus kingii*) erects a large frill on its neck when faced with a predatory threat. However, in this species, the frill is most commonly erected in interactions between conspecifics

(Shine, 1990), and it may be that the evolution of some startle signals can be understood as co-option of signals originally selected through social or sexual signalling. Another possible evolutionary pathway to a startle signal is that the signal functions aposematically against some predatory types faced by the focal prey species, and as a startle defence against other predators; its evolution can, thus, be understood via the pathways discussed in Chapter 6.

Even when the startle signal seems unrelated to other functions, its evolution can be understood relatively simply. It is easy to imagine how a predator might show a startle response to stimuli generated by the prey inadvertently during the process of fleeing from an attack or resisting attack. Such a situation would offer an immediate survival benefit to the prey. Incremental changes in behaviour and/or anatomy might be selected if they increased the frequency and/or salience of such stimuli and did not incur a prohibitive cost (see section 10.5). Over time, the original act of fleeing may not be required to produce the stimulus, which shifts from being inadvertent to being under voluntary control. Differential exposure to predation appears to have impacted the development of acoustic defence, for example, in *Poecilimon ornatus* tettigoniids, where the higher proportion of male startling stridulation, relative to females, is consistent with a male-biased predation risk (Kowalski et al., 2014).

A startle display that is triggered too easily by proximity of a predator might incur costs in attracting the attention of that predator (or other predators nearby) when it had not previously detected the prey. However, this cost must be traded off against avoiding a situation where the predator has already damaged the prey before the startle display is triggered.

There may also be conflict between secondary defences; for example, conflict between some aspects of startle signal production and simultaneous ability to flee. For some taxa it may be better (in terms of survival) simply to flee on first detecting a predator, rather than engage in a startle defence that may be effective on some occasions but on other occasions allows the predator to approach sufficiently closely that risk of capture is much higher if the startle signal is not effective.

Lastly, it may be that there are costs associated with investment in startle defences that are paid regardless of whether attacks take place or not. These might be the physiological cost of construction of signal structures, or compromises made in morphology to accommodate signalling structures. However, in general, these costs may often be relatively small, since in most cases startle signals involve relatively minor physical modification to existing structures: clearly the prime function of lepidoptera wings are for flight and not for startle displays.

10.4 Ecological aspects of startle defences

Links between the startle defences and the ecologies of the species involved have not been subject to any significant research, but it is possible to make a number of predictions that should be open to empirical testing. Since being startled is detrimental to the predator in an encounter with prey, we would expect that predators should habituate to startle signals. Thus, it seems reasonable to predict that startle defences will most often be seen, and therefore most often be successful, when utilized by prey that the predator does not encounter frequently. Thus, we would expect startle displays to be less commonly used and to be less effective when the prey are locally common, relative to alternative prey types of the predator concerned. Thus, startle defences may be uncommon in outbreaking species and those for which aggregation is an important aspect of life history. Similarly, we might expect that susceptibility to startle defences is a cost of dietary generalism for a predator, with startle signals being most effective when presented to predators with a broad diet. Again to counteract predator habituation, we might predict that prey with relatively high local densities would be selected to show within-population variation in the nature of any startle signals used, and where two species share a predator, there would be selection for between-species variation in startle signalling.

The susceptibility of a predator to startle defences is likely to rest on the relative costs of missed opportunities to feed versus increased risk of predation (see section 10.5). We would predict that when prey are particularly valuable to predators (either because of a high level of hunger in the predator, high nutritional value of the prey, and/or low availability of alternative prey), then startle defences will be less effective. Conversely, where predators are vulnerable to predators of their own, then startle defences should be more effective. Often these two aspects will covary with the relative size of predator and prey, and so making general predictions about body-size trends in the effectiveness and/or prevalence of startle displays is difficult. For example, a mantid's startle display might be expected to be less effective against a chimpanzee (*Pan troglodytes*) that has few potential predators relative to a small lizard. However, conversely, a single mantid is likely to be a much more valuable food resource for the lizard.

We can also predict that startle defences should be more common in situations where other secondary defences are less available to the prey concerned. For example, one common trait of the species of prey that have been shown to use startle defences is that none of them make use of retreat into a burrow or similar physical refuge from predation. We might expect that startle will be more common in physically complex environments, where prey are unable to detect predators a long distance away and flee; conversely, startle defences will be ineffective if the predator is able to completely surprise the prey, such that the predator can subdue the prey before the prey has had opportunity to detect the predator and mount a startle defence. This may be open to experimental investigation. For example, at high ambient temperatures butterflies may be able to escape terrestrial predators—like lizards—by taking flight, whereas at colder temperatures startle defences may be more attractive. Similarly, we predict that where the undergrowth is physically complex (and thus potentially concealing) a mantid in vegetation might escape from an avian predator simply by dropping from the plant, whereas startle might be more attractive where ground cover is less, or when the same mantid encounters the bird on open ground and dropping is not a behavioural option.

10.5 Co-evolutionary considerations in startle defences

If a predator is startled by a prey type such that its success in capturing that prey type is reduced, then one would expect this situation to produce a selection pressure to reduce this cost in the predator. This selection pressure might take the form of changes to the cognitive processes associated with prey recognition, to reduce the effectiveness of this startle signal and/or learned habituation to the signal. However, such selection pressure may not be strong if the predator is a dietary generalist and the prey is infrequently encountered. Essentially, this is a signal detection problem for the predator, and selection pressure to reduce one type of error (being startled by harmless prey) will depend on: the relative frequency with which this type of error can occur compared to errors where dangerous animals are misidentified as innocuous; the relative costs of these two types of errors; and how any change in focal predator cognition influences the relative frequencies of the two types of error. The life–dinner principle can be used to argue that where the individual value of the prey items lost to inappropriate startle responses is low compared to any increase in predation risk associated with changes in responses to ambiguous stimuli, then prey lost to startle effects may be a price worth paying. As a corollary of this, we predict that startle defences will be less effective in high-value prey items. Thus, we might expect startle defences in caterpillars to be more effective against avian predators (where a single prey item might only meet a tiny fraction of the predator's daily energetic requirements) than, for example, digger wasps (where a single caterpillar might be sufficient to provision an entire brood of the wasp's offspring). By similar reasoning, we might expect predators to be less susceptible to startle defences when particularly hungry and more susceptible at times of heightened expectation of predation. We know of no tests of these hypotheses.

We would expect predators to evolve mechanisms that allow them to habituate to startle signals from harmless prey. That is, we would expect predators to react to startle signals not by completely fleeing the scene, but by retreating to a safer distance from which they may be able to assess the potential threat. Fleeing may deprive a predator of the ability to learn about the other organism (except in as much as the other organism did not successfully pursue it). Thus, the behaviour of a predator towards a startle signal should not be fixed but, rather, should be responsive to that individual's experiences subsequent to previous exposures to similar signals.

Habituation by birds to initially startling displays has been demonstrated under laboratory conditions (Ingalls, 1993; Vaughan, 1983; Schlenoff, 1985). However, the relevance of this to natural conditions, where birds are likely to encounter a range of different stimuli between interactions with a particular startle display, and where those startle displays are encountered in a range of different environmental settings (e.g. different relative orientations of predator and prey), remains to be explored.

Experiments of Merilaita et al. (2011) into the importance of eyespot patterns in startle signals are insightful in a number of respects. Firstly, since they demonstrated that eyespot patterns induced latency to attack in naive hand-reared pied flycatchers (*Ficedula hypoleuca*), they demonstrate clearly that startle behaviours in predators can have an innate component. However, by using an experimental design that involved serial exposures to prey, they found that eyespots not only caused latency to attack the focal prey, but also caused increased latency to attack the next prey item offered, even when the next prey had no eyespots. This shows that there can be a carry-over effect as well as innate aspects of startle responses. In Merilaita et al.'s experiment, the treatment without eyespots still had many aspects of visual similarity to the prey with eyespots. It would be interesting to explore how predators generalize from the aversive experience of startle responses. That is, we currently have little idea of the aspects of a startling experience that predators remember in the context of subsequent experiences.

10.6 Unresolved issues and future challenges

Essentially, there is much current debate in the literature as to whether startle effects should best be seen only as a response to unexpected strong sensory stimulation, or whether they can in some instances additionally represent misidentification of the prey as a predator. It should be possible to manipulate the experience of captive birds so as to explore the relative strengths of these alternatives.

There are established protocols for manipulating the perceived predation risk of small passerines. If startle effects are primarily about misidentification of prey as a predatory threat, we might expect that startle defences should be more effective against avian predators whose perception of predation risk has been experimentally raised. Similarly, it should be interesting to either manipulate birds' propensity to neophobia (by manipulating their experience of novel stimuli) and/or to screen individuals in terms of individual variation in neophobia. If startle is essentially a special case of neophobia, then we would predict that it would be less effective against those with an innate or experimentally induced lower propensity to neophobic reaction.

Exploration of eye mimicry (see Stevens & Ruxton, 2014 for a review) and satyric mimicry rests on how relevant non-human detectors assemble aspects of an image to achieve localization and identification of a particular object. Howse (2013) argues that *Gestalt* perception is absent or greatly reduced relative to humans in non-mammalian vertebrates, which focus more on individual aspects of detail than on integration of the pattern as a whole. This issue is fundamental to Howse's theory of satyric mimicry as an explanation for why mimicry can be effective even if very imperfect to human observers. Investigation of this issue would be valuable, more generally, to understanding of how non-humans separate specific objects from the background and, thus, illuminate the effectiveness of countershading and crypsis through disruption.

It is clearly difficult to resolve in non-humans the extent to which a startle reaction is based on misidentification of innocuous prey as a predatory threat. Unlike humans, we cannot ask evolutionarily relevant predators 'what they thought they saw'. However, developments in functional magnetic resonance imaging (fMRI) and other brain activity mapping techniques may hold potential for illuminating this issue in the not-too-distant future. It would be particularly interesting to compare patterns of brain activity when relevant animals are exposed to sensory stimulation caused by real predators and when caused by startling prey and suitable controls.

CHAPTER 11

Deflecting the point of attack

11.1 Overview

Deflection involves the prey influencing the position of the initial contact of the predator with the prey's body, in a way that benefits the prey. As we discussed in the introduction, the anti-predatory mechanisms covered in this book vary greatly in current understanding of their taxonomic distribution, and deflection is an extreme example of this. It has been postulated to occur in a sparse and eclectic group of organisms, and the evidence for its existence is quite variable among members of this group. We spend the bulk of this chapter exploring this evidence. We argue that the evidence currently available allows some speculation on the evolutionary ecology of this anti-predatory strategy, and we develop hypotheses that aim to broaden the scope of research into deflective traits.

11.2 How deflecting traits work

Deflection involves traits that influence the initial point of contact of the predator on the prey's body in a way that benefits the prey. These traits might be behavioural, involve morphological structures, or involve pigmentation and other appearance traits, or combinations thereof. The benefit to the prey is normally considered to be an increased likelihood of escaping the attack, and so the benefit to the prey comes at a cost to the predator. Thus, deflection may involve biasing attacks to areas of the body that can be broken off without catastrophic damage to the prey, or which are particularly difficult for the predator to grasp. However, for chemically defended prey, both predator and prey may benefit if the predator's point of contact can be biased to positions on the prey's body that allow for accurate assessment of these defences by predators, without incurring significant damage to the prey. Additionally, spiders may offer a tantalizing example of the extended phenotype in deflection, where decorations on the spider's web, that are thus entirely physically separate from the body of the spider, serve to draw predatory attacks away from the spider. Since this involves deflecting the predator away from the body of the prey entirely, this does not fit our definition of deflection. We choose to call the putative mechanism in this case *distraction* rather than *deflection*, but cover it in this chapter because of the strong conceptual similarity.

As with startle, it is important to differentiate between perceptual exploitation and mimicry in deflection. That is, there are two conceptually different mechanisms by which the predator's point of attack might be influenced by prey appearance. In perceptual exploitation, the point of attack may be drawn to particular areas of the body simply because they are the most conspicuous or salient, stimulating the senses of the predator. However, in mimicry, the cognitive systems of the predator are fooled and an attack is drawn to, for example, a false head structure, because the predator intended to attack the prey's head and has misidentified another part of the body as the head. Sensory exploitation is a more parsimonious explanation of observed deflection than specific mimicry (in terms of making fewer assumptions about cognitive complexity in the decision-making processes of the predator). Hence, we should be careful not to assume the latter is occurring without due consideration of the former.

Avoiding Attack: The Evolutionary Ecology of Crypsis, Aposematism, and Mimicry. Second Edition. Graeme D. Ruxton, William L. Allen, Thomas N. Sherratt, & Michael P. Speed, Oxford University Press (2018). © Graeme D. Ruxton, William L. Allen, Thomas N. Sherratt, & Michael P. Speed 2018. DOI: 10.1093/oso/9780199688678.001.0001

It also seems likely to us that in cases where deflection by mimicry is occurring, deflection through sensory exploitation was a likely precursor to this situation. These two mechanisms are not dichotomous, but are best seen as descriptions of ends of a continuum of a cognitive underpinning of behaviour; see Schaefer & Ruxton (2009) for a fuller discussion of these concepts.

Some conspicuous animal markings are described as having a 'distractive' effect that hinders visual detection. The suggestion is that such markings draw the attention of the viewer in a way that takes attention away from the outline of the animal (or other salient aspects of morphology) in a way that hinders visual detection of the animal (Dimitrova et al., 2009, Stevens et al., 2012) In this chapter we focus on markings that manipulate the point of first physical contact on the body; markings that influence detection are covered fully in Chapters 1–4, with Chapter 2 on disruptive camouflage being particularly relevant to distraction of attention from salient features.

Notice that this chapter deals entirely with situations where it is assumed that the predator's visual sense is the key sensory system involved in determining the point of attack. We can think of no physical reason why deflection must be confined to this modality, and suspect cases of deflection in other modalities await discovery. There have been suggestions that the morphology and flight dynamics of moths may function to deflect the point of attack of acoustically mediated bat predators towards harder-to-grasp body parts (Barber et al., 2015), but whether the errors in targeting by moths might be better seen as an acoustic version of the dazzle mechanism discussed in Chapter 12 remains an open question (Lee & Moss, 2016).

11.3 The taxonomic distribution of deflecting traits

In contrast to crypsis, aposematism, and mimicry, we do not have an extensive body of empirical evidence in support of the existence of deflecting traits in natural organisms, so in this section we need to consider the evidence that is available. We start with eyespots in butterflies and moths, as the potential deflecting effect of these has been most intensively studied. However, it is worth first pondering why deflection is much less commonly observed than the other anti-predatory traits mentioned above. We suspect that this apparent paucity of examples is a genuine reflection of its rarity in nature, rather than simply neglect or oversight by scientists. Deflection will only be successful in mobile prey that have the ability to escape their predator, even after contact between the two has been initiated, and that feature at least some body parts that are highly resistant to, or tolerant of, damage inflicted by contact with a predator. We suspect that taxa that meet both these requirements will be relatively uncommon.

11.3.1 Adult lepidopteran eyespots

Spot-shaped markings on the wings of adult Lepidoptera can have functions that are nothing to do with predation—for example, they can influence mate-choice. As discussed in Chapter 10, larger spots may have a startling effect on would-be predators; here we explore whether (as is often postulated) smaller spots can have a deflecting function.

Perhaps not surprisingly, it is certainly possible for markings on the body of prey to influence the point of attack. For example, Vallin et al. (2011) exposed blue tits (*Cyanistes caeruleus*) to artificial triangular prey objects that were uniformly coloured except that some were made conspicuously asymmetric by placing a contrastingly coloured mark on one half. Birds were more likely to peck the side of the prey with the marking when the marking was small, but there was no statistically significant effect when spots were larger. These results suggest that some forms of patterning have the potential to influence the point of attack.

In order to explore the effect of lighting conditions on deflection, Olofsson et al. (2010) presented dead specimens of the woodland brown butterfly (*Lopinga achine*) to blue tits. These butterflies show numerous white spots near the peripheral edge of their wings which reflect strongly in UV wavelengths. The authors found that the point of attack on the body of the butterflies was strongly affected by ambient light conditions. If the light contained a UV component, then at high light intensities attacks were directed to the butterfly's head, whereas at low light intensities the attacks were directed more often towards the peripheral edge of the wing. In another treatment with low light intensity but the UV component removed, attacks were again directed towards the head area. It seems difficult to explain these results other than by concluding that the interaction of the natural appearance of this butterfly species with ambient light has the potential to influence the point of attack of these avian predators. In the absence of experimental manipulations of the butterflies' appearance or comparative studies with other species, we cannot identify the specific appearance attributes that influenced predators under the experimental conditions. But, it does seem as if those attacks not focussed on the head often seemed directed towards the peripheral spot markings. Following on from this study, Olofsson et al. (2013) investigated whether the background against which butterflies are concealed may deceive birds so that they err in similar ways to the low UV-containing light intensity condition. They presented speckled wood butterflies, decorated with or without eyespots, on oak and birch bark backgrounds to blue tits. Their observations suggest that, firstly, eyespots, independent of background, were effective in deflecting attacks; secondly, the time elapsed between a bird landing and the attack depended on the background and whether the butterfly had an eyespot or not; and, thirdly, faster-to-attack birds were more prone to errors than slower birds. Backgrounds, therefore, play some role in the anti-predatory effectiveness of marginal eyespots, and predators can experience speed–accuracy trade-offs.

While larger size and conspicuousness may make individual eyespots serve an intimidating, startling effect; smaller, sometimes serial, eyespots located closer to the periphery of wings can be argued to serve a deflective role (Kodandaramaiah et al., 2013). A considerable number of older studies have exploited variation in the number and/or size of spots on the periphery of the wings between different populations, morphs, or sexes within one species, or variation between different species (summarized by Stevens, 2005). These studies are based on the premise that, if a greater fraction of the type with more prominent spots have wing-damage associated with failed predation, then this can be seen as evidence of a survival advantage of spots through a deflec-

tive effect. As an example, the beak mark data collected in Pinheiro et al.'s (2014) capture-recapture study of *Junonia evarete* butterflies in Brazil, indicated to the authors that birds do focus on the eyespots when making attacks on butterflies The interpretation of these sorts of studies hinges on (i) the ability to differentiate damage associated with failed predation from other types of damage, and (ii) confidence that there is no reason to expect the type with the spots to be exposed to higher levels of predatory attack. There has been little empirical effort invested in demonstrating the first assumption. However, the second assumption is particularly problematic, given that different species and different sexes of one species have different behaviours (e.g. activity levels and microhabitat preferences), and comparison between different morphs of the same species generally requires comparison between locations or across different time periods in the same location (both of which are likely to lead to variation in predator community composition).

Similarly, interpretation of older studies (again well summarized in Stevens, 2005) that attempt to interpret position of wing damage relative to the position on the wing of eyespots also require consideration of the validity of a number of assumptions. These assumptions assert (i) that damage resulting from predatory attack can be unambiguously identified, and (ii) that unsuccessful attacks on different parts of the wing are equally likely to cause breakage. Further, if these assumptions hold, it might be possible to infer whether spots on the wing influence the placement of unsuccessful attacks. However, this would not allow inference to be made about the fraction of attacks that are successful (and so the selective forces—if any—associated with the deflective nature of the spots). Hence, this does not seem a particularly fruitful line of evidence. However, related work by Hill & Vaca (2004) is illuminating. They demonstrated in the laboratory that the strength needed to tear the outer region of the hindwing of *Pierella astyoche* (which has putatively deflective spot patterns on its wings) was less than that required to make equivalent damage to wings of species of the same family that lack this wing patterning. This observation is consistent with the deflection theory, as less prey investment in outer hindwing strength is perhaps to be expected if these areas are adapted to be the targets of attacks. Prey butterflies would benefit from their outer hindwings tearing easily at 'intended' areas when attacked there, as the predator will perhaps more readily pull back from the assault after breaking a region and allow the prey to escape quicker. Given the relatively simple nature of the experiment, it would be straightforward to evaluate whether this effect can be found consistently across a range of species.

Paul Brakefield and colleagues have intensively studied selection pressures on the appearance of the African satyrine butterfly *Bicyclus anynana*. The species has two morphs, termed *dry season* and *wet season*, which differ in appearance, mainly in that the *wet season* form has prominent spots on the wing margin that are absent in the *dry season* form. These researchers have discovered that individuals of similar genotype (e.g. full-sibs) can be made to develop into one form or the other depending on larval-rearing temperature. Thus, the two forms have been used to explore the possible deflective effect of marginal spots in a series of carefully controlled laboratory experiments (Lyytinen et al., 2003, 2004; Vlieger & Brakefield, 2007) that have, so far, failed to produce strong support for this mechanism against vertebrate predators. They do, however, suggest a cost to developing into a wet season rather than a dry season form in terms of reduced crypsis (see section 11.4), and so a balancing selective benefit likely exists, but is currently unknown. However, recent evidence suggests that the spots may have a deflective effect against invertebrate mantid predators (Prudic et al., 2015).

Many Lycaenid butterflies seem to show a complex of traits that, together, are strongly suggestive of the false-head mechanism: a pattern at the anal angle of the hindwings combining eye-like markings, filamentous antenna-like extensions to the edge of the wing, movement of the false-antennae whilst keeping the real antennae still, other extensions giving the appearance of legs, a resting attitude with the real head pointed downward (in contrast to the normal resting position of related species), and a tendency to walk backwards (again contrary to the norm)—see Stevens (2005) and references therein. However, not all species show all of these features, and systematic exploration of how often they appear together in one species would be welcome. It is dif-

ficult to propose plausible alternative explanations to deflection through misleading predators as to the position of the false head—especially for the false antennae. However, there has not been particularly vigorous empirical investigation of the effect of these traits on predators. The most thorough exploration has been that of Wourms & Wasserman (1985) on the response of captive avian predators (blue jays *Cyanocitta cristata*) to modified forms of the cabbage white butterfly (*Pieris rapae*). When birds were presented with dead butterflies marked in a number of different ways (see Figure 11.1), only the addition of a 'false eye' affected the target of initial strikes, with strikes being directed towards this marking. In another experiment reported in the same paper with live butterflies, those with the false eye markings were more often mishandled by the jays and escaped relative to unpainted controls. These results could be strengthened if repeated with one or more further controls for the paint applied, either applying the paint near to the real head, or applying paint that matches the natural wing colour in the false-head position. At the moment we cannot rule out that painted butterflies were less palatable to birds and/or that the application of paint influenced prey behaviour that reduced their risk of predation, regardless of any change in predator behaviour.

Sourakov (2013) presented a single individual jumping spider of the species *Phidippus pulcherrimus* with a succession of butterfly individuals. Two of those individuals were of the species *Calycopis cecrops* that sports a false head; the other 13 belonged to 11 different species that do not appear to have such false heads. The two *C. cecrops* were the only individuals that the spider failed to capture, despite making 14 attempted attacks in one case. Attacks on other species were always directed towards the head region, whereas those on *C. cecrops* were directed towards the false-head region. These results deserve development with more predatory individuals and a wider range of false-head bearing butterflies.

There are two hypotheses regarding how false heads may offer anti-predatory defence. One is that butterflies can break free and escape after being grabbed in the false head region (Robbins, 1981). The other is due to Cordero (2001) and suggests that prey sensory organs are concentrated on the head,

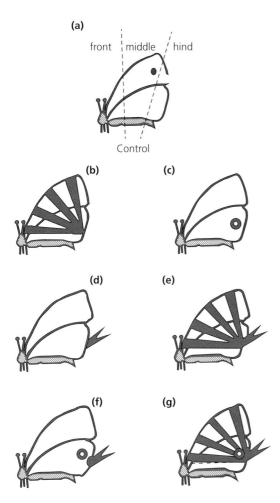

Figure 11.1 Seven types of butterfly markings presented to blue jays by Wourms & Wasserman (1985). (a) Control showing front, middle, and hind regions used to identify strike points. (b) Lines converging in the anal angle of the hindwing. (c) Eyespot 0.5 cm in diameter. (d) Tails consisting of 1.0 cm wing segments painted and dried in place. (e) Convergent lines and tails. (f) Tails and anal spot. (g) Convergent lines, tails, and anal spot.

can often be found with damage in the false-head region that is indicative of failed attack, and that the wing is particularly fragile in this region to facilitate breaking off in the grip of a predator, and the butterflies' escape. Quantification of these issues would be very helpful.

Many butterfly species with false-head wing structures also perform a typical hind wings movement (HWM) back-and-forth along the sagittal plane with their tails while perching, a movement that apparently mimics antennal movement. Interestingly, this movement has sometimes been seen to change depending on how close a predator is. One hypothesis proposes that, when butterflies are observed at close range, the HWM distorts the shape of the false head so that its deceiving effect is reduced; and butterflies that stop moving their wings when a predator is close by will be selected. A second hypothesis suggests that increasing the frequency of HWM improves its deflective effect at close range. A study by López-Palafox et al. (2015) presented 33 individuals from 18 species of Lycaenidae to a stuffed bird to test these hypotheses, and found that half of the butterflies started to move their hind wings or increased the rate of HWM when the stuffed bird was at close range; however, a substantial proportion of butterflies (30 per cent) stopped moving their hind wings or decreased the rate of HWM. Future studies could investigate the effects of HWM on predator deflection in more naturalistic settings. However, the results produced here also demonstrated that there are alternative ways butterflies can produce 'vivid' movement of the hind wing tails (the 'false antennae') in the absence of HWM, and so perhaps more fine-scale behavioural studies concerning prey responses to approaching predators may reveal more subtle effects. Additionally, as López-Palafox et al. note, it would be interesting to perform similar experiments using other visually oriented predators that are also thought to have been selective pressures in the evolution of false heads.

11.3.2 Lizard's tails

Several species of lizard have the ability to shed their tail when in the grasp of a predator (an ability called 'autotomy'). Such tails are often coloured in a

and so predators attempt surprise attacks by approaching from the rear. In such circumstances, the false head can fool the predator into approaching from an angle that makes detection and flight easier for the butterfly; here, our definition of deflection would not be met, as it may be that no contact is made between predator and prey. Sourakov suggests that his observations of staged attacks by his spider give support for the first rather than the second of these mechanisms. He also suggests that wild-caught individuals of species with false heads

way that contrasts strongly with both the rest of the lizard's body and with typical substrates. In experiments with predatory snakes, Cooper et al. (1985) and Cooper & Vitt (1991) demonstrated clearly that the appearance of tails can have a deflective effect; increasing the chance of the predator grasping the lizard by the tail, and increasing the probability of the lizard's escape from the predator's grasp following autotomy. The effect of such distinctively coloured tails is often enhanced by dramatic tail-waving behaviours that draw further attention to them. This is an unusually unambiguous case of deflection, and has been extensively studied; hence we will return to it later in the chapter.

11.3.3 Eyespots on fish

Many tropical fish feature a dark spot at the posterior end of their bodies; these 'eyespots' or 'ocelli' have sometimes been suggested as having a deflective function. Direct evidence for this is, however, limited. McPhail (1977) outlines some potentially interesting experiments, without providing detailed methods or quantitative results. In these experiments, he experimentally added dark spots to some members of a species that does not feature them (*Hyphessobrycon panamensis*), and compared the fate of manipulated fish with those of unmanipulated controls in laboratory predation experiments. He reported that fish with artificial caudal spots escaped predators more often than the same species without spots, but the difference was not significant. However, in fish captured, the place on the body where the fish were seized was shifted towards the posterior by the presence of a spot.

Dale & Pappantoniou (1986) explored the deflective function of eyespots using the cutlips minnow (*Exoglossum maxillingua*) as a predator. This species exhibits eye-picking behaviour, where it removes and consumes the eyes of attacked fish. As with the previously discussed paper, experimental details were not provided, but they compared 20 trials with single goldfish (*Carassius auratus*) as prey with 20 trials using single goldfish with an artificial eyespot (created by injecting India ink under the scales in the tail region). For fish with eyespots, 49 per cent of strikes were directed to the head; compared to 68 per cent of strikes to fish given a sham injection

that did not alter their appearance. Further, the mean time taken to remove both of the prey fish's eyes was 1.2 minutes for unmanipulated fish and 2.3 minutes for fish with eyespots. In a further series of similar experiments, unmanipulated goldfish were compared with two species that naturally feature eyespots (rosy barb *Barbus conchonius* and flag cichlid *Cichlasoma festivum*). The results were, again, suggestive of an anti-predatory role for eyespots. Finally, the authors report that a 'series of experiments with wooden models having various combinations of eye, eye camouflage, and eyespot markings, have further corroborated these results', although no details were given. We can see very strong ethical reasons for making maximal use of inanimate models in any future laboratory work related to this phenomenon.

Empirical work in the wild is very scarce. Winemiller (1990) observed that '*Astronotus ocellatus* and several other large cichlid fishes of South America exhibit bright ocelli, or eyespots, near the base of the caudal fin. *Astronotus ocellatus* sympatric with fin-nipping piranhas shows less extensive fin damage than sympatric cichlids of similar size that lack distinct caudal ocelli.' Winemiller suggested that this two-species comparison supports the hypothesis that eyespots reduce piranha attacks by confounding visual recognition of the prey's caudal region. Since the two species involved differ quite considerably in behaviour and ecology, we find the strength of evidence in support of this argument to be low.

The most extensive study of this issue is reported by Gagliano (2008). She used a mark-recapture experiment on juvenile coral reef fish *Pomacentrus amboinensis*, which features a prominent black eyespot on the posterior dorsal fin. Although she argues that these fish experience strong predation from visually hunting fish, no differential survivorship on the basis of natural variation in size of the eyespot was detected (despite three-fold variation in her sample in the ratio of eyespot area to area of the real eye). Further, none of the recaptured juveniles showed bitemarks on the posterior of their body.

Some insights can also be gained from recent studies of attacks by predatory fish on fish-like artificial prey (Kjernsmo & Merilaita, 2013; Kjernsmo et al., 2016). Eyespots on such artificial prey could be designed such that they drew attacks towards them. It can be common for fish (and other taxa) to have a stripe running through their eye as well as sometimes sporting an eyespot. These experiments showed that adding a stripe through an artificial eyespot on these artificial prey reduced the tendency of predators to attack the eyespot. The inference of this in real prey fish is that the combination of an eyespot and a stripe marking through the real eye may maximize the effectiveness of deflection of the point of attack towards the eyespot.

11.3.4 Tadpole tails

Tadpoles often have patterning on their tail that has long been suggested to have a deflective function. Touchon & Warkentin (2008) reared tadpoles of the Neotropical treefrog *Dendropsophus ebraccatus* subject to cues from either fish predators, or from dragonfly nymph predators, or under control conditions. Tadpoles reared with dragonfly cues developed larger and redder tails than controls; those reared with fish cues had shallower achromatic tails. This is argued to be adaptive because fish are long-range cruising predators, against which crypsis is the best defence; whereas dragonfly larvae are ambush predators, for which deflection may be a more effective defence because (unlike fish) the dragonfly has little ability to pursue the tadpole through open water after a failed attack.

Van Buskirk et al. (2004) presented model tadpoles to dragonfly larvae. The models either had a dark body and a pale tail, a dark spot in the middle of the tail, or a dark spot near the tip of the tail. For this first model type, almost all attacks were directed to the body, whereas for the second model about half of all attacks were directed to the mid-tail, whereas for the final model type around 2/3 of attacks were to the body, but the other 1/3 were near the tip of the tail (with very few strikes to other parts of the tail). Previous work (Van Buskirk et al., 2003) with dragonfly larvae preying on *Rana temporaria* tadpoles demonstrated that tadpoles were around three times as likely to survive attacks to the tail as attacks directed to the body. Taken together, this recent body of work has considerably strengthened the evidence for deflection by traits in tadpole tails.

11.3.5 Weasels' tails

Powell (1982) hypothesized that the black tip to the tail of a stoat (*Mustela erminea*, sometimes called an ermine) acted to draw attacks from potential avian predators towards the tail, which is a smaller target than the stoat's body and more easily missed by the predator. He further hypothesized that weasels (*Mustela nivalis*, sometimes called 'least weasels'), being smaller than stoats and having shorter tails, do not have a black tip to their tail because this would be too close to the body to provide any advantage. These hypotheses were tested using three captive hawks trained to attack targets that moved rapidly across the floor of their enclosure. Six types of targets were used, three of similar size and tail length to a stoat and three similar to a weasel. One of each set of three was all white; one was white with a black tip to the tail and one white with a black band on the body. Each hawk was presented with each of the six models on twelve occasions, in random order. On every occasion the hawk attacked the moving target, but sometimes missed (see Table 11.1). Of the larger, more stoat-like, long-tailed targets, each hawk was much more likely to miss the model with the tail spot than the ones with no spot or a band on the back. Conversely, for the smaller, more weasel-like, short-tailed models, they were each much more likely to miss the spotless model than either of the other two. These clear-cut results very much warrant further investigation, especially with regard to other mammals with contrastingly coloured tips to their tails. We would welcome a cross-species comparison among mammals to explore whether there were any morphological or ecological variables that could be related to contrastingly-coloured tail tips. Powell's hypothesis predicts that contrasting tips would be more prevalent in species with longer tails and in those facing greatest predation pressure.

11.3.6 Caterpillars

Many caterpillars have markings that have generally been considered to have a startling effect on predators (see Chapter 10). However, Hossie & Sherratt (2012) provide some evidence, from models exposed to free-living birds, that some spot markings

Table 11.1 Results of Powell (1982).

			Number of times a hawk missed a fake weasel of the designated size-spot morph		
			Hawk		
	Size-spot morph	1	2	3	Total
	Long-tail no spot	1	1	0	2
	Long-tail tail spot	11	4	9	24
	Long-tail black spot	2	0	2	4
	Short-tail no spot	9	7	9	25
	Short-tail tail spot	1	0	1	2
	Short-tail black spot	3	0	0	3

Note: each kind of fake weasel morph was presented 12 times to each hawk

may influence the point of birds' attacks on a caterpillar's body. Given the body-plan of caterpillars, any such deflection is unlikely on its own to increase the likelihood of the caterpillar's escape. However it may be that the position of the caterpillar within the mouth of a bird may influence the ability of birds to detect chemical defences deployed by the caterpillar and, thus, increase the chance of taste rejection. Alternatively, or additionally, it may be that some areas of the body are more resistant to damage incurred prior to taste rejection, or that damage in some areas of the body can be more easily tolerated than in others. Empirical evaluation of the survival and growth of chemically defended invertebrates following handling and rejection by predators is needed, not just in the context of potential deflection but also to improve our understanding of the evolution of chemical defences and associated signalling (see Chapter 6).

11.3.7 Web decorations in orb spiders functioning in distraction

Many orb-web spiders add so-called decorations (called *stabilimenta*) to their webs, made variously from silk, prey remains, egg sacs, and plant material. There has been much debate on the function of these structures, and it seems highly likely that they serve a variety of functions in different species (see general reviews by Herberstein et al. (2000) and Théry & Casas (2009)). However, Tseng & Tso (2009) provide evidence suggestive that in the Taiwanese species *Cyclosa mulmeinsis* their function may be to act as decoys that distract the attacks of predatory wasps from the spider itself. The webs built by these spiders are vertically orientated and spiders characteristically rest in the middle. Over time, a series of decorations are added in a line above and below the normal resting place of the spider. These decorations are made from prey remains, egg sacs, and silk. At any one time there is natural variation in the numbers of decorations, because spiders are sometimes forced to relocate due to damage by wind, falling leaves, or large animals, and then the number of decorations increases slowly over the lifetime of a web at a particular location. These authors compared webs that contained one or no decorations with those having more than one. Wasp attacks were twice as frequent on webs in the latter category. However, of 12 attacks observed on webs with one or no decorations, nine (75 per cent) were directed towards the spider with the others being directed at the decoration; whereas, of 22 attacks observed on more-decorated webs only seven (~32 per cent) were directed towards the spiders. It is unsurprising that decorations increased observed attack rates, since both decorations and spiders appear to be conspicuously coloured in comparison to the background. However, distraction also seems plausible, since web decorations appear similar in appearance (as measured by reflectance spectra) to spiders. Further, decorations on a particular web were similar in size to the associated individual spider, despite five-fold variation in linear measures of spider size being observed. Other studies have offered evidence from other species of apparent strong similarity of appearance between spiders and decorations, as seen through the visual systems of relevant predators (Gan et al., 2010). In this system it would be interesting to explore the consequence of attacks initially directed to decorations for spider survival. In particular, it would be of value to have confirmation that attacks misdirected at decorations enhance survival, by allowing spiders to flee to hiding places in the vegetation around the web before being attacked themselves. Further, it would be valuable to have a closer exploration of whether such reduced success rates provide a net mortality benefit, given the enhanced predation rates that seem associated with higher numbers of decorations. It is also important to exercise care in interpreting results that take advantage of natural variation in decoration number with respect to the potential for confounding factors. It might be instructive to undertake experiments in this species which manipulate the appearance of the decorations, to reduce their visual similarity to the spiders, and explore the effects that this has on attack targeting by wasps.

11.3.8 Distractive behaviour by breeding adult vertebrates

It is well known that many species of ground-nesting birds exhibit conspicuous displays in response to approaching predators. These displays are generally interpreted as functioning to distract the predator, reducing the chance of the bird's offspring being discovered and predated. This area has been thoroughly reviewed by Gochfeld (1984), who commented that it was difficult to think of another field of enquiry where the ratio of anecdotal accounts to scientific study is as great as in distraction behaviour. This remains true today, although the lack of manipulative studies seems to us less problematic for this topic than for others discussed in this book. It is difficult to interpret the wealth of information collected by Gochfeld and others using any other explanation than the conventional one of distracting the predator away from the animal's offspring. Gochfeld demonstrates that individuals use such behaviour flexibly, not adopting it when a nearby predator is moving on a course that will cause them to pass harmlessly by. Similarly, the behaviour is suppressed if the predator has approached so closely that the bird could not creep away from the nest undetected before initiating the display. Gochfeld lists several instances where the predators substantially modify their trajectory in order to move towards such a displaying bird. Displays are often apparently designed to make the bird look like an easier target than usual: e.g. by apparently feigning injury or mimicking a small rodent, in either case potentially suggesting that the adult bird is an easy target that cannot simply fly away if attacked. All these points are strongly supportive of the classical distraction interpretation.

Behaviour that distracts predators away from offspring has also been reported in fish. Sticklebacks can show parental care, with males constructing a nest that they defend against conspecifics, which can show egg cannibalism. There is a body of evidence (e.g. Foster, 1988; Ridgway & McPhail, 1987; Whoriskey & Fitzgerald, 1985; Whoriskey, 1991) suggesting that males use deception rather than aggression when approached by a group of conspecifics. This deception involves moving away from the nest and simulating feeding in the substrate, such apparent feeding behaviour attracting the nearby group away from the proximity of the nest. Whoriskey (1991) demonstrates that males with empty nests as well as those defending eggs use this behaviour. He thinks it unlikely that nesting materials were being defended; hence this aspect of behaviour requires further consideration; especially since Ridgway & McPhail (1987) found that males with empty nests did not show distraction behaviour. Long (1993) reported cases of distraction behaviour in the American red squirrel (*Tamiasciurus hudsonicus*), and we would not be surprised to find reports of it in other mammals whose offspring go through a relatively immobile stage in nests accessible to predators.

11.3.9 Summary of current knowledge of the distribution of deflective traits

The clearest evidence we currently have regarding deflection comes from coloration and behaviour of the tails of lizards that can show autotomy when grasped by a predator. Evidence for a deflective function to tail coloration in tadpoles has strengthened considerably in the last decade. The possibility that the contrasting coloration of the tip of some mammals' tails can be deflecting still rests on one intriguing 30-year old study. Despite being the focus of much historical interest in deflection, evidence for its importance in butterflies and in fish has not strengthened greatly in recent years, and it may be time to shift research focus to more promising groups.

Autotomy occurs across quite a wide spectrum of animals: reptiles, salamanders, both terrestrial and sea slugs, octopuses, crabs, brittle stars, lobsters, and spiders (see Fleming et al., 2007 for distribution among invertebrates). In animals that have this ability, there would seem to be a strong benefit to deflecting attacks towards the relevant body part; hence we would not be surprised to find that there are further examples of deflective traits associated with autotomy other than lizards' tails.

11.4 The evolution of deflective traits

11.4.1 Evolutionary history

In the experiments with blue tits attacking artificial prey discussed by Vallin et al. (2011), the existence or size of the spot on the prey did not affect the willingness of the birds to attack. Birds took longer to attack prey when the background closely matched the colour of the prey than when a contrasting background was used, but this effect was unaffected by the presence or size of the spot. These results imply that the blue tits did not use the spot to aid them in prey detection and identification, even in circumstances where the matching background made prey location challenging. This may suggest that if deflection of attack selected for the introduction of non-

background matching spots, this need not necessarily impose a cost on prey that are also selected to be cryptic to their predators. However, Lyytinen et al. (2004) found that the spotted wet-season form of *Bicyclus anynana* (which features wing spots suggested by some to have a deflective function) was less cryptic against brown leaf litter than the spotless dry-season morph, suggesting a potential cost to some spot patterns in terms of higher discovery rates by predators.

Autotomy of a lizard's tail may buy escape from the clutches of a predator, but this escape comes at a cost. The tail acts as a fat store in many lizards, and the loss of this fat store may make an individual more at risk from starvation. Forfeiture of the tail can also affect future foraging, as autotomized salamanders have been found to have a significantly greater latency to strike at prey and to make fewer predatory strikes than intact salamanders (Gildemeister et al., 2017). Further, after autotomy, the lizard must invest in regrowing the tail, and until regrowth is complete this anti-predator technique is unavailable to the individual concerned. Exploratory movements, escape distance, and temperature preference can also be affected. Interestingly, while feeding rates do not increase, salamanders compensated for tail loss by preferring warmer microenvironments that might accelerate tail regeneration or healing (Bliss & Cecala, 2017). During some lizards' tail regeneration, digestive performance is also affected, as protein income needs to be maximized (Sagonas et al., 2017). There is also likely to be a cost in increased conspicuousness to predators associated with bright tail coloration (see Husak et al., 2006 for empirical support). Thus, not all lizards show autotomy and associated coloration and tail-waving behaviours that likely cause predators to deflect their point of attack towards the tail. It is quite common in those species that show brightly coloured detachable tails that this coloration is lost over ontogeny. This is likely to be associated with a greater ability to outrun or outfight predators in older, larger individuals. This shifts the trade-off in the costs and benefits of alternative strategies away from autotomy. Telemeco et al. (2011) studied hatchling Australian three-lined skinks, *Bassiana duperreyi*, and found that individuals with less ability to run fast when exposed to a predatory threat were more likely to use tail-waving behaviours. Starostová et al. (2017) looked at juvenile males of the Madagascar ground gecko (*Paroedura picta*) with and without regenerating tails, and found that tail regeneration had a negligible influence on metabolic rate; this suggests to the authors that fast-growing juveniles with unrestricted food can largely compensate for costs of tail loss and regeneration in their somatic growth without significant metabolic costs. Another reason why juveniles might have greater need of deflection would be if they are more active foragers, but adults switch to a more sit-and-wait foraging style, for which cryptic coloration may be more effective. Cryptic coloration is often compromised by movement. As conspicuous coloration is lost over ontogeny, so too are associated waving behaviours (Hawlena, 2009), again suggesting that coloration and behaviour work synergistically in deflective defence, but that the costs of such defences begin to outweigh the benefits as individuals develop. Cooper & Perez-Mellado (2004) compared three populations of lacertid lizard of the genus *Podarcis* that they suggest differed in predation risk. They argued that physiological and behavioural traits associated with improving the effectiveness of autotomy as an anti-predatory defence were more pronounced as predation threat increased. This work deserves further exploration with a wider comparative data set. It suggests that there are substantial costs associated with having the ability to use autotomy, which are paid even when the ability is not used. Such costs mean that it is unattractive for individuals that are unlikely to face predation threats to invest in having this ability. However, researchers performing such studies face the considerable challenge of quantifying the pressure from relevant predators in different situations. Cooper et al. argued that islands offer ideal study systems, since (they suggest) smaller islands have lower predator densities than larger islands. However, Itescu et al. (2017) suggest that intraspecific competition may be more important than predation in imposing selection on autotomy on islands. An added complication is that predator diversity may also be lower on smaller islands, and it is conceivable at least that those types of predators more likely to be exploited by deflective traits may not be a constant fraction of predator populations of different sizes.

One potential drawback to brightly coloured deflective signals is increased detection by predators. Cooper & Vitt (1991) explored this with a simple model. Imagine that an individual with a cryptically coloured tail is detected by a predator with probability P_d. After detection, the predator attacks and the probability of the prey escaping this attack is P_e. If these two probabilities are independent, then the prey individual's probability of being captured by a predator is:

$$P_d(1-P_e). \tag{11.1}$$

We assume that having a conspicuously coloured tail increases the probability of detection by α (subject to the condition that $\alpha + P_d < 1$), but also increases the probability of escaping by an amount β (again assuming $\beta + P_e < 1$). Conspicuous tail coloration will be favoured if the inconspicuous type is more likely to be captured by the predator: i.e. if:

$$P_d(1-P_e) > (P_d + \alpha)(1 - P_e - \beta). \tag{11.2}$$

This rearranges to the condition:

$$\beta > \frac{\alpha(1-P_e)}{P_d + \alpha} \tag{11.3}$$

Hence, conspicuousness will be favoured if β (the advantage that is gained from conspicuousness) is large or α (the increased probability of detection due to conspicuousness) is small compared to β and P_d (the probability of being detected in the absence of the conspicuous signal). Also unsurprisingly, a large probability of detection in the absence of conspicuous coloration favours selection of the conspicuous signal, as an anti-predatory deflective tactic should offer some benefit to the prey that is almost certain to be detected anyway. More interestingly, a high probability of escape following detection in the absence of conspicuous coloration (high P_e) also favours the evolution of the conspicuous deflective coloration. This suggests that tail autotomy (increasing P_e)—and perhaps associated behaviours that draw attention to the tail—probably developed before the conspicuous coloration of these body parts in some species. This prediction could be explored in a comparative survey across the reptiles. More generally, this model demonstrates that if deflective markings cause an increase in the rate at which their bearer is attacked, this need not necessarily mean that such markings will not be selected for. They can still be selected for providing their enhancement of probability of escape from an attack is sufficient to compensate for this potential cost.

The use of comparisons across populations, species, or ontogeny in the presence or prominence of putative deflective markings is greatly hampered by the potential for these markings to sometimes fulfil other (perhaps simultaneous) functions. Let us take fish as an illustrative example. Considering first predation, Meadows (1993) suggested an entirely novel anti-predatory function for eyespots. His investigation of these spots on the foureye butterflyfish (*Chaetodon capistratus*) found that they were generally oval in shape, being longer perpendicular to the main (head to tail) axis of the fish's body. Meadows suggests that predators are more successful when attacking perpendicular to their prey's longitudinal axis. Thus, at this angle, a predator might expect to see a round eye. An oval 'eye' may suggest to the predator that its angle to the prey is other than perpendicular. As a result, the predator may misdirect its strike or may delay its attack while it tries to manoeuvre into the 'correct' position. This idea is currently without an empirical foundation, but is worthy of further investigation. Misleading viewers as to the prey's direction was one of the key objectives of many camouflage patterns painted on warships during the twentieth century (Williams, 2001), discussed in depth in the next chapter on dazzle camouflage. Alternatively, Karplus & Algom (1981) suggested that fish use distance between the eyes to evaluate whether another fish is a predator, and suggest that false eyes may function to make relatively benign fish appear to be more fearsome to potential competitors or predators. Again, this idea remains untested. Experiments described by Paxton et al. (1994) suggest that caudal eyespots on predatory fish influence the inspection behaviour of potential prey. Additionally, or alternatively, to these varied putative predation-related effects, eyespots may have functions unrelated to predation such as species recognition (Uiblein & Nielsen, 2005) or mediating within-species social interactions (Gagliano, 2008).

The situation is perhaps more tractable for tadpoles than for fish. Using models, it should be possible to identify the specific traits that seem effective against ambushing dragonfly larvae through deflec-

tion, and then it should be possible to test how closely these traits correspond to morphological changes caused by exposure to cues associated with this predator in the laboratory in different species. It should also be possible to predict, and then test, the relative effectiveness of different morphs, or different species, of tadpole in terms of these trait values.

11.4.2 Evidence for costs to deflective traits

In the study discussed in section 11.3.3, Gagliano (2008) reported a particularly interesting experiment on juvenile coral reef fish of the species *Pomacentrus amboinensis* that feature a prominent dorsal eyespot. She kept some fish in laboratory conditions for a month and compared them with fish allowed to develop in their natural environment. The laboratory-reared fish had smaller eyespots. She argued that the explanation for this was unlikely to lie in any dietary deficiency or food shortage in her laboratory condition. Rather, she speculated that there was some cost to eyespot production, but also a benefit (either in reduced predation or reduced aggression from conspecific adult males), and that the juveniles have developmental plasticity that allows them to invest less in eyespot production in the absence of both types of potential antagonist in the laboratory. As discussed previously, she found no evidence for an anti-predatory function of eyespots, but exploration of the eyespot development in different laboratory populations experiencing different manipulated predatory cues might be very instructive in fish, as it has been in tadpoles.

Tadpole tails also present an attractive group for exploring the costs of deflection. The study of Touchon & Warkentin (2008) suggests that tadpoles exposed to cues from dragonfly larvae predators develop a larger, more colourful tail, but that these changes came at a cost of reduced body size. It would be useful to quantify costs and benefits more fully. It would also be interesting to explore whether this induced defence affects the timing and/or size at metamorphosis, and how effective the induced change was in affecting survival rate in as close to a natural environment as possible. Although there is currently good evidence that tadpoles can survive some attacks by dragonfly larvae especially when grabbed by the tail (Van Buskirk et al., 2003), and larvae can readily be found with tail damage (Blair & Wassersug, 2000), the longer-term fortunes of surviving tadpoles remain ripe for exploration.

Importantly, in terms of costs for lizards losing their tails, a recent 7-year, capture-mark-recapture study by Lin et al. (2017) found that, following autotomy in *Takydromus viridipunctatus*, the survival rate of tailless individuals over the next month was significantly reduced; however, the risk of mortality returned to baseline after the tails were fully grown. So, despite initial costs, this goes some way to explaining the maintenance of this trait as an evolutionarily beneficial adaption to predator–prey interactions on a longer-term timescale.

11.4.3 Linkage with other anti-predatory defences

If deflection works to free prey from the grasp of predators, this will only be of fitness value if such momentary release can be converted into longer-term escape from the predator. This might be achieved in a number of ways. In the cases of autotomic lizard tails, the predator retains the nutritionally valuable tail and so its motivation to pursue the rest of the lizard may be reduced, especially if such pursuit (with uncertain outcome) requires the predator to relinquish the tail. In other circumstances, the prey may be able to flee to a refuge. This may be what happens in orb spiders that characteristically sit in the middle of their web to allow rapid closing on caught prey. Wasp attacks misdirected at web decorations may allow the spider to flee into a crevice in the surrounding vegetation that hinders discovery and/or access by the wasp. Otherwise, escape may require simply the ability to move out of the reach of relatively immobile predators, as may be the case for tadpoles attacked by dragonfly larvae that are essentially benthic. This line of argument may explain why extensive research on deflection of birds by butterfly markings and of one fish species by another have failed to provide clear evidence of effective deflection. It may be that in these cases, it would be difficult for prey to turn momentary escape from the predator's grasp into long-term escape, although an exception to this might be in freshwater and coral-reef fish that have access to nearby physical refuges. In complex vegetation,

butterflies may be able to escape simply by dropping to the ground, if complex vegetative structure makes it inefficient for the bird to attempt to search for it.

Regarding possible deflection in caterpillars, our argument was that in this case the release from the predator's grip may be voluntary on the predator's part, and that the predator would have no motivation to repeat the attack. This argument is built on the assumption that deflection works to enhance the likelihood of taste rejection in chemically defended species, and as such provides a synergistic enhancement to such defence. However, this remains our speculative hypothesis.

It is clear that, in general, the momentary freedom that deflection can provide must be converted to longer-term escape through the use of other anti-predatory tactics, but at present this is entirely unexplored in the literature.

11.4.4 Phylogenetic studies

Kelley et al. (2013) took a comparative approach to the evolution of spots and stripes across 95 species of the butterflyfish (family *Chaetodontidae*). They suggested that if eyespots were evolutionarily labile, this might imply that they are primarily a sexually selected trait, whereas if they experience strong and consistent pressure as an anti-predatory adaptation, then they should be much more conserved. In fact, they found that such spots had evolved relatively recently and independently on at least 12 occasions. Further, they predicted that correlations between spot traits such as size, colour, and position might be suggestive of a defensive mechanism, but no such correlations were apparent. They further suggest that spots intended to misdirect predator attacks might be expected to be preferentially located in the posterior part of the body, but there was no evidence for this in their comparative data set.

Phylogenetic studies in Lepidoptera are similarly unsupportive of a defensive function. Specifically across the junoniine butterflies, eyespots have evolved multiple times, and the total number of eyespots on the wings had both increased and decreased over evolutionary time (Kodandaramaiah, 2009). Further investigation across 13 butterfly species suggested that species with similar-looking patterning do not necessarily use the same genetic pathways to express these spots (Shirai et al., 2012).

11.5 Ecology

One important aspect of the ecology of a species that will affect selection pressure for (and use of) deflection is the mixture of predatory threats faced. Not all predators will be vulnerable to potential deflection. In the extreme case, deflection would be ineffective against predators, such as baleen whales, that engulf the whole prey. One of the most promising models for the study of deflection seems, currently, to be tadpole tails. It appears, from our earlier discussion (section 11.3.4), that coloration in the tail can be effective against dragonfly larvae predators but not fish predators. We would welcome systematic comparison of variation in tail morphology between populations of tadpoles exposed to different relative threats from these two groups. Except for the case of chemically defended prey, we would expect that—as a generality—deflection will be much more appealing to prey that have some advantage in mobility relative to their predators, that allows momentary escape to be capitalized on. Against more mobile predators, we would expect that other anti-predatory traits might be relatively more effective.

Since deflection relies on the predator's visual representation of the prey, it will be affected by ambient light levels. This has already been demonstrated by Olofsson et al. (2011) for deflection induced by butterfly wing patterning. It seems very plausible to us that some traits may only be effective in deflection under certain light conditions. Anecdotally, it appears to us that the contrasting tip to the tail of, for example, the red fox (*Vulpes vulpes*) is much more salient when the animal is viewed under low-light conditions. Expansion of the work of Olofsson et al. to explore how variation in natural lighting conditions affects deflection would be valuable.

It has recently been found that the shades of blue colour in the tails of juvenile *Plestiodon latiscutatus* lizards vary across island populations with different predator assemblages. Kuriyama et al. (2016) found that tail coloration varied with the colour vision of specific predators: vivid blue reflectance occurred in communities with either weasel or snake predators, both groups of which can detect blue wavelengths; UV reflectance was much higher in populations with only snake predators—snakes can detect UV, but weasels cannot; and cryptic brown tails occurred independently on islands where birds were the primary predators, likely because birds have keen visual acuity and so a camouflaged phenotype may be more advantageous. This adaptation of different levels of tail conspicuousness indicates a deflective function of the tails against specific predators.

11.6 Co-evolutionary considerations

The empirical evidence above is suggestive that some prey may be able to reduce their risk of being captured in a predatory attack by inducing the predator to attack specific parts of their body. It seems logical to ask why predators 'allow' such deflection to occur, if it costs them prey items. Firstly, it is important to be certain that deflection is, in fact, costly to the predator. We speculated, in the example of chemically defended prey, whether in such cases deflection may enhance ease of taste rejection and thus be to the benefit of both prey and predator. However, in most cases we do expect that the deflection is costly to the predator, and thus should be selected against. However, predators of reptiles that shed their tails upon attack may not be under strong selection pressure to stop 'falling for this trick', since they do end up with a substantial meal from the tail, particularly as tails are often used as fat stores. More generally, it seems possible that for deflection involving autotomy, there may be selection on the prey to make the consolation prize of the automized body part sufficiently valuable to prevent predators being selected to stop responding to deflective traits. In the case of some reptile tails, autotomy can occur at a variety of different positions along the length of the tail. The 'consolation-prize' hypothesis would predict that sometimes breakage would occur nearer to the body than the predator's point of contact with the tail in order to offer a higher reward for allowing prey escape. This, however, remains entirely our own speculation and has not been empirically explored.

The issues discussed above notwithstanding, we expect that in most situations deflection is costly to the predator and should be subject to counter-selection to ignore the deflective traits. That deflection still exists suggests that the predator concerned has not experienced strong and consistent selection pressure to change its behaviour when its senses detect deflective traits. Following this line of argument, we hypothesize that deflection occurs because of lack of familiarity with the prey type. Hence, we would predict that deflective markings will be relatively unsuccessful when used by prey species that the predator attacks frequently, compared to prey items that it attacks infrequently. Specialist predators should not be fooled by deflective markings, whereas generalist predators have to accept such costs as a by-product of having evolved to be able to handle diverse prey types. Associated with this, we might expect that given sufficient practice, predators may be able to learn to ignore deflective traits, and thus we might expect that deflective traits are less common in species that have life-history traits (e.g. aggregation of individuals, outbreaking population dynamics) that would allow predators repeated experience of being deflected within a concentrated time interval. This line of argument is based on the assumption that we expect to see predators habituating, such that their probability of being fooled by deflective marking declines with increased exposure. Whilst this has been demonstrated repeatedly for startle signals (see Chapter 10), it has not been explored for deflective signals.

Restricting the ability of predators to counteract deflective traits (either evolutionarily or behaviourally) may have impact on aspects of the design of the deflective traits and on the wider life history of the prey. Specifically, a generalist predator may find deflective marking difficult to combat in one species encountered infrequently if similar visual cues are useful when attacking a different species, individuals of which are encountered more frequently. This argument may provide a theoretical framework for consideration as to why some styles of signal will be more effective at deflecting than others. It also raises the testable hypothesis that prey that use deflective signals will generally not be the main prey of predatory species that they successfully deflect, and that the success of deflection will be affected by predator exposure to other prey types. Specifically, we predict that deflective traits work because they provoke an out-of-context response that is effective in other circumstances. That is, we consider deflection to involve some sensory and/or cognitive traits in the predator that are retained despite the costs to the predator associated with deflec-

tion. They may be retained because they exploit some constraint of the sensory system, or because there is counter-selection because changes that reduce the risk of deflection in this context have a greater cost to the predator than the benefit of reduced deflection. These costs might manifest as reduced ability to capture other prey, or detect another valuable resources, or detect its own predators. Again, we know of no empirical investigation of the fundamental idea that deflection only occurs because a similar response to similar cues benefits the predator in another context. Admittedly, identifying all the potentially relevant contexts for any given predator would be difficult, but this idea is so fundamental to the concept of deflection that we feel deeper exploration is warranted. Much of the argument above is analogous to the frequency dependence inherent in Batesian mimicry, and consideration of the extensive empirical literature on that subject (see Chapter 9) may provide a useful guide when designing studies on deflection.

For distraction of predators away from offspring by avian parents, we should again ask why predators allow themselves to be fooled in this way if, indeed, it is a matter of fooling; the adult, if it can be caught, may make a better meal than eggs or chicks—Gochfeld (1984) cites several cases where the predator was able to successfully capture a bird exhibiting a distraction display. Of course, not all birds behaving as if they have a broken wing are faking it, and a predator that ignored such signals would forgo easy meals. However, there will be instances where a fox has a number of such breeding birds in its territory and encounters distraction displays much more commonly than genuinely injured birds. We might expect such a fox to become habituated to such displays and ignore them, or even use them to alert it to the close proximity of a nest. In support of this, Sonerud (1988) reports observations where foxes were not drawn to displaying grouse, but did change their behaviour in a way that was interpreted as enhanced searching for a nest.

11.7 Future challenges

We have suggested a number of potentially advantageous lines of further study in deflection traits throughout the chapter. We will not repeat these here, but instead highlight some other areas that we have not had opportunity to raise earlier where we feel substantial progress can be made.

The experiment of Vallin et al. (2011), discussed previously (section 11.3.1), suggests that it might be possible to introduce markings that function in deflection to cryptic prey, without the prey suffering a loss in crypsis. However, this experiment involved artificial prey presented to captive predators in a cafeteria-style format; it would be interesting to explore whether this effect can be replicated in a more realistic setting. However, we currently lack a clear, simple, and effective methodology for detecting deflection occurring with wild-living predators. A useful step forward would be to take the prey types used by Vallin et al. and modify them for presentation as artificial moths on the trunks of trees (as discussed at length in Chapter 10; see Stevens et al. (2008a) for an example of the use of this methodology), to explore whether cryptic prey with spots of the type predicted to have a deflective effect were removed at the same rate as those without (as Vallin et al.'s experiment would predict). Now that Olofsson et al. (2010) have identified different lighting conditions under which spots on a butterfly's wing can have a deflective function, it would also be useful to repeat these experiments, but with light more carefully calibrated to match naturally occurring light spectra in order to identify the ecological conditions where deflection appears most potent.

If it can be demonstrated that evidence of failed predatory attack can be reliably obtained from inspection of captured butterflies, then a capture-mark-recapture experiment may be of value where the size and or number of contrasting spots on the periphery of wings of a species are manipulated. Such manipulation would resolve concerns about confounding effects of varying exposure to predation,

provided it could be convincingly argued that the nature of the change in appearance caused by different types of markings might influence the point of attacks but would not influence the rates at which attacks occur. This might be plausible given that such naturally occurring eyespots take up a relatively small fraction of the wing area in comparison to putatively startling spot markings discussed in Chapter 10.

As discussed earlier, use of among-population, among-species, or across-phylogeny comparisons to explore factors associated with deflective marking are currently stymied by the complex of possible functions that putatively deflective eyespots might have. Hence, identification (probably in laboratory experiments) of specific features of such markings that are effective in deflection of predators would be very beneficial in allowing comparative work to focus particularly on these features. Although the strongest evidence for deflection comes from autotomic lizards' tails, there are non-trivial ethical and practical challenges in exploring anti-predatory traits manipulatively in such large vertebrates. For this reason, we should fully exploit artificial model prey when studying the traits that may cause deflection of predators. It has also been suggested that 'redirection' may work in combination with deflective autotomy in lizards, such that longitudinal-striped patterns on anterior body parts may redirect attacks towards less vulnerable posterior parts during motion, for example, the autotomous tail (Murali & Kodandaramaiah, 2016); this perhaps warrants further study. Recent evidence that the web decorations of at least one orb spider appear to have a distractive function is worth further exploration, because the deflective structure being separate from the body of the animal may make experimental manipulations much more practical than other systems.

CHAPTER 12

Dazzle camouflage

12.1 Camouflage in motion?

Camouflage using background matching, disruptive, and countershading principles can be highly effective strategies for motionless animals, but experiments have shown that movement 'breaks' camouflage (Hall et al., 2013; Ioannou & Krause, 2009). Reflecting on our own experience of crossing roads, playing sports, or observing wildlife, it is clear that moving objects attract attention and 'pop out' from the background. Computationally, the task of segregating a moving object from a textured background is easy because its edges are readily defined. From a prey's point of view, the salience of movement is helpful for detecting approaching predators, but the prey's own movement can cost it in apparency to predators (Skelly, 1994).

The simplest solution to mitigating this risk is to remain motionless as much as possible, or to move only when it is dark, and indeed these are common tactics. However, movement is vital for most animals, necessary to perform essential functions such as finding food and mates. So how do animals reduce predation risk while moving if standard camouflage strategies are ineffective? Anti-predator adaptations proposed to be less dependent on remaining motionless include living in groups, chemical defence, fighting ability, or fleeing ability (Chapter 5), and signals that advertise these (Chapters 6 and 8), as well as perhaps startle and deflection defences (Chapters 10 and 11). These give an animal increased freedom to roam, collect resources, and find mates compared to strategies that work by preventing detection

(Speed et al., 2010). However, they do not exploit movement itself to gain protection, nor are they necessarily dependent on movement to function.

This chapter discusses a putative anti-predator adaptation that works to reduce predation by camouflaging movement itself. Dazzle camouflage comprises coloration that interferes with predator perception of prey speed and trajectory. Accurately estimating speed and trajectory is essential for any predator that needs to know where its prey is likely to be in the near future so that it can position itself to intercept. The dazzle hypothesis is that prey coloration can interfere with these judgements. The idea is appealing because physical models of pursuit hunting show that relatively small errors in estimates of future position can dramatically adversely affect capture success (R. Wilson et al., 2015). In instances where the predator has limited ability to alter its attack path, the prey may well not be there when they arrive. Even if the attack path can be updated as better estimates of prey position are obtained, the longer curving pursuit path that results, and potential decrease in speed as a consequence of having to turn, makes the physical task of catching the prey harder for the predator.

The idea that coloration may be able to affect judgements of speed and trajectory to facilitate predator evasion was first suggested by Thayer (1909) in tandem with what is now known as disruptive camouflage under the banner of 'ruptive' coloration. These two camouflage strategies share important similarities; both aim to generate errors in specific aspects of predator perceptual processing, and maximizing pattern contrast is thought to be

Avoiding Attack: The Evolutionary Ecology of Crypsis, Aposematism, and Mimicry. Second Edition. Graeme D. Ruxton, William L. Allen, Thomas N. Sherratt, & Michael P. Speed, Oxford University Press (2018). © Graeme D. Ruxton, William L. Allen, Thomas N. Sherratt, & Michael P. Speed 2018. DOI: 10.1093/oso/9780199688678.001.0001

important for both (section 12.3.2). However the function of both strategies is quite different; preventing detection and recognition of prey outline and shape for disruptive coloration vs preventing estimation of prey speed and trajectory for dazzle coloration, so they are now treated as separate strategies. Some definitions of dazzle camouflage include errors of target size or shape created by moving patterns under the category of dazzle (Scott-Samuel et al., 2011). We feel that despite the role of prey movement, these effects would be better classified as disruptive unless they lead to misperception of speed or trajectory (which may commonly be the case, see Figure 12.1). The likely form of dazzle and disruption, as well as the ecological circumstances in which dazzle evolves compared to disruption are also thought to be quite different (sections 12.3 and 12.6).

While research questions, experimental paradigms, and response measures vary, experimental investigation of the dazzle camouflage hypothesis has thus far almost exclusively utilized human 'predators' tasked with capturing artificial prey on computer screens. Only two studies have investigated dazzle effects in non-humans (Hämäläinen et al., 2015; Santer, 2013). Overall, the basic idea that animal coloration can impact perceived speed and trajectory is quite controversial. Several studies find support (Hughes et al., 2014; Scott-Samuel et al., 2011; Stevens et al., 2011; Stevens et al., 2008b), while others find no difference between dazzle and non-dazzle targets (exp 1. Hughes et al., 2015; von Helversen et al., 2013), or even a disadvantage for moving targets designed to dazzle relative to uniformly coloured controls (exp. 2, Hughes et al., 2015). Other experiments find decreased capture success for patterns designed to dazzle compared to other similar patterns designed without dazzle principles, but also find that uniformly coloured background-matching controls survive in motion just as well as dazzle targets (Hughes et al., 2014; Stevens et al., 2008b). Thus currently the experimental evidence for the effectiveness of dazzle is suggestive but still requires further work. In section 12.3 of this chapter we aim to identify where firm and less firm conclusions can be made on the role of pattern contrast, pattern orientation, and internal object motion, before moving on to consider the evolution (12.4) and ecology (12.6) of

dazzle camouflage. First though we discuss illustrative examples of putative disruptive effects in biological and military design.

12.2 Examples of dazzle

The only objects we know that have been selected to create dazzle-type effects are warships operating in World Wars I and II that were painted with high-contrast geometric patterns as a defence against enemy torpedoes. One of the difficulties faced by dazzle theory is identifying and confirming examples of dazzle-type effects operating in nature. The groups displaying the most likely examples of dazzle are zebra, snakes and lizards, and cephalopods. These examples are discussed in turn along with the evidence for and against their dazzle function.

12.2.1 Dazzle ships

In World War I German submarines posed a major threat to Allied shipping, especially in the later stages of the war following the German campaign of unrestricted submarine warfare. Strategies were sought to minimize shipping losses, leading to the involvement of artist Norman Wilkinson in designing ship camouflage and the formation of the British Dazzle Section in 1917 (Behrens, 1999). Since effectively camouflaging large objects that operate in open but very varied lighting conditions is near impossible using conventional background matching, disruptive, and countershading strategies, ships were painted with high-contrast dazzle patterns that aimed to mislead the enemy about a ship's speed, heading, and range. At the time submarines sought to fire torpedoes that travelled at around 15m/s from between 300 and 1000 m distance from the target, aiming at a point ahead of the ship's course on an interception path. To navigate to a good position and then fire torpedoes in the right direction, submariners needed to estimate how far away the ship was currently, how fast it was travelling, and on what bearing. To make these calculations, range finding equipment at the time relied on matching images of the target acquired from two different horizontal locations. Any effect of coloration on submariners' ability to do this might have led to adoption of suboptimal firing positions and targeting errors that resulted in torpedo misses.

Thousands of ships were painted with dazzle patterns by the UK and US. As well as high-contrast markings designed to distort perception of speed and trajectory, designs included other deceiving features such as false bow-waves, attempts to make the stern appear like the bow, and efforts to make backward-leaning funnels appear forward-leaning (Figure 12.1).

It is a topic of contemporary debate whether these efforts were effective in reducing shipping losses (Forbes, 2011). Most pre-implementation tests of model ships found some supporting evidence, for example an average trajectory error of between 20° and 30° for twelve ship models (Blodgett, 1919), but experimental protocols were rudimentary and difficult to draw firm conclusions from. There is plentiful anecdotal supporting evidence, including that of the King of the United Kingdom, George V, that

Figure 12.1 Images of WW1 dazzle ships: USS Wilhelmina (1918) and USS Nebraska (1918). The repeating pattern along the side of USS Wilhelmina reduces in size in order to exploit size constancy mechanisms. If the observer makes the incorrect perceptual assumption that the patterns are all the same actual size, then the larger patterns will appear closer than the smaller ones, affecting perceived trajectory.

observers found the patterns distorting. However some German officers reported being unfazed as they generally used ship's masts rather than hulls for targeting. Post-war analysis of the effects of dazzle camouflage on shipping showed some suggestive effects of dazzle, such as dazzle ships being missed by torpedoes more, hit towards fore or aft rather than amidships, and overall sunk at a reduced rate (Behrens, 1999). However, these trends may have been confounded by the larger average size of dazzle ships compared to non-dazzle ships, or to changes in military tactics throughout the deployment period (Behrens, 2012; Forbes, 2011). Furthermore, ships were often painted with dazzle patterns when being overhauled and fitted with other modifications and improvements, making separating out the effect of the dazzle camouflage even more challenging. Recent work has aimed to assess whether 3D computer models of dazzle ships are any harder to recognize than uniformly coloured ships using machine learning techniques, a quantitative approach that has promise for the future if it is able to include models of motion processing (Bekers et al., 2016). Dazzle camouflage continued to be applied to ships in World War II but its use faded out as developments in radar technology and targeting systems made the visual effects of dazzle redundant (Williams, 2001). In modern asymmetric wars where adversaries may use low-tech targeting or bomb-triggering devices, military dazzle may find a new use (Scott-Samuel et al., 2011).

12.2.2 Zebras

The idea that zebra stripes work to 'confuse' predators while the zebra is in motion is a relatively early theory (Kruuk, 1972) for a phenotype that attracts many theories (Caro, 2016). This is a plausible possibility; zebras have the high-contrast patterning generally thought to promote dazzle effects (section 12.3.2), and as herd animals they might benefit from an interaction between dazzle effects and confusion effects created when targeting individuals in groups (section 12.6). They also have visually oriented lions and hyena as their main predators, and as large animals that live in the open, much like ships they might find camouflaging via other mechanisms difficult. The motion

dazzle hypothesis has been studied in zebras by modelling the motion signals created by moving zebra stripes (How & Zanker, 2014). This analysis, discussed in more detail in section 12.3.1, shows that zebra stripes create strong motion signals in directions other than the true direction of movement, supporting the hypothesis that zebra stripes make movement processing difficult. While most recent work focusses on exploring whether zebra stripes function as protection against biting insects rather than mammalian carnivores (Caro, 2016), this idea is not necessarily mutually exclusive with the dazzle hypothesis; rather than (or in addition to) lions and hyenas, stripes may target the motion processing of much smaller but perhaps no less fitness-reducing horse (tabanid) and tsetse (glossinid) flies instead.

12.2.3 Squamate reptiles

Many species of snake feature high-contrast striped patterns that have been suggested to confuse predator perception of speed and trajectory (Brodie, 1992; Jackson et al., 1976). In support of this hypothesis, a comparative analysis of 171 species of snake by Allen et al. (2013) found high-contrast patterns were more common in fast-moving species, that stripes running longitudinally were associated with small species that had rapid escape behaviours, and that stripes running transverse to the body were associated with erratic movements. These behaviours link colour and movement in the way predicted by dazzle camouflage theory (Hogan et al., 2016a). The very simple body shape of snakes and their undulating movements create potential for movement camouflage effects that may not be effective in animals with more complex body forms. Tracking forward motion of snakes may be difficult because of the lack of suitable reference points on a snake's cylindrical body. Longitudinal stripes do not add reference points that track forward motion, but may create local motion signals in directions different to the global heading of the snake as a result of undulating movements. This may be particularly effective if the snake is viewed through a partially occluding foreground, effectively creating a version of the barber pole illusion (section 12.3.1; Wallach, 1935).

As in snakes, lizard longitudinal stripes have a negative relationship with body size at the interspecific level, potentially because dazzle effects might only be large enough to displace attacks to less vulnerable tails when lizards are small, an idea further suggested by an association between stripes and caudal autotomy across species (Murali & Kodandaramaiah, 2017). In this respect the outcome of dazzle is similar to deflection, though it is achieved via a different mechanism. Matching results in snakes, in lizards longitudinal stripes are more common on mobile foragers that operate in the open (Halperin et al., 2017). A similar pattern has been observed at the intraspecific level in the Be'er Sheva fringe-fingered lizard *Acanthodactylus beershebensis* (Hawlena, 2006). Overall there are clear associations between patterning and the movement ecology of squamate reptiles, with the similarities between lizards and snakes suggesting that the effectiveness of longitudinal stripes in motion is not necessarily dependent on snakes' simple body plan.

12.2.4 Cephalopods

The rapid colour-changing ability of many cephalopod species allows for different patterns to be displayed whilst still and in motion. This ability to, in principle, overcome any disadvantage dazzle camouflage might have whilst motionless, makes cephalopods a good candidate for displaying dazzle camouflage.

However current evidence suggests that this is not what cephalopods do. Zylinski and colleagues (Zylinski et al. 2009a, b) found that common cuttlefish *Sepia officinalis* used relatively low-contrast patterns while in motion compared to the high-contrast disruptive patterns they are capable of displaying. This is contra to the prediction that dazzle works best when pattern contrast is high, though this is a contentious issue (section 12.3.2). A further complicating factor is that this investigation was done in the absence of predator cues. Additionally colour change during movement may work to improve the camouflage of the animal once it resettles rather than during movement, as the predator has to find an object with a different appearance to the one it saw whilst it was moving.

Cephalopods are one of the few groups capable of displaying the internally moving 'dynamic dazzle' patterns (see section 12.3.4) investigated by

Hall et al. (2016) and Hughes et al. (2017), where waves of moving pattern on an integument can strongly influence the perceived motion of the overall object. The function of these 'passing clouds' in cephalopods is currently unclear; they are used in the vicinity of conspecifics, predators, and prey, especially following capture attempts. Thus they may or may not simultaneously work as either a conspecific signal, motion camouflage, an elusiveness signal to predators, or a signal to startle prey (Laan et al., 2014; Mather & Mather, 2004).

12.2.5 Other groups

Dazzle may well operate in other aquatic systems. In cichlids horizontal stripes are associated with shoaling behaviour (Seehausen & Van Alphen, 1999) and while other groups have not been systematically investigated, there are several other examples that conform to this relationship (e.g. bluestripe snapper, *Lutjanus kasmira*). Whether these are examples of adaptive convergence on the zebra stripe phenotype in group-living animals, and whether they have dazzling effects on underwater predators, is as yet unknown.

In plants a creative but untested hypothesis proposes leaf patterns blowing in the wind produce dazzle effects, making it hard for large herbivores to target individual leaves and insects to land on them (Lev-Yadun, 2014b).

The relationship between elongated patterning and movement speed observed in squamate reptiles has also been observed at the intraspecific level in a frog *Dendrobates tinctorius* (Rojas et al., 2014a); it will be interesting to find out how general this rule is.

Little investigated is the potential for dazzle effects in insect coloration (Théry & Gomez, 2010). Flying insects might be ideal beneficiaries of dazzle camouflage as at relevant attack distances they move through predators' visual fields rapidly, behaviour thought to be necessary for dazzle effects to occur to a meaningful degree (Scott-Samuel et al., 2011). However, while research has shown in locusts that responses of neurons tuned to moving stimuli are affected by dazzle-type patterns (Santer, 2013), dazzle camouflage in insect prey has not yet been investigated. An interesting suggestion made by Hall et al. (2016) is that insect multilayer iridescence, which produces a striped pattern that would be perceived as moving when the predator and prey are in motion, perhaps creates a dynamic dazzle effect akin to cephalopod moving clouds (section 12.3.4).

12.3 How does dazzle camouflage work?

12.3.1 How can coloration affect perception of speed and trajectory?

Very simplistically, object movement is perceived following the perceptual interpretation of edges moving across the visual field, after accounting for the observer's body, head, and eye movements. Dazzle effects will occur if edges formed by colour patterns on an object can create incorrect interpretations of object movement. This section discusses how examples of motion illusions and our understanding of motion-processing systems illuminate how this can take place.

The direction a line moving across a visual receptive field is going in is inherently ambiguous; many combinations of speed and trajectory can create the same responses in neural cells tuned to moving stimuli (Adelson & Movshon, 1982). To resolve this ambiguity the mammalian visual system utilizes additional information, especially the changing position of terminal points on lines (end-stops), relative movement of non-moving foreground and background objects, and the orientation of patterning on the moving object. The importance of these factors was demonstrated in a series of experiments that introduced the aperture problem, most famously illustrated by the barber pole illusion, where diagonal stripes moving horizontally through a vertically elongated aperture are generally perceived as moving vertically (Wallach, 1935; Wuerger et al., 1996). Wallach's investigations, supplemented by more recent results, show how the orientation of high-contrast striped patterns passing through an aperture can influence the perceived speed and trajectory of object movement, i.e. create dazzle effects (Fisher & Zanker, 2001; Sun et al., 2015).

Another form of motion illusion resulting from object pattern is known as the wagon wheel effect, named after the Western movies it is sometimes noticed in. Here a slowly rotating stimulus is perceived as moving in the correct direction, but as speed increases the direction of rotation is perceived

to reverse. This results in the position of one spoke of the wagon wheel correlating with the position of a different spoke across movie frames (though it can also occur in continuously presented stimuli; Purves et al., 1996), creating interpretations of wheel motion consistent with rotation in the opposite direction. The effect is created by antagonistically oriented motion detectors generating rival perceptual interpretations of motion direction as a result of object pattern (Kline et al., 2004; VanRullen et al., 2005).

The barber-pole and wagon wheel effects show that linear and rotational shifts in patterned stimuli can create incorrect perceptions of motion, and give insight into how this happens as a result of a failure of motion detection mechanisms to correctly interpret the changing position of contours as the retinal image changes. How & Zanker (2014) demonstrate that both illusions can be replicated by a simple computational model of biological motion detection

(a) Original frame

(b) Motion map

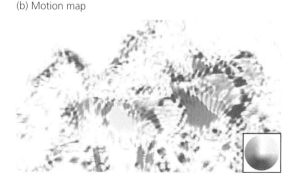

Figure 12.2 Motion vector distribution of a herd of zebras viewed by the 2DMD computational model of motion detection (How & Zanker, 2014). The colour indicates the trajectory of the motion signal and the saturation the strength of the motion signal. A herd of zebra moving in the same direction creates a patchwork of motion signals in different directions (see Plate 19).

systems (Borst et al., 2010). They go on to show that similarly misleading motion signals are created when viewing images of zebra stripes that shift position on the retina following either zebra movement or saccadic eye movements. Movement of the rump generates motion signals 50–60° offset clockwise from the true direction of movement, similar to the size of displacement generated by the barber-pole illusion. Results further show that different regions of the body of zebras such as the flank and rump would be perceived as moving in different directions, creating further confusion about overall movement (Figure 12.2). In contrast the same movements of plainly patterned horses were observed to create no misleading or incoherent motion signals.

In summary, there is reasonable evidence that coloration can affect the accurate estimation of speed and direction by exploiting features of how motion is processed locally and globally. In the next two sections we discuss specific features of coloration thought to be particularly important in influencing these judgements: pattern contrast, texture, and orientation.

12.3.2 The form of dazzle camouflage: pattern contrast

Dazzle is often defined as a high-contrast pattern that interferes with perceived speed and/or trajectory in order to reduce predation. Thus the definition includes both phenotype and function (Hogan et al., 2016b). Both theoretical frameworks which promote definitions based on function (Caro & Allen, 2017; Merilaita et al., 2017), and empirical evidence demonstrating either reduced capture success for low-contrast moving patterns (Stevens et al., 2011), or no difference between high- and low-contrast patterns (Hall et al., 2016; Hogan et al., 2016b), suggest the 'highcontrast' element of the definition may be unnecessary. Of studies testing effects of contrast in dazzle camouflage, only Scott-Samuel et al. (2011) find high contrast to be important.

There is some possibility that conflicting results are due to features of experimental design. While all studies used psychophysical techniques that tasked human predators with observing simplified virtual prey on a monitor, they differed in their response measures: Scott-Samuel et al. (2011) asked subjects which of two targets was moving faster without

requiring physical 'capture', whereas Hogan et al., (2016b) recorded tracking errors made by subjects as they followed targets around a screen, and Stevens et al. (2011) recorded target capture success. Thus it may be that perception of speed is affected by pattern contrast, but that the effect is obscured when using measures that require production of a gross motor response to capture the target.

Alternatively it may be that dazzle is only effective at moderate levels of contrast with no advantage of increasing contrast further. Only Scott-Samuel et al. (2011) tested very low-contrast (6.25 per cent) patterns against black-and-white (100 per cent contrast) patterns, whereas Hogan et al. tested at 50 per cent and 100 per cent and Stevens et al. tested 'intermediate grey' and 100 per cent contrast patterns. Clearly more work is needed on resolving this issue as uncertainty presents a risk for stimulus design in experiments investigating other aspects of dazzle camouflage, which for now seem to have settled on using high-contrast patterns as dazzle stimuli (e.g. Hogan et al., 2017; Hughes et al., 2017; Murali & Kodandaramaiah, 2017).

The evidence against an advantage of high pattern contrast in creating dazzle effects is surprising. Most theoretical models of motion detection depend on the contrast of the input (Adelson & Bergen, 1985; Simoncelli & Heeger, 1998; Van Santen & Sperling, 1985), thus, all else being equal, maximizing contrast should maximize the strength of misleading motion signals created by patterning (although see Georgeson & Scott-Samuel, 1999). It is fairly well established in human psychophysical experiments outside a dazzle context that low-contrast patterns are perceived as moving slower than high-contrast patterns at low speeds (< 4°/s), but faster at higher speeds (Stone & Thompson, 1992; Thompson, 1982). However, in dazzle camouflage experiments these results have not been replicated. Scott-Samuel et al. (2011) measured the perceived speed of a range of geometric patterns and found that the only targets that differed in perceived speed compared to uniform controls were fast-moving (20°/s) high-contrast zig-zags and checkerboards. However, these patterns were perceived as moving *slower* than low-contrast and uniform patterns, opposite to the results of Thompson (1982). Overall current evidence suggests that the relationship between pattern contrast and perceived speed is not a simple positive correlation consistent across tasks. The issue needs further investigation as resolution has important implications for the predicted form of dazzle camouflage, both in terms of whether contrast is important at all, and whether dazzle camouflage has been selected to cause over- or under-estimation of speed.

12.3.3 The form of dazzle camouflage: pattern texture and orientation

The examples of motion illusions and the species suggested as utilizing dazzle camouflage indicate that repeating patterns, especially striped patterns, should be the prime candidate for inducing dazzle effects. Studies have focussed on investigating the effects of stripes running parallel, orthogonal, or diagonal to the direction of movement.

The theoretical background for this focus is based on our understanding of the architecture of receptive fields and motion-processing mechanisms. Neurons sensitive to directional movement are also often only sensitive to certain stimuli, usually edges orthogonal to the direction of movement (Henry et al., 1974; Maunsell & Van Essen, 1983). Thus for a given stimulus movement, stimuli textured with parallel, orthogonal, or diagonal stripes should create different motion signals.

Several experiments have aimed to establish what sort of pattern makes the most effective dazzle camouflage. Overall no consistent result emerges, so here we briefly review the current state of knowledge. Von Helversen et al. (2013, exp 1) and Hughes et al. (2015, exp. 1) find targets with stripes parallel to the direction of motion are captured more easily than orthogonal or diagonal stripes or uniformly coloured controls. However, other dazzle experiments find no difference between parallel or orthogonal stripes (Hughes et al., 2015, exp. 2; Scott-Samuel et al., 2011; Stevens et al., 2008b; von Helversen et al., 2013, exp. 2), while others find parallel stripes hardest to track (Hogan et al., 2016a, b, 2017). This mirrors mixed findings from human observers in non-camouflage contexts, with some results indicating that errors in speed are greatest when patterns are orthogonal to the direction of movement and others finding the opposite result (Castet et al., 1993; Georges et al., 2002; Scott-Brown & Heeley,

2001). Indeed, it may be that stripes of any sort are not particularly effective dazzle camouflage. As previously mentioned, Scott-Samuel et al. (2011) found that only check and zig-zag treatments affected speed perception (Figure 12.3).

The study reporting the clearest effect of stripe orientation is Hughes et al. (2017), which focusses on whether dazzle patterns produce trajectory errors rather than speed errors. In these experiments diagonally striped stimuli move across a screen in a straight line on different trajectories before being occluded. The observer's task is to indicate at which point along a vertical line the target will intersect, with the response measure being the error from the true intersection point. Thus this task reflects the judgement a pursuit predator has to make about its prey's future position. Results show that when the target is moving at moderate to high angular speeds ($> 10°/s$) there is a remarkably strong average effect of about $5°$ in heading for obliquely striped patterns compared to vertically striped patterns, with errors in the direction that the leading diagonal is pointing towards (Figure 12.4). How & Zanker's model also finds that diagonal stripes are most the most likely pattern orientation to create a consistent error in perceived direction of movement; however, the error is in the opposite direction, perhaps as a result of measuring local rather than global motion perception. A further reason to predict an advantage for diagonal stripes is the higher proportion of edge detectors tuned to vertical and horizontal orientations compared to oblique orientations (Appelle, 1972). Compared to the weak or mixed results in studies focussing on speed errors, or a combination of speed and trajectory errors, the results of Hughes et al. suggest that the main effect of pattern orientation is on estimation of trajectory. Confirming this finding with a broader variety of oriented stimuli and in non-human animals should be a priority. How & Zanker's model suggests that it is likely, and that a further consequence of orientation-dependent trajectory errors is that stripes at a range of orientations over the whole object should create confusing multidirectional motion signals.

Recent discussion suggests that processing dazzle patterns at different orientations may be achieved by fundamentally different motion detection mechanisms, with stimuli patterned with orthogonal stripes

Figure 12.3 Experimental stimuli used in Scott-Samuel et al. (2011). The perceived speed of stimuli (a)–(e) were compared to that of (f), a 1D Gaussian: (a) horizontal, (b) vertical, (c) zigzag, (d) check, (e) plain. Stimuli (c) and (d) were perceived as moving slower than (f) at high speeds. All stimuli were seen to move at the same speed at low speeds.

processed by classical inference on the 'intersection of constraints' of two or more motion detectors tuned to different orientations, but the motion of stimuli with stripes parallel to the direction of movement being processed using motion streak processing (Hughes et al., 2015, 2017). Motion streak processing, an example of which can be seen by waving a sparkler around, is achieved by temporal integration of samples of a moving object and has been demonstrated for a range of species (Geisler et al., 2001). While orthogonally striped stimuli increasing in speed will lead to light and dark stripes being integrated (blurred), which only has the effect of reducing pattern contrast rather than producing smeared motion streak signals, parallel striped stimuli should create 'speedlines' that indicate direction and vel-

ocity of motion (Burr & Ross, 2002). Hughes et al. (2017) concluded that utilization of motion streak cues may underlie the target trajectory errors they observed, as the direction of errors reversed at low target velocities when motion streak signals are not produced. Interestingly, this means that at low velocities the trajectory errors found by Hughes et al. match those predicted by How & Zanker, suggesting that different dazzle effects may operate at different velocities.

12.3.4 Dynamic dazzle

Although few animals have the capability of creating moving patterns on their integuments, and current evidence suggests that those that can (e.g.

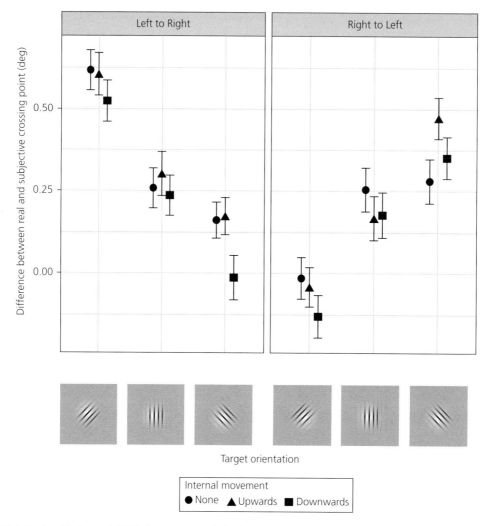

Figure 12.4 Results of Hughes et al. (2017) Observers were tasked with estimating the point at which the target would cross a line. Results show how perceived trajectory is affected significantly by stripe orientation, but minimally by internal movement of stripes. Error bars are ± 1 bootstrapped standard errors.

cephalopods) do not do so to dazzle, the question of whether internally moving patterns can create distortions of perceived speed and trajectory has been the topic of recent research interest, though the focus is more on potential military rather than biological systems.

In a series of experiments that follow the same basic paradigm as Scott-Samuel et al. (2011), but using stimuli with internally moving striped patterns, Hall et al. (2016) show that these 'dynamic dazzle' patterns can create much larger distortions of perceived speed than stimuli with internally

static patterns, up to about 18 per cent error compared to the 7 per cent maximum error observed in Scott-Samuel et al. The basic finding is that internal stripes moving in the same direction as the target increase perceived speed, while stripes moving in the opposite direction decrease perceived speed. The size of error depends on the relative speeds of target and stripe motion, with the largest effects found when they match.

The study by Hughes et al. (2017) introduced in the previous section also examines targets with internal pattern movement. This complements Hall et al. by measuring the impact of dynamic dazzle on perception of target trajectory rather than speed. The effect of moving stripes was to create trajectory errors in the same direction as the motion of internal stripes (e.g. perceiving a SW trajectory of a W moving target when internal stripes were moving to the SW); however, the magnitude of the effect was quite small (0.1°), especially compared to the effect of stripe orientation discussed in the previous section (0.5°). The authors speculated that the effect results from failures in how local and global motion signals are integrated (Tse & Hsieh, 2006).

12.4 The evolution of dazzle

If we make the (almost certainly incorrect) assumption that all the examples identified in section 12.2 are of dazzle camouflage, and that there are not many more examples of dazzle to discover, then we can draw two tentative conclusions. First that dazzle camouflage is a relatively rare anti-predator adaptation, and second that it has evolved multiple times independently, but when it does so it generally remains quite restricted. Of course these conclusions may lead from the fact that identifying dazzle camouflage and how it is perceived by relevant observers in nature is difficult, and attempts to do so are in their infancy. Establishing dazzle effects created by iridescence in insects would, for example, alter this perspective. Similarly, the pool of potentially dazzling species would broaden if it became possible to make a firm conclusion that high contrast is unnecessary to produce dazzle effects. It would also be instructive to attempt to identify the evolutionary pathways that have led to potential instances of dazzle, as has been done for zebra (Caro et al., 2014).

It is straightforward to imagine scenarios for the initial evolution of dazzle. For example, disruptive patterns could be selected for increased regularity as a result of predation against moving individuals. Another unexplored possibility is that conspicuous patterns used by mimics of aposematic species could be co-opted for dazzle following a change in the model's presence or phenotype. For now though we can only assume that the benefits of dazzle outweigh the costs for relatively few species, and therefore that the conditions that select for dazzle are

quite narrow. We explore what these might be in the next two sections.

12.5 The costs and benefits of dazzle

The major potential upside of dazzle camouflage is that it can offer protection against predation while moving without having to invest in defences or suffer predation from uneducated predators, as with aposematic strategies, or restricting movement to night time when vision is less important (Merilaita & Tullberg, 2005).

While there is conflicting evidence over the role of high-contrast and regular patterning in creating dazzle effects, assuming that dazzle does select for these phenotypes then it is likely to be balanced by the cost of decreasing camouflage performance while the animal is not moving (Stevens et al., 2011). In terms of background matching, natural backgrounds do not often feature regular patterns, nor do they generally contain very high contrast, thus these traits are likely to reduce crypsis. Like dazzle, disruptive camouflage is also thought to select for high contrast (but not pattern regularity); however, maximum contrast is constrained by the range of tones present in the background (Fraser et al., 2007) and this is possibly also the case for dazzle. Another solution to this problem is minimization of the time spent motionless. For many species movement is expensive and rest essential, but for others, such as pelagic fish, costs of movement may be relatively low, and there are many other advantages to keeping moving too. A final solution is utilization of hiding places where visual camouflage is not relevant, such as caves and burrows. In these circumstances dazzle may be effective when the animal is active and irrelevant when it is hiding. Exploration of these issues really requires resolution of the role of pattern contrast in dazzle as a prerequisite (Hogan et al., 2016b).

12.6 The ecology of dazzle

In section 12.2 we described several examples of potential dazzle camouflage and in doing so highlighted a number of ecological factors that are commonly implicated in species that might pursue a dazzle strategy. The main ones include living in groups and rapid movement. We have already discussed how dazzle effects have generally been found to increase with increasing target speed (e.g. Scott-Samuel et al., 2011), so it makes sense that they will only be effective on animals that can move across the visual field of predators sufficiently rapidly, an association that has been demonstrated in squamates (Allen et al., 2013; Halperin et al., 2017), but not yet established as a more general pattern in other taxa.

Zebra and shoaling fish are two groups of potentially dazzling prey that live in dense groups. Is this a coincidence? While the issue has not yet been extensively investigated in a comparative context, the motion modelling of How & Zanker (2014) suggests that the effect of complex motion signals created by moving zebra stripes should be enhanced when zebra are seen against a background of other moving zebra, as they might be from the viewpoint of a hunting lion. This might make it difficult to pick out a single zebra as the selected prey, both because of uncertainty about the direction of each zebra's individual motion, but also because different parts of separate but adjacent bodies may mistakenly be grouped together if they are perceived as moving in the same direction via the Gestalt principle of common fate. Several experiments have begun to follow up on this issue, aiming to find out if dazzle effects interact with confusion effects experienced by predators attempting to target individuals in groups, a well-established phenomenon where capture success decreases with increasing group size (Krause & Ruxton, 2002) or density (Ioannou et al., 2009).

Hogan et al. (2016a, b, 2017) have shown that in the errors observers make while tracking targets, dazzle patterning (which in these experiments are striped patterns) interacts with the group size of moving distractors with the same appearance, with larger groups leading to more marked tracking errors for dazzle patterns. This suggests a specific benefit of dazzle when moving in groups. However the result was not replicated by Hughes et al. (2015) who found instead that striped prey in groups of 6 or more were caught *more* often than uniform prey. One difference in experimental setup that may underlie differences between results is that in the Hogan et al. studies targets move unpredictably,

whereas in Hughes et al. targets move linearly which might make following a single target among distractors more straightforward.

Extending the basic finding of dazzle facilitating survival of individuals living in groups, Hogan et al. (2017) looked at the situation where individuals in a group vary in speed—something that occurs in natural groups and potentially undermines the protection afforded by the confusion effect because it allows individuals to be identified and separated on the basis of their unique speed (the oddity effect). Dazzle, because it interferes with speed and trajectory perception, might in turn counter the oddity effect. However, their experiments found no evidence for this; the oddity effect made tracking individuals significantly easier, and the size of effect was equal for striped and non-striped patterns.

Examining this issue from a comparative perspective, Seehausen & Van Alphen (1999) find stripes parallel to the direction of general motion in cichlid fish are correlated with feeding on fish and shoaling behaviour, suggesting a phenotype that works well in groups of moving animals; however, a similar association between stripes and sociability in butterflyfish was not supported (Kelley et al., 2013), so it is unclear how general this this pattern is.

12.7 Future challenges in dazzle camouflage research

A message that should have come through clearly in this chapter is that the dazzle camouflage hypothesis is still some way from being well established as a mechanism by which animals can achieve camouflage while in motion. While some experiments present clear effects of target patterning on perceptions of speed or trajectory, other similarly designed and powered experiments do not. What's more, these experiments take place in carefully controlled lab situations, generally using human predators and highly artificial stimuli. We have little idea whether dazzle effects occur in any natural situations, or even if they plausibly could. This is a moderately concerning situation which may be resolved by further systematic experiments examining outcomes over a wide range of conditions: however, we suggest that a current priority should be experimental investigation of potential natural dazzle systems in non-human animals.

There is a disconnect between the large errors that the modelling of How & Zanker (2014) predicts, and the relatively small errors observed in behavioural studies. One explanation for this is that human observers use higher-level object-tracking mechanisms to overcome the illusions predicted by the lower-level modelling of How & Zanker. Future work may be able to investigate this, and examine whether there are ways coloration can disrupt object-tracking processes.

The issue of prey speed requires urgent attention. It may be that to produce meaningful dazzle effects in natural predator–prey interactions, many prey simply cannot move across their predator's visual fields fast enough. Using their observed effect sizes to calculate the 'real-world' impact of dazzle camouflage, Scott-Samuel et al. (2011) estimated that a rocket-propelled grenade fired from 70 m away at a vehicle painted with dazzle patterns and moving at 90 km/h would make a targeting error of about 90 cm compared to firing at a uniformly painted vehicle, indicating some military application. It would be highly beneficial to extend this analysis and formally model different dazzle effects in natural predator–prey interactions, incorporating prey pattern phenotype, movement speed, typical attack distances, predator contrast sensitivity and acuity, and the magnitude of predator error that would give prey a chance to escape. The results should give experimental investigations in the lab and field indication of whether dazzle effects observed are likely to translate to predator–prey interactions in the wild, and whether observational studies of predator attacks are likely to involve dazzle effects.

If high-quality examples of dazzle in prey species could be established, then examining the coevolution of predator counter-adaptations may yield potentially fascinating insights. Possibilities include strategies to target and capture dazzle prey while they are motionless, or changes to motion detection mechanisms. Future work should also look to investigate the relationship between dazzle camouflage and other anti-predator strategies. For example, it may be possible that aposematic patterns featuring repetitive patterns might also function as motion dazzle.

We still see a benefit of using the controlled setting of humans targeting artificial prey in virtual computer spaces to explore some outstanding questions. If an experimental situation demonstrates a benefit to a given pattern in producing a dazzle effect, then we think it would be useful to explore the robustness of this effect following perturbations to the experimental setting. This would let us explore whether dazzle patterns can be designed so that they work effectively over a range of prey speeds, prey motion types, prey sizes, prey shapes, background types, and orientations of the prey to the predator. Such knowledge might identify likely candidate situations for finding naturally occurring dazzle markings, and also help identify likely forms for these markings. One outstanding issue in particular is the role of colour in dazzle camouflage; some dazzle ships and potentially dazzling animals are coloured, but thus far experimental studies have mainly used achromatic stimuli (Hall et al., 2016 is the exception).

Previous computer-based explorations have involved the artificial prey being in motion but the predator being effectively stationary relative to the background against which the prey are seen. Relaxing this restriction would not only be biologically realistic for mobile predators but might allow exploration of whether predators can ameliorate the effects of dazzle patterns in their prey by control of their own speed and search path trajectory.

A final thought is on the potential for dazzle camouflage in taxa where it has not yet been investigated. Flocking birds should be ideal beneficiaries of camouflage in motion, being fast, highly mobile, living in an open environment, and pursued by visually oriented predators. Examples of patterns commonly implicated in dazzle, such as high-contrast repeating stripes, are not common but do exist in a few species such as the red-bellied woodpecker *Melanerpes carolinus*. Alternatively, given the lack of clarity on what make a dazzle phenotype, it may be that we are looking for the wrong thing. Similarly, it will also be worth exploring links between visual dazzle and camouflage of movement in other sensory modalities. Acoustic camouflage of movement in moths is likely a common adaptation: for example, the shape of the luna moth's tails has been demonstrated to reflect incoming bat echolocation calls in a complex way that depends on where in the cycle the moth's wingbeat is (Lee & Moss, 2016). Averaging the centre of the echo over a period of flight, a likely heuristic used by bats to calculate where the moth is, pro- duces an estimated point behind the abdomen of the moth, in other words, a dazzle effect. This is likely to lead to targeting errors; indeed brown bats capture 47 per cent more luna moths when they have had their tails ablated (Barber et al., 2015).

CHAPTER 13

Thanatosis

When physically restrained, many animals adopt a relatively immobile state that can last after the physical constraint has been released. Sometimes in vertebrates such a behaviour involves tongue protrusion, and setting the eyes wide open, very reminiscent of a dead individual. This phenomenon has therefore been variously called *death feigning, animal hypnosis, tonic immobility, playing dead*, and *playing possum* (after the exemplary practitioners, opossums—see Gabrielsen & Smith, 1985). But the most common term used in the recent scientific literature to describe this behaviour is *thanatosis*.

This phenomenon is widespread taxonomically, and has been well known for a long time: indeed Darwin discussed it in a posthumously published essay (see Romanes, 1883). However, for reasons we discuss later, it has not been the subject of intensive study, and so there was insufficient material to justify more than a few paragraphs in our previous edition. Happily, this has changed, and there has been more purpose-designed study of this phenomenon in the last 15 years than in all the previous time since Darwin. As such, we feel thanatosis justifies a full chapter in this edition, although, unsurprisingly, we also highlight substantial areas of uncertainty in the current state of knowledge.

13.1 Introduction and overview of the chapter

It is quite difficult to obtain a clear and unambiguous definition of thanatosis. Thanatosis is an unlearned reflex action, not a proxy for 'fearfulness' (Rogers & Simpson, 2014). In general, we see it as distinct from immobility used to reduce the risk of detection or tracking by a predator, since it acts later in

the sequence of a predation event (generally after the prey has been detected by the predator, and most often after the predator has actually made physical contact with the prey). However, physical contact does not seem an absolute requirement for the triggering of thanatosis; sometimes simple detection of a predator in circumstances that suggest strong and imminent threat of attack, and lack of avenue for escape, can be sufficient (see Vogel, 1950, for reports of this in various birds). The onset of thanatosis is rapid and animals can remain in an immobilized state from seconds to hours (Rogers & Simpson, 2014). Although some instances of thanatosis are strongly suggestive of mimicry of death, some are less so and (as we will explore in section 13.3) there is potential for thanatosis to offer protection from predators through other mechanisms than feigning death. However, except for some very specialist situations considered in section 13.3.2, the function of thanatosis does appear to be employed as a means of protection from predators. Thanatosis is widely interpreted as a last-resort defence against predation after active physical resistance has proved unsuccessful and fleeing is impossible. The lack of movement is suggested to inhibit further attack by the predator, and reduce the perceived need of the predator to continue to attack or constrain the prey.

13.2 Distribution

Thanatosis is very widely but sparsely reported: in the vertebrates it has been reported in mammals, birds, reptiles (e.g. *Lygosoma* skinks, Patel et al., 2016, and *Erythrolamprus miliaris* water snakes, Muscat et al., 2016), amphibians (e.g. *Incilius* toads,

Avoiding Attack: The Evolutionary Ecology of Crypsis, Aposematism, and Mimicry. Second Edition. Graeme D. Ruxton, William L. Allen, Thomas N. Sherratt, & Michael P. Speed, Oxford University Press (2018). © Graeme D. Ruxton, William L. Allen, Thomas N. Sherratt, & Michael P. Speed 2018. DOI: 10.1093/oso/9780199688678.001.0001

Sanchez Paniagua & Abarca, 2016) and fish (recently, e.g. the Brazilian seahorse *Hippocampus reidi*, Freret-Meurer et al., 2017); in the invertebrates it has been suggested to occur in at least crabs, lobsters, stick insects, spiders, butterflies, stoneflies, water-scorpions, cicadas, crickets, mites, beetles, damselfly larvae, ants, bees, and wasps (see Cassill et al., 2008 for a partial list and references). The precise distribution is difficult to be certain of, even within a well-studied vertebrate group. For example, it seems relatively prevalent amongst snakes but even the number of families involved remains unclear (Gregory et al., 2007). One problem here is that (unlike most of the other tactics discussed for avoiding attack in this volume) there are normally no specialist morphological adaptations, so the behaviour needs to be observed as it occurs. The behaviour is very much a last line of defence after the predator has made physical contact with the prey. As such, its observation is logistically challenging in the field (with natural predation being generally unpredictable in time and space, and often deterred by human presence) and ethically challenging for staged encounters involving vertebrate prey. Finally, the use of this tactic often seems variable within even a local population, and very patchily distributed at higher levels of organization (so failure to observe it in most species of a given family would not inspire confidence in its likely absence from the remaining species).

The widespread but sparsely reported (and likely under-reported) prevalence of thanatosis makes it very difficult to speculate on underlying patterns to its taxonomic distribution. It is tempting to suggest that it might be more prevalent in snakes, for example, because snakes in general are not especially fleet of movement (reducing the efficacy of fleeing as an alternate strategy), and having an elongate body perhaps facilitates contact by a predator. However, such speculations are at this stage no more than 'just so' stories. Because thanatosis requires no anatomical specializations there seems likely often to be little cost to carrying the potential to use this last-ditch defence. As will be discussed in section 13.6, we expect the likelihood of prey displaying this trait to be more closely linked to traits of the predator than the prey—specifically, we would expect it to be

more commonly implemented against generalist predators with a broad diet, and predators that encounter prey simultaneously or in quick succession (or which are otherwise constrained in how much time they can devote to ensuring that contacted prey are dead or permanently immobilized).

The widespread but sparse nature of its taxonomic distribution suggests that it has evolved a large number of times. This seems more plausible to us than explanations based on one evolution and multiple losses of the trait. Also, thanatosis seems relatively simple to evolve, in that it is defined in the main by a lack of behaviour and does not involve specialist anatomical adaptations. Given multiple independent evolutions, one might expect considerable variation in the form and usage of thanatosis. There does seem strong evidence for such variation and this (together with the issues of almost certain under-reporting discussed above) makes generalizations challenging. However, there is evidence (discussed in later sections) that thanatosis is partly under genetic control, and is often consistent within an individual but variable within and between populations in the form and frequency of the behaviour.

Individual consistency can often be linked to other traits. As an example of this, Krams et al. (2014) found high repeatability in aspects of thanatosis in a laboratory-reared population of the mealworm *Tenebrio molitor*. They found that individuals with a higher metabolic rate had both a longer latency to enter thanatosis and a tendency to remain in that state for a shorter time. This complex of traits was repeatable within an individual, not only over time but also across environmental contexts. Similarly, Edelaar et al. (2012) exposed individuals of two avian species (yellow-crowned bishop *Euplectes afer* and tree sparrow *Passer montanus*) to mounts of predatory birds placed near their home cage and to repeated induction of thanatosis by human handling. Duration of thanatosis was found to be consistent within an individual, and low durations were correlated with higher activity rates in the presence of the mount. The authors interpret these two traits as being indicative of general 'boldness'.

Suzuki et al. (2013) found lower thanatosis responses to physical restraint in Bengalese finches

(*Lonchura striata* var. *domestica*) compared to the white-backed munia (*L. striata*), from which they were domesticated some 250 years ago. Bengalese finches also had lower corticosterone levels. The authors interpret these results as being suggestive that artificial selection during domestication has led to reduced fear responses. Evidence that thanatosis is linked to fear response also comes from the study of Fijian ground frogs (*Platymantis vitiana*) by Narayan et al. (2013). Frogs were exposed to a range of experimental manipulations. Those manipulations that experimenters expected to simulate the highest levels of predatory threat induced both higher levels of circulating corticosterone and stronger thanatosis responses in response to standardized handling.

There is evidence of variation in thanatosis over ontogeny. As a generality, those life-history classes less able to use fleeing or active resistance as defences are more likely to use thanatosis. Thus, there is a tendency in some taxa at least for it to be associated with early life-history stages. Cassill et al. (2008) report that day-old workers of the fire ant *Solenopsis invicta* responded to encounters with workers from another nest by exhibiting thanatosis, whereas week-old workers responded with aggression. This was interpreted in terms of the still-soft exoskeleton of the day-old workers rendering their ability to inflict damage on alien workers very low, but their vulnerability to damage being high. Thanatosis appears effective in this case, as day-old workers were four times more likely to survive such an encounter than week-old individuals encountering similar aliens. Interestingly, the authors speculate that death feigning is effective because aggression between workers from two different colonies often involves multiple individuals, and there may be selection to attack as many as possible, but this may come at the expense of failing to ensure that immobile individuals are actually dead.

A recent study by Matsumura et al. (2017) using red flour beetles, *Tribolium castaneum* and *T. confusum*, demonstrated that a behavioural correlation between thanatosis duration and walking distance can decouple across life stages. They found a negative correlation of traits in adults of both the species, but not in larvae of either species. The authors suggest that the negative correlation between tonic immobility and walking is decoupled across life

stages, contrasting with the phenotypic correlations between behavioural traits that have been found to be maintained from larvae to adults in insects that do not have a pupal stage. Metamorphosis is, therefore, suggested to have the potential to change trade-offs between behavioural traits, including the duration of thanatosis.

Linkages with other life-history traits have also been observed. For example, Gregory & Gregory (2006) reported that gravid garter snakes (*Thamnophis elegans*) were much more likely to show thanatosis than similar-sized non-gravid females. The authors put this down to impaired locomotive ability and flexibility of gravid individuals making fleeing or resisting less effective as alternative tactics. Other traits that have been shown in some species to influence thanatosis include sex, size, conditions during rearing, and sexual history; these effects can generally be interpreted in terms of shifting costs and benefits of the tactic and so will be discussed more fully in section 13.4. However, in keeping with an emerging theme in thanatosis, these life-history correlates can be explained in individual species or populations, but are not seen universally across species.

13.3 Form: the mechanisms involved

There are a number of mechanisms by which thanatosis might confer protection from predators; these are enumerated by Miyatake et al. (2009), and are potentially complementary.

Firstly, it might actually be death feigning, where the predator has an aversion to long-dead prey (which might be adaptive because of risk from toxins produced by microbial spoilage). Here thanatosis causes the predator to mistakenly reject a live individual because it is misidentified as a long-dead individual.

Secondly, it might be that thanatosis reduces the predator's ability to localize the prey individual, relative to a moving individual. This may be at least part of the explanation in cases where prey first of all drop from a plant to the ground (or otherwise physically distance themselves from a potential predator) and then adopt thanatosis.

Thirdly, thanatosis might serve as a physical defence against predators that swallow prey whole.

For example, this might work by thanatosis involving assuming a posture that makes swallowing more difficult or impossible. Alternatively, it could be combined with production of post-ingestion aversive secretions that trigger regurgitation, if the posture minimizes injury during the processes of ingestion and regurgitation. There is evidence in support of the first of these in Honma et al. (2006). Evidence for the second of these comes from reports in Toledo et al. (2010) of frogs swallowed whole by snakes and surviving after subsequent regurgitation.

The study of Honma et al. is particularly interesting; it involved staged interactions between the pygmy grasshopper *Criotettix japonicus* and potential predators. When faced with a gape-limited frog predator the grasshopper adopted a pose that effectively increased its functional body size and exposed spines. These adaptations make the grasshopper impossible to swallow for smaller frogs that could manage to swallow it in the prey's normal posture. For slightly larger frogs, ultimate swallowing was only possible after time-consuming manipulation to position the prey correctly in the mouth, hence making eating this prey less attractive from a foraging rate efficiency perspective. In this regard, it is significant that a field-based aspect of Honma et al.'s study found that larger frogs that could consume the grasshoppers undergoing thanatosis in the lab did not consume them in a field situation where such prey were available in abundance. In another interesting aspect, the grasshoppers never showed this behaviour in response to bird, spider, or mantid predators that do not need to consume their prey whole. In an earlier study, Moore & Williams (1990) observed stonefly nymphs *Pteronarcys dorsata* curl up into a ball in a way that projected spines and prevented ingestion by fish predators. Curling into a ball removes vulnerable body parts from exposure to the predator. Examples of prey postures where predators are presented only with an armoured shell are widely known, from insects such as pill-bugs all the way to large vertebrates like the armadillos.

Returning to possible underlying mechanisms of thanatosis, it may be that some predators have a relatively hard-wired sequence of actions that they need to perform during prey capture. In this case, thanatosis might subvert this by denying the predator successful completion of a subjugation action.

This, in turn, might prevent them from initiating later actions of the predation sequence; and, indeed, cause them to break off and abandon the interaction with that particular prey item.

It could be that thanatosis is an aposematic signal that the prey is chemically or otherwise well defended, and thus unattractive as prey. However, the one study to test this possibility found no evidence that this was the case (Miyatake et al., 2009).

If moving prey are more salient to a predator's senses, then it may be that thanatosis is a means of deflecting a predator's attention to alternative prey in a situation where prey are exposed simultaneously to the predator. Strong evidence for this was presented by Miyatake et al. (2009), who also pointed out that this mechanism was not the only one acting in their study system, since thanatosis also offered protection when prey were offered singly to the predator (see fuller discussion of this study in section 13.3.1). More generally, thanatosis might be effective in any circumstances where there is a cost to predators in time invested in making sure that a captured individual is definitely dead (or at least immobilized). The most obvious circumstance for this to occur is when a predator has potential to capture several prey items but only a short period of time when those prey are available; for example, when a predator discovers a group of prey that take some time to flee to cover.

In terms of what goes on in the prey animal itself, a number of physiological mechanisms can be involved in controlling thanatosis. In vertebrates, reduced breathing rates, bradycardia, salivation, defecation, and urination are common during thanatosis, and are consistent with mediation by the parasympathetic nervous system (Rogers & Simpson, 2014). In insects, ventilatory movements of the abdomen strongly decrease during thanatosis, but see Rogers & Simpson (2014) for some species-specific mechanisms.

13.3.1 What is the evidence that thanatosis offers protection from predators?

Krams et al. (2013a) found the resting metabolic rate (RMS) was consistent within individuals of a laboratory population of yellow mealworm beetles *Tenebrio molitor*. High metabolic rate was correlated with both longer latency to show thanatosis in response to experimental jarring with the substrate (caused by experimenters striking the experimental arena) and reduced duration of thanatosis. After RMR and thanatosis were measured, beetles were moved to another arena into which an insectivorous bird was introduced. High-RMS individuals were preferentially consumed by the bird. The authors interpret these results in terms of high RMS being part of a 'bold' personality type that will experience increased predation risk (because they exploit thanatosis less strongly) but obtained benefits in terms of higher activity and thus increasing mating opportunity. These authors subsequently observed essentially similar results with a nocturnal predator, the brown rat (*Rattus norvegicus*; Krams et al., 2013b). For both predators, the beetles could likely use vibrations through the substrate to warn of danger of predation. Further evidence of thanatosis offering insects protection from predators is discussed in later sections. Specifically, Miyatake et al. (2009) and Nakayama & Miyatake (2010b) bred lines of the red flour beetle *Tribolium castaneum* for either longer duration and higher frequency thanatosis (the L strain) or shorter and less frequent (the S strain), and demonstrated that L strain individuals suffered less from predation in an experimental arena with an Adanson's house jumping spider *Hasarius adansoni.*

13.3.2 Does thanatosis ever function in contexts other than against predators?

Hansen et al. (2008) demonstrated that males of the spider *Pisaura mirabilis* that entered a state of thanatosis during mating were more successful in gaining copulations and gained longer copulations than those males that did not. Sexual cannibalism is rare in this species, and when it does happen it occurs before the part of the mating sequence when thanatosis is manifest—so it is unlikely that this risk is the main driver of thanatosis. Males begin mating by offering the female a nuptial gift. While the female consumes this gift, the male initiates sperm transfer. The female can, however, interrupt copulation, after which she can either return to eating or depart with the gift (ending the copulation). Some males responded to such interruption by entering a state of thanatosis, which was associated with higher

probability of the female returning to feeding, after which the male left thanatosis and returned to copulating. This can happen several times during a copulation session. The authors speculate that there must also be a cost to thanatosis to explain variation in its expression between males. They speculate that this cost might be heightened vulnerability to attack by females, energetic cost of maintaining the pose, or risk of injury if they are dragged across the substrate in this pose by a fleeing female; however, as yet there has been no investigation of these alternatives. Previous studies had briefly reported that male mantids of the species *Mantis religiosa* freeze after mating to avoid post-copulatory cannibalism by the female (Lawrence 1992), and robber fly females of the species *Efferia varipes* behave similarly to avoid harassment by males (Dennis & Lavigne 1976).

Another instance of thanatosis in mating behaviours is described by Shreeve et al. (2000) in their observations of European Lepidoptera. They found that death feigning was used in the final part of a mate-rejection behavioural sequence by non-receptive female butterflies of the Satyrinae of the Palaearctic. The females closed their wings and released themselves from the underlying substrate, following extreme male persistence. Shreeve et al. suggest that this behaviour might possibly be restricted to the tribes Elymiini and Maniolini, and that it could be related to female mating frequency, male mate-locating mechanisms, and the physical structure of habitats where attempted mating occurs.

There is also highly suggestive evidence that at least two species of cichlid adopt a pose that appears subjectively to mimic a corpse falling through the water column and then lying inert on the substrate. The cichlid then appears to attract other fish that are interpreted as motivated to scavenge from the 'corpse'; such fish are then attacked by the cichlid that 'comes to life' when a would-be scavenger swims close to their mouth (McKaye, 1981; Tobler, 2005). In both studies, this hunting tactic was only shown by adults, which McKaye speculates may be because juvenile fish of these species would be sufficiently small to be vulnerable to predation if they lay immobile. A similar tactic involving apparently feigning illness or dying to draw potential prey closer has been reported in a shallow-water predatory grouper fish *Mycteroperca acutirostris* (Gibran, 2004).

Van Veen et al. (1999) discuss the use of thanatosis by gynes in the stingless bee *Melipona beecheii*. They discuss how surplus queens are continually produced but are subject to aggression from the workers and usually killed. It was noted that thanatosis appeared effective in stopping instances of worker aggression, although it was also noted that no gynes which showed this behaviour went on to successfully be accepted by the colony.

13.4 Evolutionary function: a cost/benefit approach

The last section outlined potential benefits to thanatosis, but there are also likely to be costs. The same is likely to be true for alternative last-ditch anti-predator behaviours, such as struggling to break free and fleeing, or aggressively attacking the would-be predator. Hence, we might expect the prevalence and strength of thanatosis to reflect conditions where the cost-benefit trade-off favours thanatosis over alternative behavioural tactics. As an example of this, Gyssels & Stoks (2005) found that thanatosis in larvae of the damselfly *Ischnura elegans* was more likely to be utilized in staged laboratory encounters with predators if an alternate anti-predator defence was negated by removal of lamellae, which are given up through autotomy when grasped by a predator. In general, variation and plasticity in thanatosis could be driven by evolution since (as we have shown previously) aspects of this behaviour are heritable. We will leave responsiveness to environmental factors to the next section, and here focus more on identifying the potential trade-offs that affect variation in the expression of thanatosis.

Nakayama & Miyatake (2010b) bred lines of the red flour beetle *Tribolium castaneum* for either longer duration and higher frequency thanatosis (the L strain) or shorter and less frequent (the S strain). They then demonstrated that males of the L strain suffered less from predation in an experimental arena with a spider, but had reduced mating success in a predator-free environment. This latter effect was interpreted by the authors as individuals of the L strain being overall less active and thus encountering females less frequently.

Nakayama & Miyatake (2010a) similarly bred L and S lines of the adzuki bean beetle *Callosobruchus*

chinensis and found that the L strain had reduced activity, which translated into reduced mating success in males but not females. This is likely because male mating success is more dependent on frequency of encounters and thus on activity levels. They suggest that activity levels might explain the fact that in Nakayama & Miyatake (2009) these authors reported that L individuals had higher longevity, larger egg size, and faster development. They suggest that reduced activity may conserve energy, increasing longevity, and allowing diversion of energy to larger eggs that take less time to develop.

Kuriwada et al. (2010) reported that mass rearing over 71 generations in the sweet potato weevil *Cylas formicarius* caused no change in the duration of thanatosis. This is surprising, since there was no predation pressure during these 71 generations. The authors speculate that one reason for lack of selection pressure on thanatosis might be because with a high abundance of both food and mates, any cost of reduced activity through thanatosis is trivial. We would also add that thanatosis may have been triggered very infrequently in any individuals during these 71 generations, and there may be little expressed cost in retaining the capacity to perform thanatosis in situations where the behaviour is seldom triggered.

Ohno & Miyatake (2007) argued that for the adzuki bean beetle flying away and death feigning are two alternative tactics that are used in response to predators. They artificially bred lines for long or short duration of thanatosis, and found that those bred for long-duration death feigning had weaker flying ability than those in the other line. Flying ability was measured in terms of how strongly an experimentally dropped beetle influenced its downward trajectory (i.e., how far it landed from directly below its release point). Lines bred for higher (lower) flight ability showed correlated lower (higher) duration of thanatosis. The exact mechanism underlying this genetic correlation was not determined. The authors speculated that there may be competition for resources between investment in flight muscles and in reproduction, and highly reproductive individuals may thus have poor flying ability and separately (for reasons that are unclear to us) increased ability to sustain thanatosis. Alternatively, they speculate that individuals that show a low propensity for thanatosis are more active and, thus, more likely to find a means of escape from the rearing facility. Finally (and most plausibly to us) they suggest that thanatosis might be indirectly selected, despite a direct cost, because it is genetically correlated with other traits that confer higher fitness. In this regard, they cite Nakayama & Miyatake (2009) who found that longevity, emergence rate, and egg size were all positively correlated with extent of thanatosis in the adzuki bean beetle.

Kiyotake et al. (2014) inferred potential costs to thanatosis of the red flour beetle, *T. castaneum*. These authors had previously demonstrated (Miyatake et al., 2009) that the frequency of predation by a jumping spider, *Hasarius adansoni*, was significantly lower among beetles from lines that had been selected to show higher frequencies and longer durations of tonic immobility (L-type) than those with lower frequencies and shorter durations of tonic immobility (S-type). However, they also found that the wild population showed a majority of S-type individuals. In the 2014 study, they demonstrated that L-type beetles were significantly more sensitive to environmental stressors such as mechanical vibration and high or low temperatures.

In the sweet potato weevil, Kuriwada et al. (2011) found that mated females showed longer duration thanatosis than virgin individuals, something the authors interpreted in terms of a shifting trade-off between avoiding predation and being active to search for mates. They also found that older females decreased investment in thanatosis, which they again interpreted as reduced relative costs to predation in older individuals. Males, however, showed no change with age. The authors argued that there should be a strong effect of increasing age (and thus mortality risk from factors other than predation) in females that must not only mate, but also collect food for investment in eggs and find oviposition sites, versus males, who need only find mates. A previous study (Kuriwada et al., 2009) had found an effect of copulation in both males and females. The authors suggest that the difference between studies was that the earlier study tested thanatosis 5 hours after the treatment (copulation or control) was imposed, increasing to 30 hours in the second study. The authors speculate that the long interval during which no females could be encountered may have caused perception of mating opportunity to

decline in males of both treatments, making a treatment effect difficult to pick up. Miyatake (2001b) had previously found that duration of thanatosis in the sweet potato weevil decreased with increased duration of starvation, and that this effect was stronger in males. This last result can be explained by males being more susceptible to mortality through starvation than females. Miyatake (2001a) also observed a sex difference in thanatosis in the sweet potato weevil. Both sexes showed a reduction in thanatosis at night, which Miyatake suggests might be because most of their predators are visual and so predation risk is generally lower at night. The effect was particularly strong in males, and the author interpreted this as a consequence of copulations occurring at night and males having more to gain from remaining active to obtain multiple matings than females.

Hozumi & Miyatake (2005) found that smaller aduki bean beetles showed shorter duration thanatosis, something the authors interpreted in terms of the shorter natural lifespan of smaller individuals. Nakayama & Miyatake (2009) bred adzuki bean beetles for either long or short duration of thanatosis. Fecundity was the same in the two lines, but those with longer-duration thanatosis produced bigger eggs, that developed more quickly, were more likely to hatch, and produced longer-lived offspring. This was interpreted as thanatosis requiring less energetic investment than fleeing, the energy saved being diverted into reproduction. However, they note that their stocks experienced an abundance of food and potential mates. In other ecological circumstances, the energetic savings of thanatosis may have to be traded off against the lost time that could be devoted to finding food and mating. We explore ecological influences of thanatosis more fully in the next section.

13.5 Ecological considerations

Having interpreted plasticity and variation in thanatosis in the framework of a cost–benefit trade-off, we here explore how that trade-off might be influenced by aspects of the environment. For example, Miyatake et al. (2008) found that thanatosis occurred more strongly in two species of seed beetle (*Callosobruchus maculatus* and *C. chinensis*) when individuals were kept at lower temperatures. The authors interpreted this as a consequence of higher temperatures allowing beetles to use fleeing as an alternative anti-predator behaviour more readily. Similarly, Saxena (1957) found that woodlice (*Armadillidium vulgare*) showed stronger thanatosis at lower temperatures (presumably when the alternative anti-predatory tactic of fleeing was less available) and when lighting was reduced (and so vulnerability to visual predators when stationary was reduced).

There is evidence that thanatosis is used flexibly according to how ecological circumstances influence its costs and benefits relative to alternative anti-predator behaviours (most obviously fleeing to a refuge). Arduino & Gould (1984) performed experiments where domestic chicks (*Gallus gallus domesticus*) were exposed to some stimulus for 15 seconds, then experienced 15 seconds of being manually restrained whilst still being exposed to the stimulus. This induced thanatosis, the chick was released, and the duration of thanatosis measured. They found in numerous paired comparisons that thanatosis lasted longer in situations where it was thought that the chick would consider the chance of escape to be lower (e.g. when the stimulus was a hawk model facing the chick), compared to cases when this chance might be considered to be higher (when the hawk model was facing away from the chick). Ewell et al. (1981) demonstrated in a laboratory that rabbits (*Oryctolagus cuniculus*) decreased the duration of thanatosis as proximity to a human decreased or as proximity to their home cage increased. Gallup et al. (1971) found that chickens immobilized by manual restraints in the presence of a hawk model exhibited longer durations of thanatosis than chickens immobilized without the presence of the model. Hennig et al. (1976) found that thanatosis was of shorter duration in anole lizards that had been immobilized near bushes than those immobilized in the absence of nearby protective cover. O'Brien & Dunlap (1975) reported that crabs immobilized on a sandy substrate (in which they could seek cover by burying themselves) exhibited shorter durations of thanatosis than individuals similarly immobilized on a hard surface. Santos et al. (2010) reported that thanatosis experimentally induced in lizards by manual handling lasted longer if the human remained in close vicinity. Social environment can also affect expression of thanatosis. Odén et al. (2005) reported that female

laying hens showed shorter duration thanatosis in response to standardized human handling if they had been kept in groups with males. The authors interpreted this as males acting as guards and lowering fear levels in the females.

Sometimes the environment is not always the best predictor of thanatosis behaviour, though. Farkas (2016) tested the hypothesis that *Timema cristinae* stick insect individuals that were poorly camouflaged against their backgrounds were more likely to feign death. This would indicate that individuals respond to attacks in a plastic fashion, wherein thanatosis is dependent on the colour pattern matching the host-plant species. Instead, Farkas found that smaller individuals were more likely to feign death than larger individuals, regardless of camouflage.

It may be that experience of previous environmental conditions can influence thanatosis, as well as the currently experienced condition. Tojo (1991) found that common cutworm *Spodoptera litura* raised in crowded conditions in their 4th and 5th instar showed reduced periods of thanatosis in their 6th instar compared to those raised in isolation. This might be seen as an adaptive response to a trade-off between avoiding predation and competing for food resources. He also reported variation between natural populations in duration of thanatosis and demonstrated, in crossed experiments, that this had a genetic component. In a study investigating the effects of human contact and intraspecific social learning on thanatosis in guinea pigs, *Cavia porcellus*, de Lima Rocha et al. (2017) found that, firstly, handling and habituation did not prevent tonic immobility responses in subjects. However, habituation increased the latency of thanatosis, while handling or habituation decreased the duration of the tonic immobility behaviour. Additionally, the cohabitation of unhabituated and habituated animals was found to reduce tonic immobility duration. The authors concluded that experience of human interactions can reduce experimenter fear in guinea pigs and that unhabituated guinea pigs may learn not to fear experimenters by cohabitating with habituated guinea pigs. Experience and social environment can clearly be important in determining the occurrence of thanatosis.

The complexity of how thanatosis, as an anti-predatory behaviour, interacts with other parts of an individual's responses to the environment is shown by the study of Kuriwada et al. (2009) on the effect of copulations in the sweet potato weevil. In both sexes, encounters with individuals of the opposite sex where copulation was experimentally prohibited did not affect duration of any subsequent thanatosis. However, an encounter leading to successful insemination led to reduced thanatosis length in females but not males. Regardless of whether copulation lead to insemination, females reduced thanatosis after a single copulation, and males after multiple copulations. The authors interpret the behaviour of males as follows. Sex recognition ability is poor in this species and so copulation rather than simple encounter may be needed to suggest to males that there is high current availability of females. In such a situation activity to find females should be prioritized over avoiding predation. For females, copulations may signal high male activity and thus high risk of sexual harassment. Having allowed successful sperm transfer once, a female's interest in the attentions of further males decreases and, thus, such females will be willing to accept higher predation risk in order to remain active to avoid unwanted male attentions.

The local predatory community is also part of the environment. Gyssels & Stoks (2005) found that thanatosis in larvae of the damselfly *Ischnura elegans* was more frequently induced experimentally in those from a pond in which fish predators are present compared to those from a pond where this important predatory group is absent. As well as demonstrating a genetic correlation in the laboratory between flight ability and duration of thanatosis discussed previously, Ohno & Miyatake (2007) also found such a negative correlation across 21 wild populations of the adzuki bean beetle *C. chinensis*. They speculate that one reason for this may be that the two alternate tactics have differential effectiveness against different predators and the predator communities differ between populations. Additionally, or alternatively, they speculate that one tactic (perhaps flying away) may be both more effective and more costly, and populations that face lower predation pressure are selected to make more use of the cheaper but less effective tactic (thanatosis). We

explore the predator's side of thanatosis more fully in the next section.

13.6 Co-evolutionary considerations

There is increasing evidence that thanatosis does have a functional basis in reducing the probability of an attack by a predator ending in the prey's death. We now turn to considering why thanatosis causes the predator to reduce its attacking. Our suggestion is that the thanatosis might be effective against predators that commonly face a situation where they can subdue more than one prey item in quick succession. The predator faces a trade-off, since the more time it spends making certain that the first prey item is dead (or sufficiently injured that it cannot escape), the more likely it is that other nearby prey items will make good their escape. In this circumstance, thanatosis works by inducing the predator to switch prematurely from one prey item to the next, and the trade-off may make it maladaptive for the predator to invest time in making absolutely sure that the first item is dead. In order to explore this, we suggest that it would be useful for a systematic study of anecdotal accounts of the observation of thanatosis to explore the percentage of occasions when the predator had the chance to attack another prey item soon after releasing the individual exhibiting thanatosis. Our prediction would be that this percentage will be high. Additionally, or alternatively, it would be worth exploring whether individuals of species exhibiting thanatosis occur commonly in groups. Thanatosis can still be effective for single prey individuals, if the predator concerned has evolved to release prey quickly because it often encounters prey in groups, against which quick release can aid in making multiple kills.

It may be that thanatosis is ineffective against predators that specialize on a given prey type and have thus been able to co-evolve effective responses to thanatosis. Gregory et al. (2007) demonstrated thanatosis in the grass snake *Natrix natrix* and hypothesized that this would only be an effective tactic against generalist predators that rarely encounter it. He thus suggested that grass snakes should less readily show thanatosis in those parts of its range shared by the snake specialist short-toed eagle *Circaetus gallicus*.

Evidence that thanatosis can be more effective when predators face multiple prey comes from the study of the red flour beetle *T. castaneum* attacked by Adanson's house jumping spider *H. adansoni* in laboratory experiments. Miyatake et al. (2009) used two strains bred for long and short duration of thanatosis. When a long-strain individual was presented to a spider 38 per cent were killed, dropping to 9 per cent when presented at the same time as a short-strain individual. When presented in a pair alongside an individual of another species that does not show thanatosis, 28 per cent of long-strain and 62 per cent of short-strain individuals were killed. When presented in pairs of a long-strain and short-strain individual, the short-strain individual was killed every time. When presented in a group of five from each strain, predation rates were 79 per cent and 21 per cent for short- and long-strain individuals respectively. These authors also reported that levels of potential chemical aversants were not strongly different between the two strains, so the differences in predation rates were likely to do with thanatosis and not to a correlated chemical defence.

13.7 Unresolved issues and future challenges

As already discussed, a substantial challenge remains to identify and then understand the taxonomic prevalence of this behaviour. We suspect that thanatosis may often be observed by fieldworkers in studies of other aspects of animal biology, and these incidents go unreported. We would see value in raising awareness of thanatosis amongst biologists generally and providing some easy means for observations to be logged and collated. We suggest that a valuable way forward in this would be for collaboration between those scientists who have an interest in thanatosis to reach a consensus on such an endeavour. This consensus might involve agreement on: 1.) a simple definition that would allow non-specialist biologists to identify incidents of thanatosis; 2.) the aspects of a thanatosis manifestation that should ideally be recorded; 3.) design and maintenance of a single website where such data could be input; and 4.) vigorous and sustained effort to publicize the initiative widely among animal biologists. Such a consensus might usefully

agree on how the data from which an effort could best be analysed. Based on the arguments presented in the last section, we would argue that it would be of great value to record information on the predators that induce the thanatosis as well as the prey that displays it. This would allow testing of predictions made in this chapter about which predators are more likely than others to induce thanatosis. However, both prey and predator traits should influence the prevalence of thanatosis, and we would expect (for example) that the behaviour is more common in aggregated prey that have some capacity to flee upon discovery by a predator which can only capture those prey one at a time.

The last 15 years have seen a huge leap in our understanding of thanatosis (at least in insects) driven by some very effective laboratory studies. However, captive animals experience a very unnatural (lack of) predation regime, have often experienced this artificial world their whole life, and indeed may often be bred from stock that have been removed from natural predation pressures for generations. In most studies thanatosis is induced by humans rather than a natural predator. There is a strong need for us to find systems in which we can study thanatosis in the wild. Ever-improving technology involving tracking animals and on-board cameras should be a powerful weapon in such an endeavour. Head-mounted cameras that allow predators of the size of a domestic cat to hunt for prey unimpeded are now very cheaply and widely available.

Laboratory studies have also highlighted behavioural, physiological, and life-history correlates of thanatosis, and we would benefit now from finding a suitable model system in which to explore whether similar relationships can be found under naturally occurring conditions.

There is still an important place for laboratory study, and here—for example—we would encourage an increased focus on how thanatosis influences predators' subsequent behaviours. It might be, again, that there is a place for advanced technology here. An attraction of remote-controlled robotic simulated prey is that the form and extent of thanatosis can be manipulated whilst controlling all potentially confounding variables, in a way that is very difficult in experiments with living prey.

Synthesis

In writing this new edition, we have thoroughly enjoyed exploring the most recent findings in the fascinating world of anti-predatory interactions and the diverse and sometimes astonishing related adaptations. The first section to this book was devoted to studies of crypsis, beginning with a consideration of background matching. Simply matching the background against which you are seen might seem at first pass to be the be all and end all of avoiding detection. The running theme throughout this chapter, however, is that costs and constraints mean that perfect background matching is often not obtained, and this explains why organisms often utilize other mechanisms of reducing their ease of detection.

Turning to disruptive camouflage, it is clear from our discussion of empirical research that it is fundamentally tricky to demonstrate that disruption is occurring; this does not mean that disruption is unlikely to be widespread, just that the tell-tale hallmarks of it are not as obvious as, for example, transparency. We emphasize how the effectiveness of disruption depends on subverting the perceptual processing of predators; however, thus far predator counter-adaptations to disruption have not been identified. It may be that given all the other design requirements of efficient visual processing, it is fundamentally difficult for a predator to evolve to stop being fooled by disruption. As observers experience such a diversity of different visual inputs, they need to have very generalized processes for detecting and recognizing objects; it is that very need for generality that disruption can exploit. We find the emerging picture of a trade-off, whereby some markings might make initial detection more tricky through disruption, but then might also be more easily learnt, making subsequent detection easier, absolutely fascinating. This might select for mass polymorphism of disruptive markings, or it might be that the range of scenarios that natural disruptively camouflaged animals occur in reduces the importance of this learning.

Considering next the class of countershading defences, we are likely on the cusp of really improving our empirical understanding. Fish that countershade adaptively over relatively short timescales in response to the local light environment are a really fruitful experimental resource and have already been used effectively to test between alternative mechanisms that might select for countershading. Our ability to generate computer images with different types of countershading that can be used to test ease of detection by humans is also growing ever more sophisticated—it would seem not too difficult to extend this to other types of predators interacting with computer screens or even with artificial 3D objects (3D printers might be a great tool for scientists in this regard). It seems unlikely to us that fixed countershading often offers high-quality camouflage, because in order to really work perfectly the organism must be suitably oriented to the direction of the light and the light levels must be just so; this is a very big ask for almost all organisms except under rather specialist circumstances. However, countershading does offer a relatively cheap defence adaptation that goes some way to preventing, or reducing the likelihood of, detection. Additionally, or alternatively, countershading might augment camouflage generated mainly by background matching and disruption—the relationships between countershading and other forms of camouflage are currently unexplored.

With regards to transparency and silvering, we believe there is still more to learn about the physiology of transparency. More investigations could

Avoiding Attack: The Evolutionary Ecology of Crypsis, Aposematism, and Mimicry. Second Edition. Graeme D. Ruxton, William L. Allen, Thomas N. Sherratt, & Michael P. Speed, Oxford University Press (2018). © Graeme D. Ruxton, William L. Allen, Thomas N. Sherratt, & Michael P. Speed 2018. DOI: 10.1093/oso/9780199688678.001.0001

explore what the exceptions to the pelagic transparency rule are, and why are they adaptive. There is still more to discover, also, about whether polarization change is ever important for detecting transparent organisms. Quantifying and exploring just how much cryptic benefit is gained by silvering would also be of great value.

The second section of this book looked at adaptations through which prey can avoid attack following detection. It is key to recognize that the defences employed by one individual not only affect the outcome of a given interaction between that individual and a predator but may also affect the subsequent behaviour of the predator—and thus the selection pressure it imposes on the defences of subsequently encountered prey. We focus on this a lot in the secondary defences chapters and in the mimicry chapters—but it is also true more widely. Prey that make themselves harder for predators to detect may induce the predators to shift their attention more to other prey. We don't dwell much on predator switching and predator–prey dynamics, but there is much to explore about population and ecosystem-level effects of anti-predator defences. We have to limit the book in some way or go mad, so we don't talk about behavioural defences that don't have a primarily sensory aspect; defences like fleeing, avoiding predators in time and space (e.g. by being nocturnal), and using protective microenvironments like burrows. A truly integrated consideration of predator–prey interactions would give more consideration to these.

In our chapter on aposematism, we discuss how there is now a suite of logically plausible scenarios that could explain the initial evolution of aposematism. Many of these scenarios also have empirical support for some of their key assumptions. What we don't have is a good idea of the relative importance of these scenarios either for particular cases or more generally across the multiple times that aposematism has evolved. Further, it seems easy to construct scenarios where a change in the local environment can cause aposematism to be outcompeted by crypsis—so it may be that aposematic traits are usually extinction-prone—but this deserves further investigation. In contrast to chemical defence, we see lots of comparative phylogenetic studies for ecological correlates of visual aposematic signals.

This is simply a function of how easy to it to score an individual as having a visual aposematic signal (or not), versus scoring it as chemically defended or not.

Müllerian mimicry is a fascinating anti-predator defence to unpick, because the advantages of having a shared warning signal rest so much on the cognition of predators. While the genetics of this defence are beginning to be understood, especially in *Heliconius*, and it has been the subject of some very large-scale field studies, there remain significant outstanding questions and potential controversies. In the future, there should be a greater consideration of whether or not Müllerian mimicry benefits predators. Essentially, we tend to assume that predators are trying to categorize potential prey into those worth eating or not—just two categories. In such a situation, if all the unpalatable prey share a common signal than this would aid the predator's decision-making—it only has to learn one signal. However, it is probably the case that predators are more sophisticated than this and that they actually want to evaluate just how bad different prey types are; some might be essentially unattractive in the present situation but would become attractive if alternative prey became scarce. If this is the case, Müllerian mimicry will no longer help the predator if it cannot easily tell two prey species that differed substantially in their unattractiveness apart, with one prey type a Batesian mimic in time of need. Whichever of these situations is truly the case, scientists should be more interested in what predators actually learn about prey, acknowledging that it might be more sophisticated than if a prey item is simply tasty or not. We have some theory on this, but we need more experiments.

When we think about the evolution of mimicry (especially Müllerian mimicry) we need to remember that there might be selection on level of investment in defences as well as selection on appearance. Too much of the theory of evolution of Müllerian mimicry treats the aversiveness of the species involved as fixed. As a corollary of this, we need to measure the costs of the mimetic signal compared to the costs of other potential signals and the costs of producing the defence that is being signalled. Additionally, we currently treat the predators as very simple agents in our models—they either have perfect knowledge or they are sampling in order to learn about the prey options available to them. But for long-lived predators their information about a prey type will be subject to forgetting as well as learning, and we should maybe look at memory jogging, at how long memories are retained, and how accurately they are retained to get a fuller picture. The fact that Müllerian mimicry is assumed to involve learning by predators (rather than innate genetically encoded knowledge) suggests that the prey community available to a predator is dynamic—to what extent do predators keep tabs on a prey type after they have had some aversive experiences with it? Future work is also needed to understand the transitions between different anti-predator strategies, both generally and between Batesian and Müllerian mimicry specifically.

Considering elusiveness signalling, there is a sound theoretical base and a number of highly suggestive case studies. However, there is currently little empirical work detailing the costs and benefits to each party of signalling and responding to the signal. For the future, both the development of a laboratory system and large-scale observations of predator–prey interactions in the wild are necessary for really drilling down into the selective pressures that sometimes lead to prey signalling their elusiveness to predators.

The final section of the book focusses around deceptive anti-predatory defences and we first looked at the roles of Batesian mimicry and masquerade in predator–prey interactions. Our understanding of Batesian mimicry is currently dominated by visual mimicry, although there are known examples of mimicry in other modalities and even multimodal mimicry; we likely have lots of examples still to uncover. We also need to consider whether we can predict which modalities will be important for mimicry in different contexts, whether there are particular advantages to multimodal mimicry, and whether we would expect mimicry to be similarly accurate in all modalities. It should not be a surprise that some field studies fail to produce results consistent with mimicry theory; it is unlikely that the selection pressure for mimicry would be consistently strong over short temporal and spatial scales, where there will be fluctuations in both the prey and predator communities. Indeed, it would not be surprising if you ran the same experiment at the same field site in two consecutive years and got quite different results, because of stochastic variation in the communities under study. There is, therefore, a need to make our field studies more sophisticated. At the moment, we don't seem to be embracing new technologies, maybe because most mimicry systems involve pretty small insects that are hard to track over long periods, but we could definitely have biologging equipment on avian predators, and maybe some of the fish mimicry systems might lend themselves to on-board tracking of prey. In the same vein of using different predatory systems, we could learn more from exploring possible examples of Batesian mimicry and masquerade employed against invertebrate predators. Further work on the genetics of masquerade and invasions of mimics and models generally would also be of value, as would progress on separating taste and toxicity.

There is much current debate in the literature as to whether startle effects are best seen only as a response to unexpected strong sensory stimulation, or whether they can in some instances additionally represent misidentification of the prey as a predator. It should be possible to manipulate the experience of captive birds (for example) to explore the relative strengths of these alternatives. Developments in functional magnetic resonance imaging (fMRI) and other brain activity mapping techniques may hold potential for illuminating this issue in the not-too-distant future. An important point surrounding startle signals, we feel, is that whereas some strategies are clearly pursued by certain kinds of species, this does not seem to be the case for startle signals, although systematic comparative work has not been done.

Deflection of the predator's point of attack on the body of a prey individual would appear to occur in a number of situations in ways that offer benefits to the prey. This deflection can also be linked to appearance traits of the prey. The key underexplored issue in this topic is what costs if any having these deflective traits have on other defences—most obviously crypsis, but also mimicry and aposematism. We hope that someone will also find ways to exploit new technologies to help us explore the operation of deflection in natural environments, since this remains a very lab-based area of study.

The defence discussed in this book with perhaps the shakiest empirical foundation currently is dazzle camouflage. The idea that animal patterns might distort the perception of speed and trajectory has only recently been tested in a dozen studies, and these report highly inconsistent findings. While overall these strongly suggest that pattern can 'camouflage in motion' via dazzle effects, we do not yet really understand how they do so, under what circumstances dazzle effects occur, and whether any animals actually employ dazzle camouflage as a defence. This situation compares with the growth of our understanding of disruptive camouflage following Cuthill et al. (2005) and Merilaita & Lind (2005). These studies reported consistent disruptive effects of, for example, patterns that intersect target outlines, and allowed progress to be made on understanding perceptual processes underlying disruption. Hopefully our understanding of dazzle camouflage will likewise develop now that its plausibility has been confirmed.

Death-feigning research has boomed in the last decade, causing us to devote an entire chapter to the topic. We expect the next decade to allow this study to move out of the lab into observation of how this mechanism operates in the natural world. We also highlighted that there would be real benefit to finding some way to more effectively collate observations of the phenomenon—since its taxonomic spread is very hard to evaluate in the absence of obvious anatomical correlates.

Identifying instances of many defensive strategies is very difficult. This is an issue we kept returning to in the chapters on disruptive camouflage, dazzle camouflage, aposematism, elusiveness, startle, and deflection. Even for defences that can readily identified, such as countershading and transparency, establishing the mechanism through which a particular case provides protection can be challenging.

It emerges as a consistent conclusion across chapters that the quality of predator counter-adaptations to prey defences is perhaps weaker than might be naively expected. If increased foraging efficiency results, then why are predators not able to detect and recognize prey in spite of coloration that attempts to disguise status as suitable prey by subverting the sensory systems that evolve in part to detect prey?

This emerged most strongly in the chapters on disruption, countershading, and Batesian mimicry, and we feel is one of the key outstanding issues in our understanding of sensory aspects of predator–prey relations.

Humans are primarily visual, and this book is primarily focussed on visual adaptations in part because researchers in this field are human (mostly ☺). We have strived to always remember other modalities throughout each chapter but there is a whole world of non-visual interaction between predators and prey yet to be fully uncovered. Sometimes our understanding based on vision can illuminate exploration of other modalities: for example, it seems highly likely that many alarm calls that are designed to incorporate the disruptive principles explored visually in Chapter 2.

If we had to distil down where we see most activity in the general field of predator–prey sensory interactions in the next decade then some of the key themes that stand out for future research on all of these intriguing defences are:

- a need to embrace emerging technologies (especially to move out from the lab to the field);
- a broadening of the perspective of predator learning and decision-making (to not only understand how predators react, but about the internal processes that lead to that behavioural reaction); and
- a general inclusion of many ecological factors in predictive models, so as to better reflect the complexity of natural ecosystems and predator–prey interactions (moving from describing patterns of defensive strategies to predicting).

As in our previous edition, we are keen to discourage compartmentalization of thinking on the issues discussed in this book. Indeed, many prey may have evolved a portfolio of defences—for example, relying on crypsis but then employing startle or death-feigning when their cover is blown. Despite our chapter structuring, this field will make the greatest progression through developing an integrated framework that takes into consideration all the complexities of natural systems without viewing individual behaviours and interactions in isolation. Only when we consider the costs, benefits, evolution, maintenance, and distribution of anti-predatory strategies can we begin to understand the true role of the sensory mechanisms involved and make predictions about how they may influence behaviours, fitness, and future evolution of the species involved. We have seen in recent years how a broadening of outlook can develop understanding, radically in some cases, of the ways in which prey avoid attack by predators, and we hope that future studies will continue in this vein to discover yet more.

References

Abbott, K. R. (2010) 'Background evolution in camouflage systems: A predator–prey/pollinator-flower game', *Journal of Theoretical Biology*, 262(4), pp. 662–78.

Abram, P. K., Guerra-Grenier, E., Després-Einspenner, M.-L., Ito, S., Wakamatsu, K., Boivin, G., and Brodeur, J. (2015) 'An insect with selective control of egg coloration', *Current Biology*, 25, pp. 2007–11.

Acuña, J. L. (2001) 'Pelagic tunicates: why gelatinous?', *The American Naturalist*, 158(1), pp. 100–7.

Adams, J. M., Kang, C., and June-Wells, M. (2014) 'Are tropical butterflies more colorful?', *Ecological Research*, 29(4), pp. 685–91.

Adelson, E. H., and Bergen, J. R. (1985) 'Spatiotemporal energy models for the perception of motion', *JOSA A*, 2(2), pp. 284–99.

Adelson, E. H., and Movshon, J. A. (1982) 'Phenomenal coherence of moving visual patterns', *Nature*, 300(5892), pp. 523–5.

Agrawal, A. A., and Fishbein, M. (2008) 'Phylogenetic escalation and decline of plant defense strategies', *Proceedings of the National Academy of Sciences*, 105(29), pp. 10057–60.

Aguilar-Argüello, S., Díaz-Fleischer, F., and Rao, D. (2016) 'Motion-triggered defensive display in a tephritid fly', *Journal of Ethology*, 34(1), pp. 31–7.

Akkaynak, D., Siemann, L. A., Barbosa, A., and Mäthger, L. M. (2017) 'Changeable camouflage: how well can flounder resemble the colour and spatial scale of substrates in their natural habitats?', *Royal Society Open Science*, 4(3), pp. 160824.

Alatalo, R. V., and Mappes, J. (1996) 'Tracking the evolution of warning signals', *Nature*, 382(6593), pp. 708–10.

Alcock, J. (1970) 'Punishment levels and response of white-throated sparrows (*Zonotrichia albicollis*) to three kinds of artificial models and mimics', *Animal Behaviour*, 18(V), pp. 733–9.

Alexandrou, M. A., Oliveira, C., Maillard, M., McGill, R. A. R., Newton, J., Creer, S., and Taylor, M. I. (2011) 'Competition and phylogeny determine community structure in Müllerian co-mimics', *Nature*, 469(7328), pp. 84–8.

Allen, J. A. (1988a) 'Frequency-dependent selection by predators', *Phil. Trans. R. Soc. Lond. B*, 319(1196), pp. 485–503.

Allen, J. A. (1988b) 'Reflexive selection is apostatic selection', *Oikos*, 51(2), pp. 251–3.

Allen, J. A. (1989a) 'Colour polymorphism, predation and frequency-dependent selection', *Genetics (Life Science Advances)*, 8, pp. 27–43.

Allen, J. A. (1989b) 'Searching for search image', *Trends in Ecology & Evolution*, 4, 361.

Allen, J. A., and Cooper, J. M. (1985) 'Crypsis and masquerade', *Journal of Biological Education*, 19, pp. 268–70.

Allen, W. L., Baddeley, R., Cuthill, I. C., and Scott-Samuel, N. E. (2012) 'A quantitative test of the predicted relationship between countershading and lighting environment.', *The American Naturalist*, 180, pp. 762–76.

Allen, W. L., Baddeley, R., Scott-Samuel, N. E., and Cuthill, I. C. (2013) 'The evolution and function of pattern diversity in snakes', *Behavioral Ecology*, 24, pp. 1237–50.

Allen, W. L., Cuthill, I. C., Scott-Samuel, N. E., and Baddeley, R. (2010) 'Why the leopard got its spots: relating pattern development to ecology in felids', *Proceedings of the Royal Society of London B: Biological Sciences*. doi: 10.1098/rspb.2010.1734

Allen, W. L., and Higham, J. P. (2013) 'Analyzing visual signals as visual scenes', *American Journal of Primatology*, 75(7), pp. 664–82.

Aluthwattha, S. T., Harrison, R. D., Ranawana, K. B., Xu, C., Lai, R., and Chen, J. (2017) 'Does spatial variation in predation pressure modulate selection for aposematism?', *Ecology and Evolution*, 7(18), pp. 7560 72.

Alvarez, F. (1993) 'Alertness signalling in two rail species', *Animal Behaviour*, 46(6), pp. 1229–31.

Alvarez, F., Sánchez, C., and Angulo, S. (2006) 'Relationships between tail-flicking, morphology, and body condition in Moorhens', *Journal of Field Ornithology*, 77(1), pp. 1–6.

Amézquita, A., Castro, L., Arias, M., González, M., and Esquivel, C. (2013) 'Field but not lab paradigms support generalisation by predators of aposematic polymorphic prey: the *Oophaga histrionica* complex', *Evolutionary Ecology*, 27(4), pp. 769–82.

Ancillotto, L., and Mori, E. (2017) 'Adaptive significance of coat colouration and patterns of Sciuromorpha (Rodentia)', *Ethology Ecology & Evolution*, 29, pp. 241–54.

Andersson, M. (2015) 'Aposematism and crypsis in a rodent: antipredator defence of the Norwegian lemming', *Behavioral Ecology and Sociobiology*, 69(4), pp. 571–81.

Anthony, P. (1981) 'Visual contrast thresholds in the cod *Gadus morhua* L', *Journal of Fish Biology*, 19(1), pp. 87–103.

Antonovics, J., and Kareiva, P. (1988) 'Frequency-dependent selection and competition: empirical approaches', *Philosophical Transactions of the Royal Society of London Series B: Biological Sciences*, 319(1196), pp. 601–13.

Appelle, S. (1972) 'Perception and discrimination as a function of stimulus orientation: the "oblique effect" in man and animals', *Psychol Bull*, 78(4), pp. 266–78.

Arbuckle, K., Brockhurst, M., and Speed, M. P. (2013) 'Does chemical defence increase niche space? A phylogenetic comparative analysis of the Musteloidea', *Evolutionary Ecology*, 27(5), pp. 863–81.

Arbuckle, K., and Speed, M. P. (2015) 'Antipredator defenses predict diversification rates', *Proceedings of the National Academy of Sciences*, 112(44), pp. 13597–602.

Archetti, M., Döring, T. F., Hagen, S. B., Hughes, N. M., Leather, S. R., Lee, D. W., Lev-Yadun, S., Manetas, Y., Ougham, H. J., Schaberg, P. G., and Thomas, H. (2009) 'Unravelling the evolution of autumn colours: an interdisciplinary approach', *Trends in Ecology & Evolution*, 24, pp. 166–73.

Arcizet, F., Jouffrais, C., and Girard, P. (2009) 'Coding of shape from shading in area V4 of the macaque monkey', *BMC Neuroscience*, 10, pp. 140.

Arduino, P. J., and Gould, J. L. (1984) 'Is tonic immobility adaptive?', *Animal Behaviour*, 32(3), pp. 921–3.

Arenas, L. M., and Stevens, M. (2017) 'Diversity in warning coloration is easily recognized by avian predators', *Journal of Evolutionary Biology*, 30(7), pp. 1288–1302.

Arenas, L. M., Troscianko, J., and Stevens, M. (2014) 'Color contrast and stability as key elements for effective warning signals', *Frontiers in Ecology and Evolution*, 2(25). https://doi.org/10.3389/fevo.2014.00025

Arenas, L. M., Walter, D., and Stevens, M. (2015) 'Signal honesty and predation risk among a closely related group of aposematic species', *Scientific Reports*, 5, 11021. doi:10.1038/srep11021

Arias, M., le Poul, Y., Chouteau, M., Boisseau, R., Rosser, N., Thery, M., and Llaurens, V. (2016a) 'Crossing fitness valleys: empirical estimation of a fitness landscape associated with polymorphic mimicry', *Proceedings of the Royal Society B: Biological Sciences*, 283(1829).

Arias, M., Mappes, J., Thery, M., and Llaurens, V. (2016b) 'Inter-species variation in unpalatability does not explain polymorphism in a mimetic species', *Evolutionary Ecology*, 30(3), pp. 419–33.

Aronsson, M., and Gamberale-Stille, G. (2008) 'Domestic chicks primarily attend to colour, not pattern, when learning an aposematic coloration', *Animal Behaviour*, 75(2), pp. 417–23.

Aronsson, M., and Gamberale-Stille, G. (2012) 'Colour and pattern similarity in mimicry: evidence for a hierarchical discriminative learning of different components', *Animal Behaviour*, 84(4), pp. 881–7.

Aronsson, M., and Gamberale-Stille, G. (2013) 'Evidence of signaling benefits to contrasting internal color boundaries in warning coloration', *Behavioral Ecology*, 24(2), pp. 349–54.

Aubier, T. G., Joron, M., and Sherratt, T. N. (2017) 'Mimicry among unequally defended prey should be mutualistic when predators sample optimally', *The American Naturalist*, 189, pp. 267–82.

Aubier, T. G., and Sherratt, T. N. (2015) 'Diversity in Müllerian mimicry: the optimal predator sampling strategy explains both local and regional polymorphism in prey', *Evolution*, 69(11), pp. 2831–45.

Auko, T. H., Trad, B. M., and Silvestre, R. (2015) 'Bird dropping masquerading of the nest by the potter wasp *Minixi suffusum* (Fox, 1899) (Hymenoptera: Vespidae: Eumeninae)', *Tropical Zoology*, 28(2), pp. 56–65.

Aviezer, I., and Lev-Yadun, S. (2015) 'Pod and seed defensive coloration (camouflage and mimicry) in the genus *Pisum*', *Israel Journal of Plant Sciences*, 62(1–2), pp. 39–51.

Ayala, F.J., and Campbell, C.A., (1974) 'Frequency-dependent selection', *Annual Review of Ecology and Systematics*, 5(1), pp. 115–38.

Badcock, J. (1970) 'The vertical distribution of mesopelagic fishes collected on the SOND cruise', *Journal of the Marine Biological Association of the United Kingdom*, 50(4), pp. 1001–44.

Bagnara, J. T., and Matsumoto, J. (2007) 'Comparative anatomy and physiology of pigment cells in nonmammalian tissues', in Nordlund, J. J., Boissy, R. E., Hearing, V. J., King, R. A., Oetting, W. S., and Ortonne, J.-P. (eds), *The Pigmentary System: Physiology and Pathophysiology*, 2nd edn, Hoboken, NJ: Wiley-Blackwell, pp. 11–59.

Bailey, I. E., Muth, F., Morgan, K., Meddle, S. L. and Healy, S. D. (2015) 'Birds build camouflaged nests', *The Auk*, 132(1), pp. 11–15.

Bain, R., Rashed, A., Cowper, V., Gilbert, F. S., and Sherratt, T. N. (2007) 'The key mimetic features of hoverflies through avian eyes', *Proceedings of the Royal Society B: Biological Sciences*, 274, pp. 1949–54.

Balogh, A. C. V., Gamberale-Stille, G., and Leimar, O. (2008) 'Learning and the mimicry spectrum: from quasi-Bates to super-Müller', *Animal Behaviour*, 76, pp. 1591–9.

Balogh, A. C. V., Gamberale-Stille, G., Tullberg, B. S., and Leimar, O. (2010) 'Feature theory and the two-step hypothesis of Müllerian mimicry evolution', *Evolution*, 64(3), pp. 810–22.

Balogh, A. C. V., and Leimar, O. (2005) 'Müllerian mimicry: an examination of Fisher's theory of gradual evolutionary change', *Proceedings of the Royal Society B: Biological Sciences*, 272(1578), pp. 2269–75.

Barber, J. R., Leavell, B. C., Keener, A. L., Breinholt, J. W., Chadwell, B. A., McClure, C. J., Hill, G. M., and Kawahara, A. Y. (2015) 'Moth tails divert bat attack: evolution of acoustic deflection', *Proceedings of the National Academy of Sciences*, 112(9), pp. 2812–16.

Barbosa, A., Mathger, L. M., Buresch, K. C., Kelly, J., Chubb, C., Chiao, C. C., and Hanlon, R. T. (2008) 'Cuttlefish camouflage: the effects of substrate contrast and size in evoking uniform, mottle or disruptive body patterns', *Vision Res*, 48(10), pp. 1242–53.

Barbosa, A., Mathger, L. M., Chubb, C., Florio, C., Chiao, C. C., and Hanlon, R. T. (2007) 'Disruptive coloration in cuttlefish: a visual perception mechanism that regulates ontogenetic adjustment of skin patterning', *J Exp Biol*, 210(Pt 7), pp. 1139–47.

Barbour, M. A., and Clark, R. W. (2012) 'Ground squirrel tail-flag displays alter both predatory strike and ambush site selection behaviours of rattlesnakes', *Proceedings of the Royal Society of London B: Biological Sciences*, 279(1743), pp. 3827–33. doi: 10.1098/rspb.2012.1112

Barnard, C. (1984) 'When cheats may prosper', *Producers and Scroungers: Strategies of Exploitation and Parasitism*. London: Chapman & Hall, pp. 6–33.

Barnett, C. A., Bateson, M., and Rowe, C. (2007) 'State-dependent decision making: educated predators strategically trade off the costs and benefits of consuming aposematic prey', *Behavioral Ecology*, 18(4), pp. 645–51.

Barnett, C. A., Skelhorn, J., Bateson, M. and Rowe, C. (2012) 'Educated predators make strategic decisions to eat defended prey according to their toxin content', *Behavioral Ecology*, 23(2), pp. 418–24.

Barnett, J. B., and Cuthill, I. C. (2014) 'Distance-dependent defensive coloration', *Current Biology*, 24(24).

Barnett, J. B., Cuthill, I. C., and Scott-Samuel, N. E. (2017) 'Distance-dependent pattern blending can camouflage salient aposematic signals', *Proceedings of the Royal Society B-Biological Sciences*, 284(1858).

Barnett, J. B., Redfern, A. S., Bhattacharyya-Dickson, R., Clifton, O., Courty, T., Ho, T., Hopes, A., McPhee, T., Merrison, K., and Owen, R. (2016a) 'Stripes for warning and stripes for hiding: spatial frequency and detection distance', *Behavioral Ecology*, 28(2), pp. 373–81.

Barnett, J. B., Scott-Samuel, N. E., and Cuthill, I. C. (2016b) 'Aposematism: balancing salience and camouflage', *Biology Letters*, 12(8). doi: 10.1098/rsbl.2016.0335

Bateman, A. W., Vos, M. and Anholt, B. R. (2014) 'When to defend: antipredator defenses and the predation sequence', *The American Naturalist*, 183(6), pp. 847–55.

Bates, H. W. (1862) 'Contributions to an insect fauna of the Amazon valley. Lepidoptera: Heliconidae.', *Transactions of the Linnean Society of London*, 23, pp. 495–566.

Bateson, W. (1892) 'The alleged "aggressive mimicry" of Volucellæ', *Nature*, 46, pp. 585–6.

Beatty, C. D., Beirinckx, K., and Sherratt, T. N. (2004) 'The evolution of Müllerian mimicry in multispecies communities', *Nature*, 431(7004), pp. 63–7.

Beccaloni, G. W. (1997) 'Vertical stratification of ithomiine butterfly (Nymphalidae: Ithomiinae) mimicry complexes: the relationship between adult flight height and larval host-plant height', *Biological Journal of the Linnean Society*, 62(3), pp. 313–41.

Beddard, F. E. (1892) *Animal coloration: an account of the principal facts and theories relating to the colours and markings of animals*. London: S. Sonnenschein & Company.

Behrens, R. R. (1999) 'The role of artists in ship camouflage during World War I', *Leonardo*, 32(1), pp. 53–9.

Behrens, R. R. (2002) *False colors: art, design and modern camouflage*. Bobolink Books.

Behrens, R. R. (2009) 'Revisiting Abbott Thayer: non-scientific reflections about camouflage in art, war and zoology.', *Philosophical Transactions of the Royal Society B: Biological Sciences*, 364, pp. 497–501.

Behrens, R. R. (2012) *Ship Shape: A Dazzle Camouflage Sourcebook; an Anthology of Writings about Ship Camouflage During World War One*. Bobolink Books.

Bekers, W., De Meyer, R., and Strobbe, T. (2016) 'Shape recognition for ships: World War I naval camouflage under the magnifying glass', *WIT Transactions on The Built Environment*, 158, pp. 157–68.

Beldale, P., Koops, K., and Brakefield, P. M. (2002) 'Modularity, individuality, and evo-devo in butterfly wings', *Proceedings of the National Academy of Sciences of the United States of America*, 99, pp. 14262–7.

Belt, T. (1874) *The Naturalist in Nicaragua*. New York: E.P. Dutton.

Benson, W. W. (1972) 'Natural selection for Müllerian mimicry in *Heliconius erato* in Costa Rica', *Science*, 176(4037), pp. 936–9.

Berenbaum, M. (1983) 'Coumarins and caterpillars: a case for coevolution', *Evolution*, 37(1), pp. 163–79.

Berenbaum, M. R., and Miliczky, E. (1984) 'Mantids and milkweed bugs: efficacy of aposematic coloration against invertebrate predators.', *American Midland Naturalist*, 111, pp. 64–8.

Berenbaum, M. R., and Zangerl, A. R. (1992) 'Genetics of physiological and behavioral resistance to host furanocoumarins in the parsnip webworm', *Evolution*, 46(5), pp. 1373–84.

Berenbaum, M. R., and Zangerl, A. R. (1998) 'Chemical phenotype matching between a plant and its insect herbivore', *Proceedings of the National Academy of Sciences of the United States of America*, 95(23), pp. 13743–8.

Berenbaum, M. R., Zangerl, A. R., and Nitao, J. K. (1986) 'Constraints on chemical coevolution – wild parsnips and the parsnip webworm', *Evolution*, 40(6), pp. 1215–28.

Bergstrom, C. T., and Lachmann, M. (2001) 'Alarm calls as costly signals of antipredator vigilance: the watchful babbler game', *Animal Behaviour*, 61(3), pp. 535–43.

Bernays, E. A., and Wcislo, W. T. (1994) 'Sensory capabilities, information processing, and resource specialization', *The Quarterly Review of Biology*, 69(2), pp. 187–204.

Bhandiwad, A., and Johnsen, S. (2011) 'The effects of salinity and temperature on the transparency of the grass shrimp *Palaemonetes pugio*', *Journal of Experimental Biology*, 214(5), pp. 709–16.

Bian, X., Elgar, M. A., and Peters, R. A. (2016) 'The swaying behavior of *Extatosoma tiaratum*: motion camouflage in a stick insect?', *Behavioral Ecology*, 27, pp. 83–92.

Binetti, V. R., Schiffman, J. D., Leaffer, O. D., Spanier, J. E., and Schauer, C. L. (2009) 'The natural transparency and piezoelectric response of the *Greta oto* butterfly wing', *Integrative Biology*, 1(4), pp. 324–9.

Bitton, P.-P., and Doucet, S. M. (2013) 'A multifunctional visual display in elegant trogons targets conspecifics and heterospecifics', *Behavioral Ecology*, 25(1), pp. 27–34.

Blair, J., and Wassersug, R. J. (2000) 'Variation in the pattern of predator-induced damage to tadpole tails', *Copeia*, 2000(2), pp. 390–401.

Blaisdel, M. (1982) 'Natural theology and nature's disguises', *Journal of the History of Biology*, 15, pp. 163–89.

Blake, R., and Wilson, H. (2011) 'Binocular vision', *Vision Research*, 51, pp. 754–70.

Blanchard, B. D., and Moreau, C. S. (2017) 'Defensive traits exhibit an evolutionary trade-off and drive diversification in ants', *Evolution*, 71(2), pp. 315–28.

Bliss, M. M., and Cecala, K. K. (2017) 'Terrestrial salamanders alter antipredator behavior thresholds following tail autotomy', *Herpetologica*, 73(2), pp. 94–9.

Blodgett, L. S. (1919) *Ship camouflage*. Cambridge, MA: Massachusetts Institute of Technology Press.

Blount, J. D., Rowland, H. M., Drijfhout, F. P., Endler, J. A., Inger, R., Sloggett, J. J., Hurst, G. D., Hodgson, D. J., and Speed, M. P. (2012) 'How the ladybird got its spots: effects of resource limitation on the honesty of aposematic signals', *Functional Ecology*, 26(2), pp. 334–42.

Blount, J. D., Speed, M. P., Ruxton, G. D., and Stephens, P. A. (2009) 'Warning displays may function as honest signals of toxicity', *Proceedings of the Royal Society of London B: Biological Sciences*, 276(1658), pp. 871–7.

Blum, M. J. (2002) 'Rapid movement of a *Heliconius* hybrid zone: for phase III of Wright's shifting balance theory?', *Evolution*, 56, pp. 1992–8.

Boevé, J. L., Sonet, G., Nagy, Z. T., Symoens, F., Altenhofer, E., Haberlein, C., and Schulz, S. (2009) 'Defense by volatiles in leaf-mining insect larvae', *Journal of Chemical Ecology*, 35(5), pp. 507–17.

Bohl, E. (1982) 'Food supply and prey selection in planktivorous Cyprinidae', *Oecologia*, 53(1), pp. 134–8.

Bohlin, T., Gamberale-Stille, G., Merilaita, S., Exnerova, A., Stys, P., and Tullberg, B. S. (2012) 'The detectability of the colour pattern in the aposematic firebug, *Pyrrhocoris apterus*: an image-based experiment with human 'predators'', *Biological Journal of the Linnean Society*, 105(4), pp. 806–16.

Bohlin, T., Tullberg, B. S., and Merilaita, S. (2008) 'The effect of signal appearance and distance on detection risk in an aposematic butterfly larva (*Parnassius apollo*)', *Animal Behaviour*, 76, pp. 577–84.

Bolton, S. K., Dickerson, K., and Saporito, R. A. (2017) 'Variable alkaloid defenses in the dendrobatid poison frog *Oophaga pumilio* are perceived as differences in palatability to arthropods', *Journal of Chemical Ecology*, 43(3), pp. 273–89.

Bond, A. B. (1983) 'Visual search and selection of natural stimuli in the pigeon: the attention threshold hypothesis', *Journal of Experimental Psychology: Animal Behavior Processes*, 9(3), pp. 292.

Bond, A. B., and Kamil, A. C. (1998) 'Apostatic selection by blue jays produces balanced polymorphism in virtual prey', *Nature*, 395(6702), pp. 594.

Bond, A. B., and Kamil, A. C. (2002) 'Visual predators select for crypticity and polymorphism in virtual prey', *Nature*, 415(6872), pp. 609.

Bond, A. B., and Kamil, A. C. (2006) 'Spatial heterogeneity, predator cognition, and the evolution of color polymorphism in virtual prey', *Proceedings of the National Academy of Sciences*, 103, pp. 3214.

Borer, M., van Noort, T., Rahier, M., and Naisbit, R. E. (2010) 'Positive frequency-dependent selection on warning color in alpine leaf beetles', *Evolution*, 64(12), pp. 3629–33.

Borst, A., Haag, J., and Reiff, D. F. (2010) 'Fly motion vision', *Annual Review of Neuroscience*, 33, pp. 49–70.

Bouwma, P. E., and Herrnkind, W. F. (2009) 'Sound production in Caribbean spiny lobster *Panulirus argus* and its role in escape during predatory attack by *Octopus briareus*', *New Zealand Journal of Marine and Freshwater Research*, 43(1), pp. 3–13.

Bowdish, T. I., and Bultman, T. L. (1993) 'Visual cues used by mantids in learning aversion to aposematically colored prey', *American Midland Naturalist*, 129(2), pp. 215–22.

Bowers, M. D. (1992) 'The evolution of unpalatability and the cost of chemical defense in insects', in Evans, D. L., and Schmidt, J. O. (eds), *Insect Chemical Ecology*, London: Chapman & Hall, pp. 216–44.

Bowers, M. D., Brown, I. L., and Wheye, D. (1985) 'Bird predation as a selective agent in a butterfly population', *Evolution*, 39(1), pp. 93–103.

Boyden, T. C. (1976) 'Butterfly palatability and mimicry - experiments with *Ameiva* lizards', *Evolution*, 30(1), pp. 73–81.

Braude, S., Ciszek, D., Berg, N. E., and Shefferly, N. (2001) 'The ontogeny and distribution of countershading in colonies of the naked mole-rat (*Heterocephalus glaber*)', *Journal of Zoology*, 253, pp. 351–7.

Bretagnolle, V. (1993) 'Adaptive significance of seabird coloration: The case of Procellariiformes', *The American Naturalist*, 142, pp. 141–73.

Britton, N., Planqué, R. and Franks, N. (2007) 'Evolution of defence portfolios in exploiter–victim systems', *Bulletin of Mathematical Biology*, 69(3), pp. 957–88.

Brockhurst, M. A., Chapman, T., King, K. C., Mank, J. E., Paterson, S., and Hurst, G. D. (2014) 'Running with the Red Queen: the role of biotic conflicts in evolution', *Proceedings of the Royal Society of London B: Biological Sciences*, 281(1797), pp. 20141382.

Brodie, E. (1992) 'Correlational selection for color pattern and antipredator behavior in the garter snake *Thamnophis ordinoides*', *Evolution*, 46, pp. 1284–98.

Brodie, E. D. (1980) 'Differential avoidance of mimetic salamanders by free-ranging birds', *Science*, 208(4440), pp. 181–2.

Brodie, E. D. (1981) 'Phenological relationships of model and mimic salamanders', *Evolution*, 35(5), pp. 988–94.

Brodie, E. D., and Howard, R. R. (1973) 'Experimental study of Batesian mimicry in the salamanders *Plethodon jordani* and *Desmognathus ochrophaeus*', *American Midland Naturalist*, 90, pp. 38–46.

Brodie III, E. D., and Brodie Jr, E. D. (1999a) 'Costs of exploiting poisonous prey: evolutionary trade-offs in a predator-prey arms race', *Evolution*, 53, pp. 626–31.

Brodie III, E. D., and Brodie Jr, E. D. (1999b) 'Predator-prey arms races: asymmetrical selection on predators and prey may be reduced when prey are dangerous', *Bioscience*, 49(7), pp. 557–68.

Brooks, J. L. (2015) 'Traditional and new principles of perceptual grouping', *Oxford handbook of perceptual organization*, pp. 57–87. Edited by Johan Wagemans. OUP

Broom, M., Higginson, A. D. and Ruxton, G. D. (2010) 'Optimal investment across different aspects of anti-predator defences', *Journal of Theoretical Biology*, 263(4), pp. 579–86.

Broom, M., and Ruxton, G. D. (2012) 'Perceptual advertisement by the prey of stalking or ambushing predators', *Journal of Theoretical Biology*, 315, pp. 9–16.

Brouwer, G. J., and Heeger, D. J. (2011) 'Cross-orientation suppression in human visual cortex', *Journal of Neurophysiology*, 106(5), pp. 2108–19.

Brower, A. V. (1995) 'Locomotor mimicry in butterflies? A critical review of the evidence.', *Philosophical Transactions of the Royal Society of London Series B: Biological Sciences*, 347, pp. 413–25.

Brower, J. V. (1958a) 'Experimental studies of mimicry in some North American butterflies .1. The monarch, *Danaus plexippus*, and Viceroy, *Limenitis archippusarchippus*', *Evolution*, 12(1), pp. 32–47.

Brower, J. V. (1958b) 'Experimental studies of mimicry in some North American butterflies .2. *Battus philenor* and *Papilio troilus*, *P. polyxenes* and *P. glaucus*', *Evolution*, 12(2), pp. 123–36.

Brower, J. V. (1958c) 'Experimental studies of mimicry in some North American butterflies .3. *Danaus gilippus berenice* and *Limenitis archippus floridensis*', *Evolution*, 12(3), pp. 273–85.

Brower, J. V. (1960) 'Experimental studies of mimicry .IV. The reactions of starlings to different proportions of models and mimics', *American Naturalist*, 94(877), pp. 271–82.

Brower, L. P. (1988) 'Avian predation on the monarch butterfly and its implications for mimicry theory', *The American Naturalist*, 131, pp. S4–S6.

Brower, L. P., Brower, J. V., and Westcott, P. W. (1960) 'Experimental studies of mimicry .5. The reactions of toads (*Bufo terrestris*) to bumblebees (*Bombus americanorum*) and their robberfly mimics (*Mallophora bomboides*), with a discussion of aggressive mimicry', *American Naturalist*, 94(878), pp. 343–55.

Brower, L. P., Brower, J. V., and Collins, C. T. (1963) 'Experimental studies of mimicry. 7. Relative palatability and Müllerian mimicry among neotropical butterflies of the subfamily Heliconiinae', *Zoologica, New York*, 48, pp. 65–84.

Brower, L. P., and Calvert, W. H. (1985) 'Foraging dynamics of bird predators on overwintering monarch butterflies in Mexico', *Evolution*, 39, pp. 852–68.

Brower, L. P., Cook, L. M., and Croze, H. J. (1967a) 'Predator responses to artificial Batesian mimics released in a neotropical environment', *Evolution*, 21(1), pp. 11–23.

Brower, L. P., Hower, A. S., Croze, H. J., Brower, J. V., and Stiles, F. G. (1964) 'Mimicry - differential advantage of color patterns in the natural environment', *Science*, 144(361), pp. 183–5.

Brower, L. P., Brower, J. V., and Corvino, J. M. (1967b) 'Plant poisons in a terrestrial food chain', *Proceedings of the National Academy of Sciences of the United States of America*, 57(4), pp. 893–8.

Browman, H., Novales-Flamarique, H., and Hawryshyn, C. (1994) 'Ultraviolet photoreception contributes to prey search behaviour in two species of zooplanktivorous fishes', *Journal of Experimental Biology*, 186(1), pp. 187–98.

Brown, C. M., Henderson, D. M., Vinther, J., Fletcher, I., Sistiaga, A., Herrera, J., and Summons, R. E. (2017) 'An exceptionally preserved three-dimensional armored dinosaur reveals insights into coloration and Cretaceous predator-prey dynamics', *Current Biology*, 27(16), pp. 2514–21. e3.

Brown, G. E., Godin, J.-G. J., and Pedersen, J. (1999) 'Fin-flicking behaviour: a visual antipredator alarm signal in a characin fish, *Hemigrammus erythrozonus*', *Animal Behaviour*, 58(3), pp. 469–75.

Brown, J. L., Maan, M. E., Cummings, M. E., and Summers, K. (2010) 'Evidence for selection on coloration in a Panamanian poison frog: a coalescent-based approach', *Journal of Biogeography*, 37(5), pp. 891–901.

Brown, S. G., Boettner, G. H., and Yack, J. E. (2007) 'Clicking caterpillars: acoustic aposematism in *Antheraea polyphemus* and other Bombycoidea', *Journal of Experimental Biology*, 210(6), pp. 993–1005.

Brownell, C. L. (1985) 'Laboratory analysis of cannibalism by larvae of the Cape anchovy *Engraulis capensis*', *Transactions of the American Fisheries Society*, 114(4), pp. 512–18.

Bura, V. L., Rohwer, V. G., Martin, P. R., and Yack, J. E. (2011) 'Whistling in caterpillars (*Amorpha juglandis*, Bombycoidea): sound-producing mechanism and function', *Journal of Experimental Biology*, 214(1), pp. 30–7.

Buresch, K. C., Mäthger, L. M., Allen, J. J., Bennice, C., Smith, N., Schram, J., Chiao, C.-C., Chubb, C. and Hanlon, R. T. (2011) 'The use of background matching vs. masquerade for camouflage in cuttlefish *Sepia officinalis*', *Vision Research*, 51(23–24), pp. 2362–8.

Burr, D. C., and Ross, J. (2002) 'Direct evidence that "speedlines" influence motion mechanisms', *Journal of Neuroscience*, 22(19), pp. 8661–4.

Cairns, D. K. (1986) 'Plumage colour in pursuit-diving seabirds: Why do penguins wear tuxedos?', *Bird Behavior*, 6, pp. 58–65.

Caley, M. J., and Schluter, D. (2003) 'Predators favour mimicry in a tropical reef fish', *Proceedings of the Royal Society of London Series B: Biological Sciences*, 270(1516), pp. 667–72.

Camazine, S. (1985) 'Olfactory aposematism: Association of food toxicity with naturally-occurring odor', *Journal of Chemical Ecology*, 11(9), pp. 1289–95.

Cardoso, M. Z., and Gilbert, L. E. (2007) 'A male gift to its partner? Cyanogenic glycosides in the spermatophore of longwing butterflies (*Heliconius*)', *Naturwissenschaften*, 94, pp. 39–42.

Carmona, D., Lajeunesse, M. J. and Johnson, M. T. (2011) 'Plant traits that predict resistance to herbivores', *Functional Ecology*, 25(2), pp. 358–67.

Caro, T. (1986a) 'The functions of stotting: a review of the hypotheses', *Animal Behaviour*, 34(3), pp. 649–62.

Caro, T. (1986b) 'The functions of stotting in Thomson's gazelles: some tests of the predictions', *Animal Behaviour*, 34(3), pp. 663–84.

Caro, T. (2005) *Antipredator defenses in birds and mammals*. University of Chicago Press.

Caro, T. (2013) 'The colours of extant mammals', *Seminars in Cell & Developmental Biology*, 24(6–7), pp. 542–52.

Caro, T. (2016) *Zebra stripes*. Chicago: University of Chicago Press.

Caro, T., and Allen, W. L. (2017) 'Interspecific visual signalling in animals and plants: a functional classification', *Phil. Trans. R. Soc. B*, 372(1724), pp. 20160344.

Caro, T., Graham, C., Stoner, C., and Vargas, J. (2004) 'Adaptive significance of antipredator behaviour in artiodactyls', *Animal Behaviour*, 67(2), pp. 205–28.

Caro, T., Lombardo, L., Goldizen, A., and Kelly, M. (1995) 'Tail-flagging and other antipredator signals in white-tailed deer: new data and synthesis', *Behavioral Ecology*, 6(4), pp. 442–50.

Caro, T., Stankowich, T., Kiffner, C., and Hunter, J. (2013) 'Are spotted skunks conspicuous or cryptic?', *Ethology, Ecology & Evolution*, 25(2), pp. 144–60.

Caro, T., Stoddard, M. C., and Stuart-Fox, D. (eds) (2017). Animal coloration: production, perception, function and application. *Philosophical Transactions of the Royal Society B: Biological Sciences*, 372 (1724), theme issue.

Caro, T. M., Beeman, K., Stankowich, T., and Whitehead, H. (2011) 'The functional significance of colouration in cetaceans', *Evolutionary Ecology*, 25, pp. 1231–45.

Caro, T. M., Izzo, A., Reiner Jr, R. C., Walker, H., and Stankowich, T. (2014) 'The function of zebra stripes', *Nature Communications*, 5. doi:10.1038/ncomms4535

Caro, T. M., and Melville, C. (2012) 'Investigating colouration in large and rare mammals: the case of the giant anteater', *Ethology, Ecology and Evolution*, 24, pp. 104–15.

Caro, T. M., and Stankowich, T. (2009) 'The function of contrasting pelage markings in artiodactyls', *Behavioral Ecology*, 21, pp. 78–84.

Caro, T. M., Stankowich, T., Mesnick, S. L., Costa, D. P., and Beeman, K. (2012) 'Pelage coloration in pinnipeds: functional considerations', *Behavioral Ecology*, 23, pp. 765–74.

Carpenter, G. D. H., and Ford, E. B. (1933) *Mimicry*. London: Methuen & Co.

Carroll, J. and Sherratt, T. (2013) 'A direct comparison of the effectiveness of two anti-predator strategies under field conditions', *Journal of Zoology*, 291(4), pp. 279–85.

Case, J., Warner, J., Barnes, A., and Lowenstine, M. (1977) 'Bioluminescence of lantern fish (Myctophidae) in response to changes in light intensity', *Nature*, 265(5590), pp. 179–81.

Cassill, D. L., Vo, K., and Becker, B. (2008) 'Young fire ant workers feign death and survive aggressive neighbors', *Naturwissenschaften*, 95(7), pp. 617–24.

Castet, E., Lorenceau, J., Shiffrar, M., and Bonnet, C. (1993) 'Perceived speed of moving lines depends on orientation, length, speed and luminance', *Vision Research*, 33(14), pp. 1921–36.

Catarino, M. F., and Zuanon, J. (2010) 'Feeding ecology of the leaf fish *Monocirrhus polyacanthus* (Perciformes: Polycentridae) in a terra firme stream in the Brazilian Amazon', *Neotropical Ichthyology*, 8, pp. 183–6.

Ceinos, R. M., Guillot, R., Kelsh, R. N., Cerdá-Reverter, J. M., and Rotllant, J. (2015) 'Pigment patterns in adult fish result from superimposition of two largely independent pigmentation mechanisms', *Pigment Cell & Melanoma Research*, 28, pp. 196–209.

Cerdá-Reverter, J. M., Haitina, T., Schiöth, H. B., and Peter, R. E. (2005) 'Gene structure of the goldfish agouti-signaling protein: a putative role in the dorsal-ventral pigment pattern of fish', *Endocrinology*, 146, pp. 1597–1610.

Chai, P. (1986) 'Field observations and feeding experiments on the responses of rufous-tailed jacamars (*Galbula ruficauda*) to free-flying butterflies in a tropical rain-forest', *Biological Journal of the Linnean Society*, 29(3), pp. 161–89.

Chai, P. (1990) 'Relationships between visual characteristics of rainforest butterflies and responses of a specialized insectivorous bird', in Wicksten, M. (ed.) *Adaptive coloration in invertebrates. Proceedings of Symposium sponsored by American Society of Zoologists.* Galveston: Texas A&M University, pp. 31–60.

Chai, P. (1996) 'Butterfly visual characteristics and ontogeny of responses to butterflies by a specialized tropical bird', *Biological Journal of the Linnean Society*, 59(1), pp. 37–67.

Chamberlain, N. L., Hill, R. I., Kapan, D. D., Gilbert, L. E., and Kronforst, M. R. (2009) 'Polymorphic butterfly reveals the missing link in ecological speciation', *Science*, 326(5954), pp. 847–50.

Charlesworth, D., and Charlesworth, B. (2011) 'Mimicry: The hunting of the supergene', *Current Biology*, 21(20), pp. R846–8.

Chatelain, M., Halpin, C. G., and Rowe, C. (2013) 'Ambient temperature influences birds' decisions to eat toxic prey', *Animal Behaviour*, 86(4), pp. 733–40.

Cheney, K. L. (2010) 'Multiple selective pressures apply to a coral reef fish mimic: a case of Batesian–aggressive mimicry', *Proceedings of the Royal Society B: Biological Sciences*, 277(1689), pp. 1849–55.

Cheney, K. L., and Côté, I. M. (2005) 'Mutualism or parasitism? The variable outcome of cleaning symbioses', *Biology Letters*, 1, pp. 162–5.

Chester, J. (2001) *The nature of penguins*. Vancouver: Greystone Books.

Chiao, C.-C., Chubb, C., Buresch, K. C., Siemann, L., and Hanlon, R. T. (2009) 'The scaling effects of substrate texture on camouflage patterning in cuttlefish.', *Vision Research*, 49, pp. 1647–56.

Chiao, C.-C., Chubb, C., and Hanlon, R. T. (2015) 'A review of visual perception mechanisms that regulate rapid adaptive camouflage in cuttlefish', *Journal of Comparative Physiology A*, 201(9), pp. 933–45.

Chiao, C. C., Chubb, C., and Hanlon, R. T. (2007) 'Interactive effects of size, contrast, intensity and configuration of background objects in evoking disruptive camouflage in cuttlefish', *Vision Res*, 47(16), pp. 2223–35.

Chiao, C. C., Ulmer, K. M., Siemann, L. A., Buresch, K. C., Chubb, C., and Hanlon, R. T. (2013) 'How visual edge features influence cuttlefish camouflage patterning', *Vision Res*, 83, pp. 40–7.

Chittka, L., and Osorio, D. (2007) 'Cognitive dimensions of predator responses to imperfect mimicry?', *PloS Biology*, 5(12), pp. 2754–8.

Chouteau, M., and Angers, B. (2011) 'The role of predators in maintaining the geographic organization of aposematic signals', *American Naturalist*, 178(6), pp. 810–17.

Chouteau, M., and Angers, B. (2012) 'Wright's shifting balance theory and the diversification of aposematic signals', *PloS One*, 7(3). https://doi.org/10.1371/journal.pone.0034028

Chouteau, M., Arias, M., and Joron, M. (2016) 'Warning signals are under positive frequency-dependent selection in nature', *Proceedings of the National Academy of Sciences*, 113(8), pp. 2164–9.

Chouteau, M., Llaurens, V., Piron-Prunier, F., and Joron, M. (2017) 'Polymorphism at a mimicry supergene maintained by opposing frequency-dependent selection pressures', *Proceedings of the National Academy of Sciences*, 114(31), pp. 8325.

Chouteau, M., Summers, K., Morales, V., and Angers, B. (2011) 'Advergence in Müllerian mimicry: the case of the poison dart frogs of Northern Peru revisited', *Biology Letters*, 7(5), pp. 796–800.

Cipollini, D., Walters, D. and Voelckel, C. (2018) 'Costs of resistance in plants: From theory to evidence', *Annual Plant Reviews*: John Wiley & Sons, Ltd. Annual Plant Reviews book series, Volume 47: Insect-Plant Interactions. Section III. Ecology and Evolution of Insect-Plant Interactions. https://doi.org/10.1002/9781119312994.apr0512

Claes, J. M., Aksnes, D. L., and Mallefet, J. (2010) 'Phantom hunter of the fjords: Camouflage by counterillumination in a shark (*Etmopterus spinax*)', *Journal of Experimental Marine Biology and Ecology*, 388, pp. 28–32.

Claes, J. M., and Mallefet, J. (2010) 'The lantern shark's light switch: turning shallow water crypsis into midwater camouflage', *Biology Letters*, 6, pp. 685–7.

Claes, J. M., Nilsson, D.-E., Straube, N., Collin, S. P., and Mallefet, J. (2014) 'Iso-luminance counterillumination drove bioluminescent shark radiation', *Scientific Reports*, 4, 4328. doi:10.1038/srep04328

Clark, R., Brown, S. M., Collins, S. C., Jiggins, C. D., Heckel, D. G., and Vogler, A. P. (2008) 'Colour pattern specification in the Mocker swallowtail *Papilio dardanus*: the transcription factor invected is a candidate for the mimicry locus H', *Proceedings of the Royal Society B: Biological Sciences*, 275(1639), pp. 1181–8.

Clark, R. W. (2005) 'Pursuit-deterrent communication between prey animals and timber rattlesnakes (*Crotalus horridus*): the response of snakes to harassment displays', *Behavioral Ecology and Sociobiology*, 59(2), pp. 258–61.

Clark, R. W., Dorr, S. W., Whitford, M. D., Freymiller, G. A., and Hein, S. R. (2016) 'Comparison of anti-snake displays in the sympatric desert rodents *Xerospermophilus tereticaudus* (round-tailed ground squirrels) and *Dipodomys deserti* (desert kangaroo rats)', *Journal of Mammalogy*, 97(6), pp. 1709–17.

Clarke, B. (1962) 'Natural selection in mixed populations of two polymorphic snails', *Heredity*, 17(3), p. 319.

Clarke, C., and Sheppard, P. M. (1975) 'Genetics of mimetic butterfly *Hypolimnas bolina* (L)', *Philosophical Transactions of the Royal Society of London Series B: Biological Sciences*, 272(917), pp. 229–65.

Clarke, C. A., Clarke, F. M. M., Gordon, I. J., and Marsh, N. A. (1989) 'Rule-breaking mimics - palatability of the butterflies *Hypolimnas bolina* and *Hypolimnas misippus*, a sister species pair', *Biological Journal of the Linnean Society*, 37(4), pp. 359–65.

Clarke, C. A., and Sheppard, P. M. (1959) 'The genetics of some mimetic forms of *Papilio dardanus*, Brown, and *Papilio glaucus*, Linn.', *Journal of Genetics*, 56, pp. 237–59.

Clarke, C. A., and Sheppard, P. M. (1960) 'Supergenes and mimicry', *Heredity*, 14, pp. 175–85.

Clay, K. (2014) 'Defensive symbiosis: a microbial perspective', *Functional Ecology*, 28(2), pp. 293–8.

Clay, K., and Schardl, C. (2002) 'Evolutionary origins and ecological consequences of endophyte symbiosis with grasses', *The American Naturalist*, 160(S4), pp. S99–S127.

Cock, M. (1978) 'The assessment of preference', *The Journal of Animal Ecology*, 47, pp. 805–16.

Cogni, R., Trigo, J. R., and Futuyma, D. J. (2012) 'A free lunch? No cost for acquiring defensive plant pyrrolizidine alkaloids in a specialist arctiid moth (*Utetheisa ornatrix*)', *Molecular Ecology*, 21(24), pp. 6152–62.

Comeault, A., and Noonan, B. (2011) 'Spatial variation in the fitness of divergent aposematic phenotypes of the poison frog, *Dendrobates tinctorius*', *Journal of Evolutionary Biology*, 24(6), pp. 1374–9.

Confer, J. L., Howick, G. L., Corzette, M. H., Kramer, S. L., Fitzgibbon, S., and Landesberg, R. (1978) 'Visual predation by planktivores', *Oikos*, 31(1), pp. 27–37.

Conner, W. E., Boada, R., Schroeder, F. C., Gonzalez, A., Meinwald, J., and Eisner, T. (2000) 'Chemical defense: bestowal of a nuptial alkaloidal garment by a male moth on its mate', *Proceedings of the National Academy of Sciences of the United States of America*, 97, pp. 14406–11.

Cook, L., Grant, B., Saccheri, I., and Mallet, J. (2012a) 'Selective bird predation on the peppered moth: the last experiment of Michael Majerus', *Biology Letters*, 8(4), pp. 609–12.

Cook, L. M., Brower, L. P., and Alcock, J. (1969) 'An attempt to verify mimetic advantage in a neotropical environment', *Evolution*, 23(2), pp. 339–45.

Cook, R. G., Qadri, M. A. J., Kieres, A., and Commons-Miller, N. (2012b) 'Shape from shading in pigeons', *Cognition*, 124, pp. 284–303.

Cooper, W. (2000) 'Pursuit deterrence in lizards', *Saudi J. Biosoc. Sci*, 7, pp. 15–28.

Cooper, W. E. (2001) 'Multiple roles of tail display by the curly-tailed lizard *Leiocephalus carinatus*: Pursuit deterrent and deflective roles of a social signal', *Ethology*, 107(12), pp. 1137–49.

Cooper, W. E. (2007) 'Escape and its relationship to pursuit-deterrent signalling in the Cuban curly-tailed lizard *Leiocephalus carinatus*', *Herpetologica*, 63(2), pp. 144–50.

Cooper, W. E. (2010a) 'Pursuit deterrence varies with predation risks affecting escape behaviour in the lizard *Callisaurus draconoides*', *Animal Behaviour*, 80(2), pp. 249–56.

Cooper, W. E. (2010b) 'Timing during predator–prey encounters, duration and directedness of a putative pursuit-deterrent signal by the zebra-tailed lizard, *Callisaurus draconoides*', *Behaviour*, 147(13), pp. 1675–91.

Cooper, W. E. (2011) 'Pursuit deterrence, predation risk, and escape in the lizard *Callisaurus draconoides*', *Behavioral Ecology and Sociobiology*, 65(9), pp. 1833.

Cooper, W. E., Caffrey, C., and Vitt, L. J. (1985) 'Aggregation in the banded gecko, *Coleonyx variegatus*', *Herpetologica*, 41(3), pp. 342–50.

Cooper, W. E., and Perez-Mellado, V. (2004) 'Tradeoffs between escape behavior and foraging opportunity by the Balearic lizard (*Podarcis lilfordi*)', *Herpetologica*, 60(3), pp. 321–4.

Cooper, W. E., and Vitt, L. J. (1991) 'Influence of detectability and ability to escape on natural-selection of conspicuous autotomous defenses', *Canadian Journal of Zoology-Revue Canadienne De Zoologie*, 69(3), pp. 757–64.

Cordero, C. (2001) 'A different look at the false head of butterflies', *Ecological Entomology*, 26(1), pp. 106–108.

Cortesi, F., and Cheney, K. (2010) 'Conspicuousness is correlated with toxicity in marine opisthobranchs', *Journal of Evolutionary Biology*, 23(7), pp. 1509–18.

Cortesi, F., Feeney, W. E., Ferrari, M. C. O., Waldie, P. A., Phillips, G. A. C., McClure, E. C., Sköld, H. N., Salzburger, W., Marshall, N. J., and Cheney, K. L. (2015) 'Phenotypic plasticity confers multiple fitness benefits to a mimic', *Current Biology*, 25(7), pp. 949–54.

Cott, H. B. (1940) *Adaptive coloration in animals.* London: Methuen.

Cox, C. L., and Davis Rabosky, A. R. (2013) 'Spatial and temporal drivers of phenotypic diversity in polymorphic snakes', *American Naturalist*, 182(2), pp. E40–E57.

Craig, J. L. (1982) 'On the evidence for a "pursuit deterrent" function of alarm signals of swamphens', *The American Naturalist*, 119(5), pp. 753–5.

Cresswell, W. (1994) 'Song as a pursuit-deterrent signal, and its occurrence relative to other anti-predation behaviours of skylark (*Alauda arvensis*) on attack by merlins (*Falco columbarius*)', *Behavioral Ecology and Sociobiology*, 34(3), pp. 217–23.

Cronin, T. W., Gagnon, Y. L., Johnsen, S., Marshall, N. J., and Roberts, N. W. (2016) 'Comment on "Open-ocean fish reveal an omnidirectional solution to camouflage in polarized environments"', *Science*, 353(6299), pp. 552.

Crook, A. C. (1997) 'Colour patterns in a coral reef fish Is background complexity important?', *Journal of Experimental Marine Biology and Ecology*, 217(2), pp. 237–52.

Crothers, L., Saporito, R. A., Yeager, J., Lynch, K., Friesen, C., Richards-Zawacki, C. L., McGraw, K., and Cummings, M. (2016) 'Warning signal properties covary with toxicity but not testosterone or aggregate carotenoids in a poison frog', *Evolutionary Ecology*, 30(4), pp. 601–21.

Crothers, L. R., and Cummings, M. E. (2015) 'A multifunctional warning signal behaves as an agonistic status signal in a poison frog', *Behavioral Ecology*, 26(2), pp. 560–8.

Cummings, M. E., and Crothers, L. R. (2013) 'Interacting selection diversifies warning signals in a polytypic frog: an examination with the strawberry poison frog', *Evolutionary Ecology*, 27(4), pp. 693–710.

Cummings, M. E., Rosenthal, G. G., and Ryan, M. J. (2003) 'A private ultraviolet channel in visual communication', *Proceedings of the Royal Society of London B: Biological Sciences*, 270(1518), pp. 897–904.

Cuthill, I. C., Allen, W. L., Arbuckle, K., Caspers, B., Chaplin, G., Hauber, M. E., Hill, G. E., Jablonski, N. G., Jiggins, C. D. and Kelber, A. (2017) 'The biology of color', *Science*, 357(6350), pp. eaan0221.

Cuthill, I. C., and Bennett, A. T. D. (1993) 'Mimicry and the eye of the beholder', *Proceedings of the Royal Society of London Series B: Biological Sciences*, 253, pp. 203–4.

Cuthill, I. C., Hiby, E., and Lloyd, E. (2006a) 'The predation costs of symmetrical cryptic coloration.', *Proceedings of the Royal Society B: Biological Sciences*, 273, pp. 1267–71.

Cuthill, I. C., Sanghera, N. S., Penacchio, O., Lovell, P. G., Ruxton, G. D., and Harris, J. M. (2016) 'Optimizing countershading camouflage', *Proceedings of the National Academy of Sciences of the United States of America*, 113(46), pp. 13093–7.

Cuthill, I. C., Stevens, M., Sheppard, J., Maddocks, T., Párraga, C. A., and Troscianko, T. S. (2005) 'Disruptive coloration and background pattern matching', *Nature*, 434(7029), pp. 72–4.

Cuthill, I. C., Stevens, M., Windsor, A. M. M., and Walker, H. J. (2006b) 'The effects of pattern symmetry on detection of disruptive and background-matching coloration', *Behavioral Ecology*, 17, pp. 828.

Cuthill, I. C., and Székely, A. (2009) 'Coincident disruptive coloration', *Philos Trans R Soc Lond B Biol Sci*, 364(1516), pp. 489–96.

Cuthill, I. C., and Troscianko, T. S. (2011) 'Animal camouflage: biology meets psychology, computer science and art', in Brebbia, C.A., Greated, C., and Collins, M.W. (eds), *Colour in art, design and nature*, Southampton: WIT Press, pp. 5–24.

Dale, G., and Pappantoniou, A. (1986) 'Eye-picking behavior of the cutlips minnow, *Exoglossum maxillingua* - Applications to studies of eyespot mimicry', *Annals of the New York Academy of Sciences*, 463, pp. 177–8.

Dall, S. R., and Cuthill, I. C. (1997) 'The information costs of generalism', *Oikos*, 80, pp. 197–202.

Darst, C. R., and Cummings, M. E. (2006) 'Predator learning favours mimicry of a less-toxic model in poison frogs', *Nature*, 440(7081), pp. 208–11.

Darst, C. R., Cummings, M. E., and Cannatella, D. C. (2006) 'A mechanism for diversity in warning signals: conspicuousness versus toxicity in poison frogs', *Proceedings of the National Academy of Sciences*, 103(15), pp. 5852–7.

Darwin, C. (1859) *On the Origin of Species by Means of Natural Selection, or the Preservation of Favoured Races in the Struggle for Life.* 1st edn. London: John Murray.

Darwin, C. (1887) *The life and letters of Charles Darwin: including an autobiographical chapter, edited by his son Francis Darwin.* London: Murray.

Davis, M. P., Sparks, J. S. and Smith, W. L. (2016) 'Repeated and widespread evolution of bioluminescence in marine fishes', *PLoS One*, 11, pp. e0155154.

Davis Rabosky, A. R., Cox, C. L., Rabosky, D. L., Title, P. O., Holmes, I. A., Feldman, A., and McGuire, J. A. (2016) 'Coral snakes predict the evolution of mimicry across New World snakes', *Nature Communications*, 7, pp. 11484.

Dawkins, M. (1971) 'Perceptual changes in chicks: another look at the 'search image' concept', *Animal Behaviour*, 19(3), pp. 566–74.

Dawkins, R. (1982) *The extended phenotype.* Oxford: Oxford University Press.

De Bona, S., Valkonen, J. K., López-Sepulcre, A., and Mappes, J. (2015) 'Predator mimicry, not conspicuousness, explains the efficacy of butterfly eyespots'. *Proceedings of the Royal Society of London B: Biological Sciences*, 282(1806). doi: 10.1098/rspb.2015.0202

De Cock, R., and Matthysen, E. (2003) 'Glow-worm larvae bioluminescence (Coleoptera: Lampyridae) operates as an aposematic signal upon toads (*Bufo bufo*)', *Behavioral Ecology*, 14(1), pp. 103–8.

de Lima Rocha, A. D., Menescal-de-Oliveira, L., and da Silva, L. F. S. (2017) 'Effects of human contact and intra-specific social learning on tonic immobility in guinea pigs, *Cavia porcellus*', *Applied Animal Behaviour Science*, 191, pp. 1–4.

de Ruiter, L. (1952) 'Some experiments on the camouflage of stick caterpillars', *Behaviour*, 4(3), pp. 222–32.

de Ruiter, L. (1956) 'Countershading in caterpillars', *Archives Neerlandaises de Zoologie*, 11, pp. 285–341.

de Ruiter, L. (1959) 'Some remarks on problems of the ecology and evolution of mimicry', *Archives Neerlandaises de Zoologie*, 13, pp. 351–68.

Dell'Aglio, D. D., Stevens, M., and Jiggins, C. D. (2016) 'Avoidance of an aposematically coloured butterfly by wild birds in a tropical forest', *Ecological Entomology*, 41(5), pp. 627–32.

Dennis, D. S., and Lavigne, R. J. (1976) 'Ethology of *Efferia varipes* with comments on species coexistence (Diptera: Asilidae)', *Journal of the Kansas Entomological Society*, 49(1), pp. 48–62.

Denny, M. W. (1993) *Air and water: the biology and physics of life's media*. Princeton: Princeton University Press.

Denton, E. (1970) 'On the organization of reflecting surfaces in some marine animals', *Phil. Trans. R. Soc. Lond. B*, 258, pp. 285–313.

Denton, E., Gilpin-Brown, J., and Wright, P. (1972) 'The angular distribution of the light produced by some mesopelagic fish in relation to their camouflage', *Proceedings of the Royal Society of London B: Biological Sciences*, 182(1067), pp. 145–58.

Deshmukh, R., Baral, S., Gandhimathi, A., Kuwalekar, M., and Kunte, K. (2017) 'Mimicry in butterflies: co-option and a bag of magnificent developmental genetic tricks', *WIREs Developmental Biology*, 7, e291.

Desjardin, D. E., Oliveira, A. G., and Stevani, C. V. (2008) 'Fungi bioluminescence revisited', *Photochemical & Photobiological Sciences*, 7, pp. 170–82.

DeVries, P. J., Lande, R., and Murray, D. (1999) 'Associations of co-mimetic ithomiine butterflies on small spatial and temporal scales in a neotropical rainforest', *Biological Journal of the Linnean Society*, 67(1), pp. 73–85.

Dial, B. E. (1986) 'Tail display in two species of iguanid lizards: a test of the" predator signal" hypothesis', *The American Naturalist*, 127(1), pp. 103–11.

Dill, L. M. (1975) 'Calculated risk-taking by predators as a factor in Batesian mimicry', *Canadian Journal of Zoology-Revue Canadienne De Zoologie*, 53(11), pp. 1614–21.

Dimitrova, M., and Merilaita, S. (2009) 'Prey concealment: visual background complexity and prey contrast distribution', *Behavioral Ecology*, 21(1), pp. 176–81.

Dimitrova, M., and Merilaita, S. (2011) 'Prey pattern regularity and background complexity affect detectability of background-matching prey', *Behavioral Ecology*, 23(2), pp. 384–90.

Dimitrova, M., Stobbe, N., Schaefer, H. M. and Merilaita, S. (2009) 'Concealed by conspicuousness: distractive prey markings and backgrounds', *Proceedings of the Royal Society of London B: Biological Sciences*, pp. rspb. 2009.0052.

Dittrich, W., Gilbert, F., Green, P., McGregor, P., and Grewcock, D. (1993) 'Imperfect mimicry: a pigeon's perspective', *Proceedings of the Royal Society of London Series B: Biological Sciences*, 251(1332), pp. 195–200.

Dixey, F. A. (1920) 'The geographical factor in mimicry', *Ecological Entomology*, 68(1–2), pp. 208–11.

Dookie, A. L., Young, C. A., Lamothe, G., Schoenle, L. A., and Yack, J. E. (2017) 'Why do caterpillars whistle at birds? Insect defence sounds startle avian predators', *Behavioural Processes*, 138, pp. 58–66.

Dowdy, N. J., and Conner, W. E. (2016) 'Acoustic aposematism and evasive action in select chemically defended Arctiine (Lepidoptera: Erebidae) species: Nonchalant or not?', *PloS One*, 11(4). https://doi.org/10.1371/journal.pone.0152981

Dressler, R. L. (1979) '*Eulaema bombiformis, E. meriana*, and Müllerian mimicry in related species (Hymenoptera, Apidae)', *Biotropica*, 11(2), pp. 144–51.

Duarte, R. C., Flores, A. A., and Stevens, M. (2017) 'Camouflage through colour change: mechanisms, adaptive value and ecological significance', *Phil. Trans. R. Soc. B*, 372(1724), pp. 20160342.

Dugas, M. B., Halbrook, S. R., Killius, A. M., Sol, J. F., and Richards-Zawacki, C. L. (2015) 'Colour and escape behaviour in polymorphic populations of an aposematic poison frog', *Ethology*, 121(8), pp. 813–22.

Dukas, R., and Ellner, S. (1993) 'Information processing and prey detection', *Ecology*, 74(5), pp. 1337–46.

Dukas, R., and Kamil, A. C. (2001) 'Limited attention: the constraint underlying search image', *Behavioral Ecology*, 12(2), pp. 192–9.

Duncan, C. J., and Sheppard, P. M. (1965) 'Sensory discrimination and its role in the evolution of Batesian mimicry', *Behaviour*, 24, pp. 269–82.

Duntley, S. Q. (1946) 'Visibility studies and some applications in the field of camouflage', *Summary Tech. Rept. Div. 16, NDRC*, 2, pp. 65.

Eacock, A., Rowland, H. M., Edmonds, N., and Saccheri, I. J. (2017) 'Colour change of twig-mimicking peppered moth larvae is a continuous reaction norm that increases camouflage against avian predators', *PeerJ*, 5, pp. e3999.

Edelaar, P., Serrano, D., Carrete, M., Blas, J., Potti, J., and Tella, J. L. (2012) 'Tonic immobility is a measure of boldness toward predators: an application of Bayesian structural equation modeling', *Behavioral Ecology*, 23(3), pp. 619–26.

Edmunds, J., and Edmunds, M. (1974) 'Polymorphic mimicry and natural selection: a reappraisal', *Evolution*, 28(3), pp. 402–7.

Edmunds, M. (1972) 'Defensive behaviour in Ghanaian praying mantids', *Zoological Journal of the Linnean Society*, 51(1), pp. 1–32.

Edmunds, M. (2000) 'Why are there good and poor mimics?', *Biological Journal of the Linnean Society*, 70(3), pp. 459–66.

Edmunds, M. (2009) 'Do nematocysts sequestered by aeolid nudibranchs deter predators? - a background to the debate', *Journal of Molluscan Studies*, 75, pp. 203–5.

Edmunds, M., and Dewhurst, R. A. (1994) 'The survival value of countershading with wild birds as predators', *Biological Journal of the Linnean Society*, 51, pp. 447–52.

Edmunds, M., and Reader, T. (2014) 'Evidence for Batesian mimicry in a polymorphic hoverfly', *Evolution*, 68(3), pp. 827–39.

Edmunds, M. E. (1974) *Defence in animals: a survey of anti-predator defences.* Harlow: Longman.

Egan, J., Sharman, R. J., Scott-Brown, K. C., and Lovell, P. G. (2016) 'Edge enhancement improves disruptive camouflage by emphasising false edges and creating pictorial relief', *Sci Rep*, 6, pp. 38274.

Ehrlich, P. R., and Raven, P. H. (1964) 'Butterflies and plants: a study in coevolution', *Evolution*, 18(4), pp. 586–608.

Eisner, H., Alsop, D., and Eisner, T. (1967) 'Defense mechanisms of arthropods. XX. Quantitative assessment of hydrogen cyanide production in two species of millipedes', *Psyche*, 74(2), pp. 10.

Eisner, T., Eisner, M., and Siegler, M. (2005) *Secret weapons: defenses of insects, spiders, scorpions, and other many-legged creatures.* Cambridge, MA: Harvard University Press.

Eisner, T., and Grant, R. P. (1981) 'Toxicity, odor aversion, and olfactory aposematism', *Science*, 213(4506), pp. 476.

El-Sayed, S. Z., Van Dijken, G. L., and Gonzalez-Rodas, G. (1996) 'Effects of ultraviolet radiation on marine ecosystems', *International Journal of Environmental Studies*, 51(3), pp. 199–216.

Elder, J. H., and Velisavljević, L. (2009) 'Cue dynamics underlying rapid detection of animals in natural scenes', *Journal of Vision*, 9(7), pp. 7–7.

Elias, M., Gompert, Z., Jiggins, C., and Willmott, K. (2008) 'Mutualistic interactions drive ecological niche convergence in a diverse butterfly community', *PloS Biology*, 6(12), pp. 2642–9.

Elias, M., and Joron, M. (2015) 'Mimicry in *Heliconius* and Ithomiini butterflies: The profound consequences of an adaptation', in Maurel, M.C. & Grandcolas, P. (eds) *Origins: Studies in Biological and Cultural Evolution, BIO Web of Conferences 4.* https://doi.org/10.1051/bioconf/20150400008

Emlen, J. M. (1968) 'Batesian mimicry: a preliminary theoretical investigation of quantitative aspects', *American Naturalist*, 102(925), pp. 235–41.

Endler, B. (1991) 'Interactions between predators and prey', in Krebs, J. R., and Davies, N. B. (eds), *Behavioural Ecology: An Evolutionary Approach,* Oxford: Blackwell Scientific Publications, pp. 169–202.

Endler, J. (1988) 'Frequency-dependent predation, crypsis and aposematic coloration', *Phil. Trans. R. Soc. Lond. B*, 319(1196), pp. 505–23.

Endler, J. A. (1981) 'An overview of the relationships between mimicry and crypsis', *Biological Journal of the Linnean Society*, 16(1), pp. 25–31.

Endler, J. A., and Mappes, J. (2004) 'Predator mixes and the conspicuousness of aposematic signals', *American Naturalist*, 163(4), pp. 532–47.

Endler, J. A., and Rojas, B. (2009) 'The spatial pattern of natural selection when selection depends on experience', *The American Naturalist*, 173(3), pp. E62–E78.

Engler-Chaouat, H. S., and Gilbert, L. E. (2007) 'De novo synthesis vs. sequestration: negatively correlated metabolic traits and the evolution of host plant specialization in cyanogenic butterflies', *Journal of Chemical Ecology*, 33, pp. 25–42.

Espinosa, I., and Cuthill, I. C. (2014) 'Disruptive colouration and perceptual grouping', *PLoS One*, 9(1), pp. e87153.

Ewell, A. H., Cullen, J. M., and Woodruff, M. L. (1981) 'Tonic immobility as a predator-defense in the rabbit (*Oryctolagus cuniculus*)', *Behavioral and Neural Biology*, 31(4), pp. 483–9.

Exnerová, A., Ježová, D., Štys, P., Doktorovová, L., Rojas, B., and Mappes, J. (2015) 'Different reactions to aposematic prey in two geographically distant populations of great tits', *Behavioral Ecology*, 26(5), pp. 1361–70.

Fabricant, S. A., Exnerova, A., Jezova, D., and Stys, P. (2014) 'Scared by shiny? The value of iridescence in aposematic signalling of the hibiscus harlequin bug', *Animal Behaviour*, 90, pp. 315–25.

Fabricant, S. A., and Smith, C. L. (2014) 'Is the hibiscus harlequin bug aposematic? The importance of testing multiple predators', *Ecology and Evolution*, 4(2), pp. 113–20.

Farkas, T. E. (2016) 'Body size, not maladaptive gene flow, explains death-feigning behaviour in *Timema cristinae*', *Evolutionary Ecology*, 30(4), pp. 623–34.

Feldman, C. R., Brodie, E. D., Jr., Brodie III, E. D., and Pfrender, M. E. (2010) 'Genetic architecture of a feeding adaptation: garter snake (*Thamnophis*) resistance to tetrodotoxin bearing prey', *Proceedings of the Royal Society B: Biological Sciences*, 277(1698), pp. 3317–25.

Feller, K. D., and Cronin, T. W. (2014) 'Hiding opaque eyes in transparent organisms: a potential role for larval eyeshine in stomatopod crustaceans', *Journal of Experimental Biology*, 217(18), pp. 3263–73.

Feller, K. D., Jordan, T. M., Wilby, D., and Roberts, N. W. (2017) 'Selection of the intrinsic polarization properties of animal optical materials creates enhanced structural reflectivity and camouflage', *Phil. Trans. R. Soc. B*, 372(1724), pp. 20160336.

Ferguson, G. P., and Messenger, J. B. (1991) 'A counter-shading reflex in cephalopods', *Proceedings of the Royal Society B: Biological Sciences*, 243, pp. 63–7.

Findlay, J. M., and Gilchrist, I. D. (2003) *Active vision: The psychology of looking and seeing*. Oxford: Oxford University Press.

Fink, L. S., and Brower, L. P. (1981) 'Birds can overcome the cardenolide defence of monarch butterflies in Mexico', *Nature*, 291, pp. 67–70.

Finkbeiner, S. D., Briscoe, A. D., and Reed, R. D. (2014) 'Warning signals are seductive: Relative contributions of color and pattern to predator avoidance and mate attraction in *Heliconius* butterflies', *Evolution*, 68(12), pp. 3410–20.

Fisher, N., and Zanker, J. M. (2001) 'The directional tuning of the barber-pole illusion', *Perception*, 30(11), pp. 1321–36.

Fisher, R. A. (1930) *The Genetical Theory of Natural Selection*. Oxford: Clarendon Press.

Fisher, R. A. (1958) *The genetic theory of natural selection*. New York: Dover Books.

FitzGibbon, C. D., and Fanshawe, J. H. (1988) 'Stotting in Thomson's gazelles: an honest signal of condition', *Behavioral Ecology and Sociobiology*, 23(2), pp. 69–74.

Flanagan, N. S., Tobler, A., Davison, A., Pybus, O. G., Kapan, D. D., Planas, S., Linares, M., Heckel, D., and McMillan, W. O. (2004) 'Historical demography of Müllerian mimicry in the neotropical *Heliconius* butterflies', *Proceedings of the National Academy of Sciences of the United States of America*, 101(26), pp. 9704–9.

Fleming, P. A., Muller, D., and Bateman, P. W. (2007) 'Leave it all behind: a taxonomic perspective of autotomy in invertebrates', *Biological Reviews*, 82(3), pp. 481–510.

Font, E., Carazo, P., Pérez i de Lanuza, G., and Kramer, M. (2012) 'Predator-elicited foot shakes in wall lizards (*Podarcis muralis*): evidence for a pursuit-deterrent function', *Journal of Comparative Psychology*, 126(1), pp. 87.

Forbes, P. (2011) *Dazzled and deceived: mimicry and camouflage*. New Haven, CT: Yale University Press.

Ford, E. (1940) 'Polymorphism and taxonomy', in Huxley, J. (ed.), *The New Systematics*, London: Oxford University Press, pp. 493–513.

Ford, E. B. (1936) 'The genetics of *Papilio dardanus* Brown (Lep.)', *Transactions of the Royal Entomological Society of London*, 85, pp. 435–66.

Ford, H. (1971) 'The degree of mimetic protection gained by new partial mimics', *Heredity*, 27, pp. 227–36.

Forsman, A., and Merilaita, S. (1999) 'Fearful symmetry: pattern size and asymmetry affects aposematic signal efficacy', *Evolutionary Ecology*, 13(2), pp. 131–40.

Foster, S. A. (1988) 'Diversionary displays of paternal stickleback: Defenses against cannibalistic groups', *Behavioral Ecology and Sociobiology*, 22(5), pp. 335–40.

Frank, S. A. (1993) 'Evolution of host–parasite diversity', *Evolution*, 47(6), pp. 1721–32.

Frank, T., and Widder, E. (2002) 'Effects of a decrease in downwelling irradiance on the daytime vertical distribution patterns of zooplankton and micronekton', *Marine Biology*, 140(6), pp. 1181–93.

Franks, D. W., and Noble, J. (2004) 'Batesian mimics influence mimicry ring evolution', *Proceedings of the Royal Society B: Biological Sciences*, 271(1535), pp. 191–6.

Franks, D. W., and Sherratt, T. N. (2007) 'The evolution of multicomponent mimicry', *Journal of Theoretical Biology*, 244(4), pp. 631–9.

Fraser, S., Callahan, A., Klassen, D., and Sherratt, T. N. (2007) 'Empirical tests of the role of disruptive coloration in reducing detectability', *Proc Biol Sci*, 274(1615), pp. 1325–31.

Freret-Meurer, N. V., Fernandez, T. C., Lopes, D. A., Vaccani, A. C., and Okada, N. B. (2017) 'Thanatosis in the Brazilian seahorse *Hippocampus reidi* Ginsburg, 1933 (Teleostei: Syngnathidae)', *Acta Ethologica*, 20(1), pp. 81–4.

Fryer, J. C. F. (1914) 'An investigation by pedigree breeding into the polymorphism of *Papilio polytes*, Linn', *Philosophical Transactions of the Royal Society of London. Series B, Containing Papers of a Biological Character*, 204, pp. 227–54.

Gabbott, S. E., Donoghue, P. C. J., Sansom, R. S., Vinther, J., Dolocan, A., and Purnell, M. A. (2016) 'Pigmented anatomy in Carboniferous cyclostomes and the evolution of the vertebrate eye', *Proceedings of the Royal Society B: Biological Sciences*, 283, pp. 20161151.

Gabrielsen, G., and Smith, E. (1985) 'Physiological responses associated with feigned death in the American opossum', *Acta Physiologica*, 123(4), pp. 393–8.

Gabritchevsky, E. (1924) Farbenpolymorphismus und Vererbung mimetischer Varietäten der Fliege *Volucella bombylans* und anderer "hummelähnlicher" Zweiflügler. *Z Indukt Abstamm u Vererb*, 32, pp. 321353.

Gagliano, M. (2008) 'On the spot: the absence of predators reveals eyespot plasticity in a marine fish', *Behavioral Ecology*, 19(4), pp. 733–9.

Galarza, J. A., Nokelainen, O., Ashrafi, R., Hegna, R. H., and Mappes, J. (2014) 'Temporal relationship between genetic and warning signal variation in the aposematic wood tiger moth (*Parasemia plantaginis*)', *Molecular Ecology*, 23(20), pp. 4939–57.

Galeano, S. P., and Harms, K. E. (2016) 'Coloration in the polymorphic frog *Oophaga pumilio* associates with level of aggressiveness in intraspecific and interspecific behavioral interactions', *Behavioral Ecology and Sociobiology*, 70(1), pp. 83–97.

Gall, B. G., Stokes, A. N., French, S. S., Brodie, E. D., and Brodie, E. D. (2012) 'Female newts (*Taricha granulosa*) produce tetrodotoxin laden eggs after long term captivity', *Toxicon*, 60(6), pp. 1057–62.

Gallup, G. G., Nash, R. F., and Ellison, A. L. (1971) 'Tonic immobility as a reaction to predation: Artificial eyes as a fear stimulus for chickens', *Psychonomic Science*, 23(1), pp. 79–80.

Gamberale-Stille, G., and Guilford, T. (2004) 'Automimicry destabilizes aposematism: predator sample-and-reject behaviour may provide a solution', *Proceedings of the Royal Society of London B: Biological Sciences*, 271(1557), pp. 2621–5.

Gamberale, G., and Tullberg, B. S. (1996) 'Evidence for a peak-shift in predator generalization among aposematic prey', *Proceedings of the Royal Society of London B: Biological Sciences*, 263(1375), pp. 1329–34.

Gamberale, G., and Tullberg, B. S. (1998) 'Aposematism and gregariousness: the combined effect of group size and coloration on signal repellence', *Proceedings of the Royal Society of London B: Biological Sciences*, 265(1399), pp. 889–94.

Gan, W., Liu, F., Zhang, Z., and Li, D. (2010) 'Predator perception of detritus and eggsac decorations spun by orb-web spiders *Cyclosa octotuberculata*: Do they function to camouflage the spiders?', *Current Zoology*, 56(3), pp. 379–87.

Geisler, W. S., Albrecht, D. G., Crane, A. M., and Stern, L. (2001) 'Motion direction signals in the primary visual cortex of cat and monkey', *Visual Neuroscience*, 18(4), pp. 501–16.

Gelperin, A. (1968) 'Feeding behaviour of praying mantis - a learned modification', *Nature*, 219(5152), pp. 399–400.

Gendron, R. P., and Staddon, J. E. (1983) 'Searching for cryptic prey: the effect of search rate', *The American Naturalist*, 121(2), pp. 172–86.

Gentry, G. L., and Dyer, L. A. (2002) 'On the conditional nature of neotropical caterpillar defenses against their natural enemies', *Ecology*, 83(11), pp. 3108–19.

Georges, S., Seriès, P., Frégnac, Y., and Lorenceau, J. (2002) 'Orientation dependent modulation of apparent speed: psychophysical evidence', *Vision Research*, 42(25), pp. 2757–72.

Georgeson, M. A., and Scott-Samuel, N. E. (1999) 'Motion contrast: a new metric for direction discrimination', *Vision Research*, 39(26), pp. 4393–402.

Getty, T. (1985) 'Discriminability and the sigmoid functional response: how optimal foragers could stabilize model–mimic complexes', *American Naturalist*, 125(2), pp. 239–56.

Getty, T. (1987) 'Crypsis, mimicry, and switching: the basic similarity of superficially different analyses', *The American Naturalist*, 130(5), pp. 793–7.

Getty, T. (2002) 'The discriminating babbler meets the optimal diet hawk', *Animal Behaviour*, 63, pp. 397–402.

Gibran, F. Z. (2004) 'Dying or illness feigning: an unreported feeding tactic of the comb grouper *Mycteroperca acutirostris* (Serranidae) from the Southwest Atlantic', *Copeia*, 2004(2), pp. 403–5.

Gibson, B. M., Lazareva, O. F., Gosselin, F., Schyns, P. G., and Wasserman, E. A. (2007) 'Nonaccidental properties underlie shape recognition in Mammalian and non-mammalian vision.', *Current Biology*, 17, pp. 336–40.

Gibson, D. O. (1974) 'Batesian mimicry without distastefulness?', *Nature*, 250, pp. 77–9.

Gibson, D. O. (1980) 'The role of escape in mimicry and polymorphism. I. The response of captive birds to artificial prey', *Biological Journal of the Linnean Society*, 14, pp. 201–14.

Gierer, A., and Meinhardt, H. (1972) 'A theory of biological pattern formation', *Biological Cybernetics*, 12, pp. 30–9.

Giguère, L., and Northcote, T. (1987) 'Ingested prey increase risks of visual predation in transparent *Chaoborus* larvae', *Oecologia*, 73(1), pp. 48–52.

Gilbert, C. D., and Li, W. (2013) 'Top-down influences on visual processing', *Nature Reviews Neuroscience*, 14(5).

Gilbert, F. (2005) 'The evolution of imperfect mimicry.', in Fellowes, M., Holloway, G., & Rolff, J. (eds) *Insect Evolutionary Ecology*. Wallingford: CABI, pp. 231–88.

Gildemeister, E. A., Payette, W. I., and Sullivan, A. M. (2017) 'Effects of size, caudal autotomy, and predator kairomones on the foraging behavior of Allegheny Mountain dusky salamanders (*Desmognathus ochrophaeus*)', *Acta Ethologica*, 20(2), pp. 157–64.

Gilman, R. T., Nuismer, S. L., and Jhwueng, D.-C. (2012) 'Coevolution in multidimensional trait space favours escape from parasites and pathogens', *Nature*, 483(7389), pp. 328–30.

Giske, J., Aksnes, D. L., Baliño, B. M., Kaartvedt, S., Lie, U., Nordeide, J. T., Salvanes, A. G. V., Wakili, S. M., and Aadnesen, A. (1990) 'Vertical distribution and trophic interactions of zooplankton and fish in Masfjorden, Norway', *Sarsia*, 75(1), pp. 65–81.

Givnish, T. J. (1990) 'Leaf mottling: Relation to growth form and leaf phenology and possible role as camouflage', *Functional Ecology*, 4(4), pp. 463–74.

Glanville, P. W., and Allen, J. A. (1997) 'Protective polymorphism in populations of computer-simulated moth-like prey', *Oikos*, 80, pp. 565–71.

Gluckman, T.-L., and Mundy, N. I. (2013) 'Cuckoos in raptors' clothing: barred plumage illuminates a fundamental principle of Batesian mimicry', *Animal Behaviour*, 86(6), pp. 1165–81.

Gluckman, T.-L., and Mundy, N. I. (2017) 'The differential expression of MC1R regulators in dorsal and ventral quail plumages during embryogenesis: Implications for plumage pattern formation', *PLoS One*, 12, pp. e0174714.

Gochfeld, M. (1984) 'Antipredator behaviour: aggressive and distraction displays of shorebirds', in Burger, J. & Olla, B. (eds.) *Shorebirds: breeding behaviour and populations*. New York: Plenum Press, pp. 289–377.

Godin, J.-G. J., and Davis, S. A. (1995a) 'Who dares, benefits: predator approach behaviour in the guppy (*Poecilia reticulata*) deters predator pursuit'. *Proceedings of the Royal Society B: Biological Sciences*, 259(1355), pp.193–200.

Godin, J.-G. J., and Davis, S. A. (1995b) 'Boldness and predator deterrence: a reply to Milinski & Boltshauser'. *Proceedings of the Royal Society B: Biological Sciences*, 262(1363), pp. 107–12.

Goldstein, E. B. and Brockmole, J. (2016) *Sensation and perception*. 10th edn. Boston: Cengage Learning.

Gomez, D., and Théry, M. (2007) 'Simultaneous crypsis and conspicuousness in color patterns: Comparative analysis of a neotropical rainforest bird community', *The American Naturalist*, 169, pp. S42–S61.

Gompert, Z., Willmott, K., and Elias, M. (2011) 'Heterogeneity in predator micro-habitat use and the maintenance of Müllerian mimetic diversity', *Journal of Theoretical Biology*, 281(1), pp. 39–46.

Goodger, J., Capon, R. J., and Woodrow, I. E. (2002) 'Cyanogenic polymorphism in *Eucalyptus polyanthemos* Schauer subsp. *vestita* L. Johnson and K. Hill (Myrtaceae)', *Biochemical Systematics and Ecology*, 30(7), pp. 617–30.

Gordon, S. P., Kokko, H., Rojas, B., Nokelainen, O., and Mappes, J. (2015) 'Colour polymorphism torn apart by opposing positive frequency-dependent selection, yet maintained in space', *Journal of Animal Ecology*, 84(6), pp. 1555–64.

Goss-Custard, J. (1996) *The Oystercatcher: from individuals to populations*. Oxford: Oxford University Press.

Götmark, F. (1987) 'White underparts in gulls function as hunting camouflage', *Animal Behaviour*, 35, pp. 1786–92.

Gottfried, M. D. (1989) 'Earliest fossil evidence for protective pigmentation in an actinopterygian fish', *Historical Biology*, 3, pp. 79–83.

Gould, S. J. (2010) *Bully for Brontosaurus: reflections in natural history*. New York: W. W. Norton & Co.

Grandcolas, P., and Desutter-Grandcolas, L. (1998) 'Successful use of a deimatic display by the praying mantid *Polyspilota aeruginosa* against the yellow-vented bulbul'. *Annales de la Société Entomologique de France*, 34, pp. 335–6.

Grant, B. S., and Clarke, C. A. (2000) 'Industrial melanism', *Nature Encyclopedia of Life Sciences*. London: Nature Publishing Group.

Greene, E. (1989) 'A diet-induced developmental polymorphism in a caterpillar', *Science*, 243(4891), pp. 643.

Greene, H. W., and McDiarmid, R. W. (2005) 'Wallace and Savage: heroes, theories and venomous snake mimicry', in Donnelly, M.A., Crother, B.I., Guyer, C., Wake, M.H. & White, M.E. (eds) *Ecology and Evolution in the Tropics: A Herpetological Perspective*. Chicago (Illinois): University of Chicago Press., pp. 190–208.

Greeney, H., Dyer, L. and Smilanich, A. (2012) 'Feeding by lepidopteran larvae is dangerous: A review of caterpillars' chemical, physiological, morphological, and behavioral defenses against natural enemies', *Invertebrate Survival Journal*, 9(1).

Greenwood, J. J. (1984) 'The functional basis of frequency-dependent food selection', *Biological Journal of the Linnean Society*, 23(2–3), pp. 177–99.

Greenwood, J. J. (1985) 'Frequency-dependent selection by seed-predators', *Oikos*, 44(1), pp.195–210.

Greenwood, J. J., and Elton, R. A. (1979) 'Analysing experiments on frequency-dependent selection by predators', *The Journal of Animal Ecology*, 48(3), pp. 721–37.

Greenwood, J. J. D., Cotton, P. A., and Wilson, D. M. (1989) 'Frequency-dependent delection on aposematic prey: some experiments', *Biological Journal of the Linnean Society*, 36(1–2), pp. 213–26.

Greenwood, J. J. D., Wood, E. M., and Batchelor, S. (1981) 'Apostatic selection of distasteful prey', *Heredity*, 47(Aug), pp. 27–34.

Gregory, P. T., and Gregory, L. A. (2006) 'Immobility and supination in garter snakes (*Thamnophis elegans*) following handling by human predators', *Journal of Comparative Psychology*, 120(3), pp. 262.

Gregory, P. T., Isaac, L. A., and Griffiths, R. A. (2007) 'Death feigning by grass snakes (*Natrix natrix*) in response to handling by human "predators."', *Journal of Comparative Psychology*, 121(2), pp. 123.

Guilford, T. (1985) 'Is kin selection involved in the evolution of warning coloration?', *Oikos*, 45, pp. 31–6.

Guilford, T. (1988) 'The evolution of conspicuous coloration', *The American Naturalist*, 131, pp. S7-S21.

Guilford, T. (1994) '"Go-slow" signalling and the problem of automimicry', *Journal of Theoretical Biology*, 170(3), pp. 311–16.

Guilford, T., and Dawkins, M. S. (1987) 'Search images not proven: a reappraisal of recent evidence', *Animal Behaviour*, 35(6), pp. 1838–45.

Guilford, T. and Dawkins, M. S. (1989a) 'Search image versus search rate: A reply to Lawrence', *Animal Behaviour*, 37(1), pp. 160–2.

Guilford, T., and Dawkins, M. S. (1989b) 'Search image versus search rate: Two different ways to enhance prey capture', *Animal Behaviour*, 37(1), pp. 163–5.

Guilford, T., and Dawkins, M. S. (1993) 'Are warning colors handicaps?', *Evolution*, 47(2), pp. 400–16.

Guillot, R., Ceinos, R. M., Cal, R., Rotllant, J., and Cerdá-Reverter, J. M. (2012) 'Transient ectopic overexpression of agouti-signalling protein 1 (Asip1) induces pigment anomalies in flatfish', *PLoS One*, 7, pp. e48526.

Gyssels, F. G., and Stoks, R. (2005) 'Threat-sensitive responses to predator attacks in a damselfly', *Ethology*, 111(4), pp. 411–23.

Haddock, S. H. D., Moline, M. A., and Case, J. F. (2010) 'Bioluminescence in the sea', *Annual Review of Marine Science*, 2, pp. 443–93.

Hall, J. R., Cuthill, I. C., Baddeley, R., Attwood, A. S., Munafò, M. R., and Scott-Samuel, N. E. (2016) 'Dynamic dazzle distorts speed perception', *PloS One*, 11(5), pp. e0155162.

Hall, J. R., Cuthill, I. C., Baddeley, R., Shohet, A. J., and Scott-Samuel, N. E. (2013) 'Camouflage, detection and identification of moving targets', *Proc Biol Sci*, 280(1758), pp. 20130064.

Halperin, T., Carmel, L., and Hawlena, D. (2017) 'Movement correlates of lizards' dorsal pigmentation patterns', *Functional Ecology*, 31(2), pp. 370–6.

Halpin, C. G., and Rowe, C. (2017) 'The effect of distastefulness and conspicuous coloration on the post-attack rejection behaviour of predators and survival of prey', *Biological Journal of the Linnean Society*, 120(1), pp. 236–44.

Halpin, C. G., Skelhorn, J., and Rowe, C. (2008a) 'Being conspicuous and defended: selective benefits for the individual', *Behavioral Ecology*, 19(5), pp. 1012–17.

Halpin, C. G., Skelhorn, J., and Rowe, C. (2008b) 'Naïve predators and selection for rare conspicuous defended prey: the initial evolution of aposematism revisited', *Animal Behaviour*, 75(3), pp. 771–81.

Halpin, C. G., Skelhorn, J., and Rowe, C. (2013) 'Predators' decisions to eat defended prey depend on the size of undefended prey', *Animal Behaviour*, 85(6), pp. 1315–21.

Halpin, C. G., Skelhorn, J., and Rowe, C. (2014) 'Increased predation of nutrient-enriched aposematic prey', *Proceedings of the Royal Society B: Biological Sciences*, 281(1781). doi: 10.1098/rspb.2013.3255

Ham, A. D., Ihalainen, E., Lindström, L., and Mappes, J. (2006) 'Does colour matter? The importance of colour in avoidance learning, memorability and generalisation', *Behavioral Ecology and Sociobiology*, 60(4), pp. 482–91.

Hämäläinen, L., Valkonen, J., Mappes, J., and Rojas, B. (2015) 'Visual illusions in predator–prey interactions: birds find moving patterned prey harder to catch', *Animal Cognition*, 18(5), pp. 1059–68.

Hamner, W. M. (1995) 'Predation, cover, and convergent evolution in epipelagic oceans', *Marine & Freshwater Behaviour & Phy*, 26(2–4), pp. 71–89.

Hancox, A. P., and Allen, J. A. (1991) 'A simulation of evasive mimicry in the wild', *Journal of Zoology*, 223, pp. 9–13.

Hanifin, C. T. (2010) 'The chemical and evolutionary ecology of tetrodotoxin (TTX) toxicity in terrestrial vertebrates', *Marine Drugs*, 8(3), pp. 577–93.

Hanifin, C. T., Brodie Jr, E. D., and Brodie III, E. D. (2008) 'Phenotypic mismatches reveal escape from arms-race coevolution', *PLoS Biology*, 6(3), pp. e60.

Hanlon, R., Chiao, C.-C., Mäthger, L., Barbosa, A., Buresch, K., and Chubb, C. (2009) 'Cephalopod dynamic camouflage: bridging the continuum between background matching and disruptive coloration', *Philosophical Transactions of the Royal Society of London B: Biological Sciences*, 364(1516), pp. 429–37.

Hanlon, R. T., Conroy, L. A., and Forsythe, J. W. (2008) 'Mimicry and foraging behaviour of two tropical sand-flat octopus species off North Sulawesi, Indonesia', *Biological Journal of the Linnean Society*, 93(1), pp. 23–38.

Hansen, L. S., Gonzales, S. F., Toft, S., and Bilde, T. (2008) 'Thanatosis as an adaptive male mating strategy in the nuptial gift-giving spider *Pisaura mirabilis*', *Behavioral Ecology*, 19(3), pp. 546–51.

Harper, G. R., and Pfennig, D. W. (2007) 'Mimicry on the edge: why do mimics vary in resemblance to their model in different parts of their geographical range?', *Proceedings of the Royal Society B: Biological Sciences*, 274(1621), pp. 1955–61.

Harper, R. D., and Case, J. F. (1999) 'Disruptive counterillumination and its anti-predatory value in the plainfish [sic] midshipman *Porichthys notatus*', *Marine Biology*, 134, pp. 529–40.

Harris, R. J., and Arbuckle, K. (2016) 'Tempo and mode of the evolution of venom and poison in tetrapods', *Toxins*, 8(7), pp. 193.

Harvey, P. H., Bull, J., Pemberton, M., and Paxton, R. J. (1982) 'The evolution of aposematic coloration in distasteful prey: a family model', *The American Naturalist*, 119(5), pp. 710–19.

Hasson, O. (1991) 'Pursuit-deterrent signals: communication between prey and predator', *Trends in Ecology & Evolution*, 6(10), pp. 325–9.

Hasson, O., Hibbard, R., and Ceballos, G. (1989) 'The pursuit deterrent function of tail-wagging in the zebra-tailed lizard (*Callisaurus draconoides*)', *Canadian Journal of Zoology*, 67(5), pp. 1203–9.

Hatle, J. D., and Faragher, S. G. (1998) 'Slow movement increases the survivorship of a chemically defended grasshopper in predatory encounters', *Oecologia*, 115(1–2), pp. 260–7.

Hawlena, D. (2006) 'Blue tail and striped body: why do lizards change their infant costume when growing up?', *Behavioral Ecology*, 17, pp. 889–96.

Hawlena, D. (2009) 'Colorful tails fade when lizards adopt less risky behaviors', *Behavioral Ecology and Sociobiology*, 64(2), pp. 205–13.

Hecht, M. K., and Marien, D. (1956) 'The coral snake mimic problem: a re-interpretation', *Journal of Morphology*, 98, pp. 335–56.

Hegna, R. H., Galarza, J. A., and Mappes, J. (2015) 'Global phylogeography and geographical variation in warning coloration of the wood tiger moth (*Parasemia plantaginis*)', *Journal of Biogeography*, 42(8), pp. 1469–81.

Hegna, R. H., Nokelainen, O., Hegna, J. R., and Mappes, J. (2013) 'To quiver or to shiver: increased melanization benefits thermoregulation, but reduces warning signal efficacy in the wood tiger moth', *Proceedings of the Royal Society of London B: Biological Sciences*, 280(1755), pp. 2012–812.

Heinrich, B. (2012) 'A heretofore unreported instant color change in a beetle, *Nicrophorus tomentosus* Weber (Coleoptera: Silphidae)', *Northeastern Naturalist*, 19(2), pp. 345–52.

Hennig, C. W., Dunlap, W. P., and Gallup, G. G. (1976) 'The effect of distance between predator and prey and the opportunity to escape on tonic immobility in *Anolis carolinensis*', *The Psychological Record*, 26(3), pp. 312–20.

Henry, G. H., Bishop, P. O., and Dreher, B. (1974) 'Orientation, axis and direction as stimulus parameters for striate cells', *Vision Research*, 14(9), pp. 767–77.

Herberstein, M. E., Baldwin, H. J., and Gaskett, A. C. (2014) 'Deception down under: is Australia a hot spot for deception?', *Behavioral Ecology*, 25(1), pp. 12–16.

Herberstein, M. E., Craig, C. L., Coddington, J. A., and Elgar, M. A. (2000) 'The functional significance of silk decorations of orb-web spiders: a critical review of the empirical evidence', *Biological Reviews*, 75(4), pp. 649–69.

Herring, P. (2001) *The biology of the deep ocean.* Oxford: Oxford University Press.

Hespenheide, H. A. (1973) 'A novel mimicry complex: beetles and flies', *Journal of Entomology Series A: General Entomology*, 48(1), pp. 49–55.

Hespenheide, H. A. (1975) 'Reversed sex-limited mimicry in a beetle', *Evolution*, 29, pp. 780–3.

Hessen, D. O. (1985) 'Selective zooplankton predation by pre-adult roach (*Rutilus rutilus*): the size-selective hypothesis versus the visibility-selective hypothesis', *Hydrobiologia*, 124(1), pp. 73–9.

Heurich, M., Zeis, K., Küchenhoff, H., Müller, J., Belotti, E., Bufka, L., and Woelfing, B. (2016) 'Selective predation of a stalking predator on ungulate prey', *PloS One*, 11(8), pp. e0158449.

Higginson, A. D., De Wert, L., Rowland, H. M., Speed, M. P., and Ruxton, G. D. (2012) 'Masquerade is associated with polyphagy and larval overwintering in Lepidoptera', *Biological Journal of the Linnean Society*, 106(1), pp. 90–103.

Higginson, A. D., Delf, J., Ruxton, G. D., and Speed, M. P. (2011) 'Growth and reproductive costs of larval defence in the aposematic lepidopteran *Pieris brassicae*', *Journal of Animal Ecology*, 80(2), pp. 384–92.

Hill, R. I. (2010) 'Habitat segregation among mimetic ithomiine butterflies (Nymphalidae)', *Evolutionary Ecology*, 24(2), pp. 273–85.

Hill, R. I., and Vaca, J. F. (2004) 'Differential wing strength in *Pierella* butterflies (Nymphalidae, Satyrinae) supports the deflection hypothesis', *Biotropica*, 36(3), pp. 362–70.

Hines, H. M., Counterman, B. A., Papa, R., de Moura, P. A., Cardoso, M. Z., Linares, M., Mallet, J., Reed, R. D., Jiggins, C. D., Kronforst, M. R., and McMillan, W. O. (2011) 'Wing patterning gene redefines the mimetic history of *Heliconius* butterflies', *Proceedings of the National Academy of Sciences of the United States of America*, 108(49), pp. 19666–71.

Hines, H. M., and Williams, P. H. (2012) 'Mimetic colour pattern evolution in the highly polymorphic *Bombus trifasciatus* (Hymenoptera: Apidae) species complex and its comimics', *Zoological Journal of the Linnean Society*, 166(4), pp. 805–26.

Hiroshi, N., Haruhisa, N., Ryozo, F., and Noriko, O. (1989) 'Correlation between body color and behavior in the upside-down catfish, *Synodontis nigriventris*', *Comparative Biochemistry and Physiology Part A: Physiology*, 92, pp. 323–6.

Hoekstra, H. E. (2011) 'From Darwin to DNA: the genetic basis of color adaptations', in Losos, J. B. (ed.), *In the Light of Evolution: Essays from the Laboratory and Field.* Greenwood Village, CO: Roberts and Co. Publishers, pp. 277–95. Hogan, B. G., Cuthill, I. C., and Scott-Samuel, N. E. (2016a) 'Dazzle camouflage, target tracking, and the confusion effect', *Behavioral Ecology*, 27(5), pp. 1547–51.

Hogan, B. G., Cuthill, I. C., and Scott-Samuel, N. E. (2017) 'Dazzle camouflage and the confusion effect: the influence of varying speed on target tracking', *Animal Behaviour*, 123, pp. 349–53.

Hogan, B. G., Scott-Samuel, N. E. and Cuthill, I. C. (2016b) 'Contrast, contours and the confusion effect in dazzle camouflage', *Royal Society Open Science*, 3(7), pp. 160180.

Holen, Ø. H. (2013) 'Disentangling taste and toxicity in aposematic prey', *Proceedings of the Royal Society Series B: Biological Sciences*, 280(1753), pp. 1–8.

Holen, Ø. H., and Johnstone, R. A. (2004) 'The evolution of mimicry under constraints', *American Naturalist*, 164(5), pp. 598–613.

Holen, Ø. H., and Svennungsen, T. O. (2012) 'Aposematism and the handicap principle', *The American Naturalist*, 180(5), pp. 629–41.

Holley, A. J. (1993) 'Do brown hares signal to foxes?', *Ethology*, 94(1), pp. 21–30.

Holling, C. S. (1965) 'The functional response of predators to prey density, and its role in mimicry and population regulation', *Memoirs of the Entomological Society of Canada*, 45, pp. 1–60.

Holm, E., and Kirsten, J. F. (1979) 'Pre-adaptation and speed mimicry among Namib Desert scarabaeids with orange elytra', *Journal of Arid Environments*, 2, pp. 263–71.

Holt, A. L., and Sweeney, A. M. (2016) 'Open water camouflage via 'leaky' light guides in the midwater squid *Galiteuthis*', *Journal of The Royal Society Interface*, 13(119), pp. 20160230.

Honma, A., Mappes, J., and Valkonen, J. K. (2015) 'Warning coloration can be disruptive: aposematic marginal wing patterning in the wood tiger moth', *Ecology and Evolution*, 5(21), pp. 4863–74.

Honma, A., Oku, S., and Nishida, T. (2006) 'Adaptive significance of death feigning posture as a specialized inducible defence against gape-limited predators', *Proceedings of the Royal Society of London B: Biological Sciences*, 273(1594), pp. 1631–6.

Honma, A., Takakura, K., and Nishida, T. (2008) 'Optimal-foraging predator favors commensalistic Batesian mimicry', *PloS One*, 3(10). https://doi.org/10.1371/journal.pone.0003411

Hoskin, C. J., Higgie, M., McDonald, K. R., and Moritz, C. (2005) 'Reinforcement drives rapid allopatric speciation', *Nature*, 437, pp. 1353–6.

Hossie, T., Hassall, C., Knee, W., and Sherratt, T. (2013) 'Species with a chemical defence, but not chemical offence, live longer', *Journal of Evolutionary Biology*, 26(7), pp. 1598–1602.

Hossie, T. J., and Sherratt, T. N. (2012) 'Eyespots interact with body colour to protect caterpillar-like prey from avian predators', *Animal Behaviour*, 84(1), pp. 167–73.

Hossie, T. J. and Sherratt, T. N. (2014) 'Does defensive posture increase mimetic fidelity of caterpillars with eyespots to their putative snake models?', *Current Zoology*, 60(1), pp. 76–89.

Hossie, T. J., Skelhorn, J., Breinholt, J. W., Kawahara, A. Y., and Sherratt, T. N. (2015) 'Body size affects the evolution of eyespots in caterpillars', *Proceedings of the National Academy of Sciences of the United States of America*, 112(21), pp. 6664–9.

Houston, A. I., Stevens, M., and Cuthill, I. C. (2007) 'Animal camouflage: compromise or specialize in a 2 patch-type environment?', *Behavioral Ecology*, 18(4), pp. 769–75.

How, M. J., and Zanker, J. M. (2014) 'Motion camouflage induced by zebra stripes', *Zoology*, 117(3), pp. 163–70.

Howard, J., Grill, S. W., and Bois, J. S. (2011) 'Turing's next steps: the mechanochemical basis of morphogenesis.', *Nature Reviews Molecular Cell Biology*, 12, pp. 392–8.

Howarth, B., and Edmunds, M. (2000) 'The phenology of Syrphidae (Diptera): are they Batesian mimics of Hymenoptera', *Biological Journal of the Linnean Society*, 71(3), pp. 437–57.

Howarth, B., Edmunds, M., and Gilbert, F. (2004) 'Does the abundance of hoverfly (Syrphidae) mimics depend on the numbers of their hymenopteran models?', *Evolution*, 58(2), pp. 367–75.

Howse, P., and Allen, J. A. (1994) 'Satyric mimicry: the evolution of apparent imperfection', *Proceedings of the Royal Society of London B: Biological Sciences*, 257(1349), pp. 111–14.

Howse, P. E. (2013) 'Lepidopteran wing patterns and the evolution of satyric mimicry', *Biological Journal of the Linnean Society*, 109(1), pp. 203–14.

Hozumi, N., and Miyatake, T. (2005) 'Body-size dependent difference in death-feigning behavior of adult *Callosobruchus chinensis*', *Journal of Insect Behavior*, 18(4), pp. 557–66.

Huang, J.-N., Cheng, R.-C., Li, D., and Tso, I. M. (2011) 'Salticid predation as one potential driving force of ant mimicry in jumping spiders', *Proceedings of the Royal Society B: Biological Sciences*, 278(1710), pp. 1356–64.

Hughes, A. E., Jones, C., Joshi, K., and Tolhurst, D. J. (2017) 'Diverted by dazzle: perceived movement direction is biased by target pattern orientation'. *Proceedings of the Royal Society B: Biological Sciences*, 284(1850). doi: 10.1098/rspb.2017.0015

Hughes, A. E., Magor-Elliott, R. S., and Stevens, M. (2015) 'The role of stripe orientation in target capture success', *Frontiers in Zoology*, 12(1), pp. 17.

Hughes, A. E., Troscianko, J., and Stevens, M. (2014) 'Motion dazzle and the effects of target patterning on capture success', *BMC Evolutionary Biology*, 14(1), pp. 201.

Hummel, J. E. (2013) 'Object recognition', in Reisberg, D. (ed.), *Oxford Handbook of Cognitive Psychology*, New York: Oxford University Press, pp. 32–46.

Husak, J. F., Macedonia, J. M., Fox, S. F., and Sauceda, R. C. (2006) 'Predation cost of conspicuous male coloration in collared lizards (*Crotaphytus collaris*): An experimental test using clay-covered model lizards', *Ethology*, 112(6), pp. 572–80.

Ihalainen, E., Lindström, L., and Mappes, J. (2007) 'Investigating Müllerian mimicry: predator learning and variation in prey defences', *Journal of Evolutionary Biology*, 20(2), pp. 780–91.

Ihalainen, E., Rowland, H. M., Speed, M. P., Ruxton, G. D., and Mappes, J. (2012) 'Prey community structure affects how predators select for Mullerian mimicry', *Proceedings of the Royal Society B: Biological Sciences*, 279(1736), pp. 2099–105.

Ingalls, V. (1993) 'Startle and habituation responses of blue jays (*Cyanocitta cristata*) in a laboratory simulation of anti-predator defenses of *Catocala* moths (Lepidoptera: Noctuidae)', *Behaviour*, 126(1), pp. 77–95.

Ioannou, C., and Krause, J. (2009) 'Interactions between background matching and motion during visual detection can explain why cryptic animals keep still', *Biology Letters*, 5, pp. 191–3.

Ioannou, C. C., Morrell, L. J., Ruxton, G. D., and Krause, J. (2009) 'The effect of prey density on predators: conspicuousness and attack success are sensitive to spatial scale', *The American Naturalist*, 173(4), pp. 499–506.

Isbell, L. A., and Bidner, L. R. (2016) 'Vervet monkey (*Chlorocebus pygerythrus*) alarm calls to leopards (*Panthera pardus*) function as a predator deterrent', *Behaviour*, 153(5), pp. 591–606.

Ishii, Y., and Shimada, M. (2010) 'The effect of learning and search images on predator–prey interactions', *Population Ecology*, 52(1), pp. 27.

Itescu, Y., Schwarz, R., Meiri, S., and Pafilis, P. (2017) 'Intraspecific competition, not predation, drives lizard tail loss on islands', *Journal of Animal Ecology*, 86(1), pp. 66–74.

Jablonski, N. G., and Chaplin, G. (2010) 'Human skin pigmentation as an adaptation to UV radiation', *Proceedings of the National Academy of Sciences*, 107, pp. 8962–8.

Jackson, J. F., Ingram, W., and Campbell, H. W. (1976) 'The dorsal pigmentation pattern of snakes as an antipredator strategy: a multivariate approach', *The American Naturalist*, 110, pp. 1029–53.

Jaenike, J., and Holt, R. D. (1991) 'Genetic variation for habitat preference: evidence and explanations', *The American Naturalist*, 137, pp. S67–S90.

Jamie, G. A. (2017) 'Signals, cues and the nature of mimicry', *Proceedings of the Royal Society B: Biological Sciences*, 284(1849). doi: 10.1098/rspb.2016.2080

Janssen, J. (1981) 'Searching for zooplankton just outside Snell's window', *Limnology and Oceanography*, 26(6), pp. 1168–71.

Janzen, D. H., Hallwachs, W., and Burns, J. M. (2010) 'A tropical horde of counterfeit predator eyes', *Proceedings of the National Academy of Sciences*, 107(26), pp. 11659–65.

Järvi, T., Sillentullberg, B., and Wiklund, C. (1981) 'The cost of being aposematic: an experimental study of predation on larvae of *Papilio machaon* by the great tit, *Parus major*', *Oikos*, 36(3), pp. 267–72.

Jeffords, M. R., Sternburg, J. G., and Waldbauer, G. P. (1979) 'Batesian mimicry - field demonstration of the survival value of pipevine swallowtail and monarch color patterns', *Evolution*, 33(1), pp. 275–86.

Jiggins, C. D. (2017) *The Ecology and Evolution of Heliconius Butterflies.* Oxford, UK: Oxford University Press.

Jiggins, C. D., Estrada, C., and Rodrigues, A. (2004) 'Mimicry and the evolution of premating isolation in *Heliconius melpomene* Linnaeus', *Journal of Evolutionary Biology*, 17(3), pp. 680–91.

Jiggins, C. D., Naisbit, R. E., Coe, R. L., and Mallet, J. (2001) 'Reproductive isolation caused by colour pattern mimicry', *Nature*, 411(6835), pp. 302–5.

Johnsen, S. (2001) 'Hidden in plain sight: the ecology and physiology of organismal transparency', *The Biological Bulletin*, 201(3), pp. 301–18.

Johnsen, S. (2002) 'Cryptic and conspicuous coloration in the pelagic environment.', *Proceedings of the Royal Society B: Biological Sciences*, 269, pp. 243–56.

Johnsen, S. (2005) 'The red and the black: bioluminescence and the color of animals in the deep sea', *Integrative and Comparative Biology*, 45(2), pp. 234–46.

Johnsen, S. (2012) *The optics of life: a biologist's guide to light in nature.* Princeton, NJ: Princeton University Press.

Johnsen, S. (2014) 'Hide and seek in the open sea: pelagic camouflage and visual countermeasures', *Annual Review of Marine Science*, 6, pp. 369–92.

Johnsen, S., Gagnon, Y. L., Marshall, N. J., Cronin, T. W., Gruev, V., and Powell, S. (2016) 'Polarization vision seldom increases the sighting distance of silvery fish', *Current Biology*, 26(16), pp. R752–4.

Johnsen, S., Gassmann, E., Reynolds, R. A., Stramski, D., and Mobley, C. (2014) 'The asymmetry of the underwater horizontal light field and its implications for mirror-based camouflage in silvery pelagic fish', *Limnology and Oceanography*, 59(6), pp. 1839–52.

Johnsen, S., Marshall, N. J., and Widder, E. A. (2011) 'Polarization sensitivity as a contrast enhancer in pelagic predators: lessons from in situ polarization imaging of transparent zooplankton', *Philosophical Transactions of the Royal Society of London B: Biological Sciences*, 366(1565), pp. 655–70.

Johnsen, S., and Sosik, H. M. (2003) 'Cryptic coloration and mirrored sides as camouflage strategies in near-surface pelagic habitats: Implications for foraging and predator avoidance', *Limnology and Oceanography*, 48(3), pp. 1277–88.

Johnsen, S., and Widder, E. A. (1998) 'Transparency and visibility of gelatinous zooplankton from the northwestern Atlantic and Gulf of Mexico', *The Biological Bulletin*, 195(3), pp. 337–48.

Johnsen, S., and Widder, E. A. (2001) 'Ultraviolet absorption in transparent zooplankton and its implications for depth distribution and visual predation', *Marine Biology*, 138(4), pp. 717–30.

Johnsen, S., Widder, E. A., and Mobley, C. D. (2004) 'Propagation and perception of bioluminescence: factors affecting counterillumination as a cryptic strategy', *The Biological Bulletin*, 207, pp. 1–16.

Johnson, N. C., Graham, J. H., and Smith, F. A. (1997) 'Functioning of mycorrhizal associations along the mutualism–parasitism continuum', *New Phytologist*, 135, pp. 575–85.

Johnstone, R. A., and Grafen, A. (1993) 'Dishonesty and the handicap principle', *Animal Behaviour*, 46(4), pp. 759–64.

Jones, B. W., and Nishiguchi, M. K. (2004) 'Counterillumination in the hawaiian bobtail squid, *Euprymna scolopes* Berry (Mollusca: Cephalopoda)', *Marine Biology*, 144, pp. 1151–5.

Jones, K. J., and Hill, W. L. (2001) 'Auditory perception of hawks and owls for passerine alarm calls', *Ethology*, 107(8), pp. 717–26.

Jones, R., Fenton, A., Speed, M., and Mappes, J. (2017) 'Investment in multiple defences protects a nematode-bacterium symbiosis from predation', *Animal Behaviour*, 129, pp. 1–8.

Jones, R. S., Davis, S. C., and Speed, M. P. (2013) 'Defence cheats can degrade protection of chemically defended prey', *Ethology*, 119(1), pp. 52–7.

Jones, R. S., Fenton, A., and Speed, M. P. (2015) '"Parasite-induced aposematism" protects entomopathogenic nematode parasites against invertebrate enemies', *Behavioral Ecology*, 27(2), pp. 645–51.

Joron, M. (2003) 'Mimicry', in Cardé, R.T., & Resh, V.H. (eds) *Encyclopedia of Insects*. New York: Academic Press, pp. 417–26.

Joron, M., Frezal, L., Jones, R. T., Chamberlain, N. L., Lee, S. F., Haag, C. R., Whibley, A., Becuwe, M., Baxter, S. W., Ferguson, L., Wilkinson, P. A., Salazar, C., Davidson, C., Clark, R., Quail, M. A., Beasley, H., Glithero, R., Lloyd, C., Sims, S., Jones, M. C., Rogers, J., Jiggins, C. D., and Ffrench-Constant, R. H. (2011) 'Chromosomal rearrangements maintain a polymorphic supergene controlling butterfly mimicry', *Nature*, 477(7363), pp. 203–6.

Joron, M., and Iwasa, Y. (2005) 'The evolution of a Müllerian mimic in a spatially distributed community', *Journal of Theoretical Biology*, 237(1), pp. 87–103.

Joron, M., and Mallet, J. L. B. (1998) 'Diversity in mimicry: paradox or paradigm?', *Trends in Ecology & Evolution*, 13(11), pp. 461–6.

Joron, M., Papa, R., Beltran, M., Chamberlain, N., Mavarez, J., Baxter, S., Abanto, M., Bermingham, E., Humphray, S. J., Rogers, J., Beasley, H., Barlow, K., Ffrench-Constant, R. H., Mallet, J., McMillan, W. O., and Jiggins, C. D. (2006) 'A conserved supergene locus controls colour pattern diversity in *Heliconius* butterflies', *PloS Biology*, 4(10), pp. 1831–40.

Joron, M., Wynne, I. R., Lamas, G., and Mallet, J. (1999) 'Variable selection and the coexistence of multiple mimetic forms of the butterfly *Heliconius numata*', *Evolutionary Ecology*, 13(7–8), pp. 721–54.

Jousset, A., Rochat, L., Péchy-Tarr, M., Keel, C., Scheu, S., and Bonkowski, M. (2009) 'Predators promote defence of rhizosphere bacterial populations by selective feeding on non-toxic cheaters', *The ISME journal,* 3(6), pp. 666–74.

Kamilar, J. M. (2009) 'Interspecific variation in primate countershading: Effects of activity pattern, body mass, and phylogeny', *International Journal of Primatology*, 30, pp. 877–91.

Kamilar, J. M., and Bradley, B. J. (2011) 'Countershading is related to positional behavior in primates', *Journal of Zoology*, 283, pp. 227–33.

Kang, C., Kim, Y. E., and Jang, Y. (2016) 'Colour and pattern change against visually heterogeneous backgrounds in the tree frog *Hyla japonica*', *Scientific Reports*, 6, pp. 22601.

Kang, C., Sherratt, T. N., Kim, Y. E., Shin, Y., Moon, J., Song, U., Kang, J. Y., Kim, K., and Jang, Y. (2017) 'Differential predation drives the geographical divergence in multiple traits in aposematic frogs', *Behavioral Ecology*, 28, pp. 1122–30.

Kang, C., Stevens, M., Moon, J.-y., Lee, S.-I., and Jablonski, P. G. (2015) 'Camouflage through behavior in moths: the role of background matching and disruptive coloration', *Behavioral Ecology*, 26(1), pp. 45–54.

Kang, C. K., Moon, J. Y., Lee, S. I., and Jablonski, P. (2012) 'Camouflage through an active choice of a resting spot and body orientation in moths', *Journal of Evolutionary Biology*, 25(9), pp. 1695–1702.

Kanizsa, G., Renzi, P., Conte, S., Compostela, C., and Guerani, L. (1993) 'Amodal completion in mouse vision', *Perception*, 22(6), pp. 713–21.

Kapan, D. D. (2001) 'Three-butterfly system provides a field test of Müllerian mimicry', *Nature*, 409(6818), pp. 338–40.

Karasov, T. L., Chae, E., Herman, J. J., and Bergelson, J. (2017) 'Mechanisms to mitigate the trade-off between growth and defense', *The Plant Cell*, 29(4), pp. 666–80.

Kareksela, S., Härmä, O., Lindstedt, C., Siitari, H., and Suhonen, J. (2013) 'Effect of willow tit *Poecile montanus* alarm calls on attack rates by pygmy owls *Glaucidium passerinum*', *Ibis*, 155(2), pp. 407–12.

Karplus, I., and Algom, D. (1981) 'Visual cues for predator face recognition by reef fishes', *Zeitschrift Fur Tierpsychologie-Journal of Comparative Ethology*, 55(4), pp. 343–64.

Kassarov, L. (1999) 'Are birds able to taste and reject butterflies based on 'beak mark tasting'? A different point of view', *Behaviour*, 136(8), pp. 965–81.

Kavanagh, P. H., Shaw, R. C., and Burns, K. C. (2016) 'Potential aposematism in an insular tree species: are signals dishonest early in ontogeny?', *Biological Journal of the Linnean Society*, 118(4), pp. 951–8.

Kawaguchi, I., and Sasaki, A. (2006) 'The wave speed of intergradation zone in two-species lattice Müllerian mimicry model', *Journal of Theoretical Biology*, 243(4), pp. 594–603.

Kazemi, B., Gamberale-Stille, G., Tullberg, B. S., and Leimar, O. (2014) 'Stimulus salience as an explanation for imperfect mimicry', *Current Biology*, 24(9), pp. 965–9.

Keeler, C. E. (1925) 'Recent work by Gabritchevsky on the inheritance of color varieties in *Volucella bombylans*', *Psyche*, 33, pp. 22–7.

Kekäläinen, J., Huuskonen, H., Kiviniemi, V., and Taskinen, J. (2010) 'Visual conditions and habitat shape the coloration of the Eurasian perch (*Perca fluviatilis* L.): a trade-off between camouflage and communication?', *Biological Journal of the Linnean Society*, 99, pp. 47–59.

Kelley, J. L., Fitzpatrick, J. L., and Merilaita, S. (2013) 'Spots and stripes: ecology and colour pattern evolution in butterflyfishes', *Proceedings of the Royal Society B-Biological Sciences*, 280(1757).

Kelley, J. L., and Merilaita, S. (2015) 'Testing the role of background matching and self-shadow concealment in explaining countershading coloration in wild-caught rainbowfish', *Biological Journal of the Linnean Society*, 114, pp. 915–28.

Kelley, J. L., Rodgers, G. M., and Morrell, L. J. (2016) 'Conflict between background matching and social signalling in a colour-changing freshwater fish', *Royal Society Open Science*, 3(6), pp. 160040.

Kelley, J. L., Taylor, I., Hart, N. S., and Partridge, J. C. (2017) 'Aquatic prey use countershading camouflage to match the visual background', *Behavioral Ecology*, 28(5), 1314–1322.

Kelley, L. A., and Kelley, J. L. (2013) 'Animal visual illusion and confusion: the importance of a perceptual perspective', *Behavioral Ecology*, 25(3), pp. 450–63.

Kellner, R. L. (2002) 'Molecular identification of an endosymbiotic bacterium associated with pederin biosynthesis in *Paederus sabaeus* (Coleoptera: Staphylinidae)', *Insect Biochemistry and Molecular Biology*, 32(4), pp. 389–95.

Kelman, E. J., Baddeley, R. J., Shohet, A. J., and Osorio, D. (2007) 'Perception of visual texture and the expression of disruptive camouflage by the cuttlefish, *Sepia officinalis*', *Proc Biol Sci*, 274(1616), pp. 1369–75.

Kelman, E. J., Osorio, D., and Baddeley, R. J. (2008) 'A review of cuttlefish camouflage and object recognition and evidence for depth perception', *J Exp Biol*, 211(Pt 11), pp. 1757–63.

Kemp, D. J., Herberstein, M. E., Fleishman, L. J., Endler, J. A., Bennett, A. T. D., Dyer, A. G., Hart, N. S., Marshall, J., and Whiting, M. J. (2015) 'An integrative framework for the appraisal of coloration in nature', *American Naturalist*, 185(6), pp. 705–24.

Kikuchi, D. W., Malick, G., Webster, R. J., Whissell, E., and Sherratt, T. N. (2015) 'An empirical test of 2-dimensional signal detection theory applied to Batesian mimicry', *Behavioral Ecology*, 26(4), pp. 1226–35.

Kikuchi, D. W., and Pfennig, D. W. (2010) 'Predator cognition permits imperfect coral snake mimicry', *American Naturalist*, 176(6), pp. 830–4.

Kikuchi, D. W., and Pfennig, D. W. (2012) 'A Batesian mimic and its model share color production mechanisms', *Current Zoology*, 58(4), pp. 658–67.

Kikuchi, D. W., and Pfennig, D. W. (2013) 'Imperfect mimicry and the limits of natural selection', *Quarterly Review of Biology*, 88(4), pp. 297–315.

Kikuchi, D. W., and Sherratt, T. N. (2015) 'Costs of learning and the evolution of mimetic signals', *American Naturalist*, 186(3), pp. 321–32.

Kiltie, R. A. (1988) 'Countershading: Universally deceptive or deceptively universal?', *Trends in Ecology and Evolution*, 3, pp. 21–3.

Kiltie, R. A. (1989) 'Testing Thayer's countershading hypothesis: An image processing approach.', *Animal Behaviour*, 38, pp. 542–4.

Kirby, W., and Spence, W. (1817) *An Introduction to Entomology, Volume 2.* 2nd edn. London: Longman, Hurst, Rees, Orme & Brown.

Kircher, B. K., and Johnson, M. A. (2017) 'Why do curly tail lizards (genus *Leiocephalus*) curl their tails? An assessment of displays toward conspecifics and predators', *Ethology*, 123(5), pp. 342–7.

Kitamura, T., and Imafuku, M. (2015) 'Behavioural mimicry in flight path of Batesian intraspecific polymorphic butterfly *Papilio polytes*', *Proceedings of the Royal Society B: Biological Sciences*, 282. doi: 10.1098/rspb.2015.0483

Kiyotake, H., Matsumoto, H., Nakayama, S., Sakai, M., Miyatake, T., Ryuda, M., and Hayakawa, Y. (2014) 'Gain of long tonic immobility behavioral trait causes the red flour beetle to reduce anti-stress capacity', *Journal of Insect Physiology*, 60, pp. 92–7.

Kjernsmo, K., Grönholm, M. and Merilaita, S. (2016) 'Adaptive constellations of protective marks: eyespots, eye stripes and diversion of attacks by fish', *Animal Behaviour*, 111, pp. 189–95.

Kjernsmo, K., and Merilaita, S. (2012) 'Background choice as an anti-predator strategy: the roles of background matching and visual complexity in the habitat choice of the least killifish', *Proceedings of the Royal Society of London B: Biological Sciences*, 279(1745), pp. 4192–8.

Kjernsmo, K., and Merilaita, S. (2013) 'Eyespots divert attacks by fish', *Proceedings of the Royal Society of London B: Biological Sciences*, 280(1766), pp. 20131458.

Kleffner, D. A., and Ramachandran, V. S. (1992) 'On the perception of shape from shading', *Attention, Perception and Psychophysics*, 52, pp. 18–36.

Kline, K., Holcombe, A. O., and Eagleman, D. M. (2004) 'Illusory motion reversal is caused by rivalry, not by perceptual snapshots of the visual field', *Vision Research*, 44(23), pp. 2653–8.

Klump, G., Kretzschmar, E., and Curio, E. (1986) 'The hearing of an avian predator and its avian prey', *Behavioral Ecology and Sociobiology*, 18(5), pp. 317–23.

Klump, G., and Shalter, M. (1984) 'Acoustic behaviour of birds and mammals in the predator context; I. Factors affecting the structure of alarm signals. II. The functional significance and evolution of alarm signals', *Ethology*, 66(3), pp. 189–226.

Knill, R., and Allen, J. A. (1995) 'Does polymorphism protect? An experiment with human 'predators'', *Ethology*, 99(1–2), pp. 127–38.

Kodandaramaiah, U. (2009) 'Eyespot evolution: phylogenetic insights from *Junonia* and related butterfly genera (Nymphalidae: Junoniini)', *Evolution & Development*, 11(5), pp. 489–97.

Kodandaramaiah, U., Lindenfors, P., and Tullberg, B. S. (2013) 'Deflective and intimidating eyespots: a comparative study of eyespot size and position in *Junonia* butterflies', *Ecology and Evolution*, 3(13), pp. 4518–24.

Kokko, H., Mappes, J., and Lindström, L. (2003) 'Alternative prey can change model–mimic dynamics between parasitism and mutualism', *Ecology Letters*, 6(12), pp. 1068–76.

Körner, H. K. (1982) 'Countershading by physiological colour change in the fish louse *Anilocra physodes* L. (Crustacea: Isopoda)', *Oecologia*, 55, pp. 248–50.

Kowalski, K. N., Lakes-Harlan, R., Lehmann, G. U., and Strauß, J. (2014) 'Acoustic defence in an insect: characteristics of defensive stridulation and differences between the sexes in the tettigoniid *Poecilimon ornatus* (Schmidt 1850)', *Zoology*, 117(5), pp. 329–36.

Kraemer, A. C., and Adams, D. C. (2014) 'Predator perception of Batesian mimicry and conspicuousness in a salamander', *Evolution*, 68(4), pp. 1197–206.

Krams, I., Kivleniece, I., Kuusik, A., Krama, T., Freeberg, T. M., Mänd, R., Sivacova, L., Rantala, M. J., and Mänd, M. (2014) 'High repeatability of anti-predator responses and resting metabolic rate in a beetle', *Journal of Insect Behavior*, 27(1), pp. 57–66.

Krams, I., Kivleniece, I., Kuusik, A., Krama, T., Freeberg, T. M., Mänd, R., Vrublevska, J., Rantala, M. J., and Mänd, M. (2013a) 'Predation selects for low resting metabolic rate and consistent individual differences in anti-predator behavior in a beetle', *Acta Ethologica*, 16(3), pp. 163–72.

Krams, I., Kivleniece, I., Kuusik, A., Krama, T., Mänd, R., Rantala, M. J., Znotiņa, S., Freeberg, T. M., and Mänd, M. (2013b) 'Predation promotes survival of beetles with lower resting metabolic rates', *Entomologia Experimentalis et Applicata*, 148(1), pp. 94–103.

Krause, J., and Ruxton, G. D. (2002) *Living in groups*. New York: Oxford University Press.

Krebs, R. A., and West, D. A. (1988) 'Female mate preference and the evolution of female-limited Batesian mimicry', *Evolution*, 42(5), pp. 1101–4.

Kronforst, M. R., Young, L. G., and Gilbert, L. E. (2007) 'Reinforcement of mate preference among hybridizing *Heliconius* butterflies', *Journal of Evolutionary Biology*, 20(1), pp. 278–85.

Kruuk, H. (1972) *The spotted hyena: a study of predation and social behavior*. Chicago: University of Chicago Press.

Kuchta, S. R., Krakauer, A. H., and Sinervo, B. (2008) 'Why does the yellow-eyed ensatina have yellow eyes? Batesian mimicry of pacific newts (genus *Taricha*) by the salamander *Ensatina eschscholtzii xanthoptica*', *Evolution*, 62(4), pp. 984–90.

Kunte, K. (2008) 'Mimetic butterflies support Wallace's model of sexual dimorphism', *Proceedings of the Royal Society B: Biological Sciences*, 275(1643), pp. 1617–24.

Kunte, K. (2009a) 'The diversity and evolution of Batesian mimicry in *Papilio* swallowtail butterflies', *Evolution*, 63(10), pp. 2707–16.

Kunte, K. (2009b) 'Female-limited mimetic polymorphism: a review of theories and a critique of sexual selection as balancing selection', *Animal Behaviour*, 78(5), pp. 1029–36.

Kunte, K., Zhang, W., Tenger-Trolander, A., Palmer, D. H., Martin, A., Reed, R. D., Mullen, S. P., and Kronforst, M. R. (2014) 'doublesex is a mimicry supergene', *Nature*, 507(7491), pp. 229–32.

Kuriwada, T., Kumano, N., Shiromoto, K., and Haraguchi, D. (2009) 'Copulation reduces the duration of death-feigning behaviour in the sweetpotato weevil, *Cylas formicarius*', *Animal Behaviour*, 78(5), pp. 1145–51.

Kuriwada, T., Kumano, N., Shiromoto, K., and Haraguchi, D. (2010) 'The effect of mass-rearing on death-feigning behaviour in the sweet potato weevil (Coleoptera: Brentidae)', *Journal of Applied Entomology*, 134(8), pp. 652–8.

Kuriwada, T., Kumano, N., Shiromoto, K., and Haraguchi, D. (2011) 'Age-dependent investment in death-feigning behaviour in the sweetpotato weevil *Cylas formicarius*', *Physiological Entomology*, 36(2), pp. 149–54.

Kuriyama, T., Morimoto, G., Miyaji, K., and Hasegawa, M. (2016) 'Cellular basis of anti-predator adaptation in a lizard with autotomizable blue tail against specific predators with different colour vision', *Journal of Zoology*, 300(2), pp. 89–98.

Laan, A., Gutnick, T., Kuba, M. J., and Laurent, G. (2014) 'Behavioral analysis of cuttlefish traveling waves and its implications for neural control', *Current Biology*, 24(15), pp. 1737–42.

Laiolo, P., Tella, J. L., Carrete, M., Serrano, D., and López, G. (2004) 'Distress calls may honestly signal bird quality to predators', *Proceedings of the Royal Society of London B: Biological Sciences*, 271(Suppl 6), pp. S513–15.

Langham, G. M. (2004) 'Specialized avian predators repeatedly attack novel color morphs of *Heliconius* butterflies', *Evolution*, 58(12), pp. 2783–7.

Langham, G. M. (2005) 'Rufous-tailed jacamars and aposematic butterflies: do older birds attack novel prey?', *Behavioral Ecology*, 17(2), pp. 285–90.

Latz, M., and Case, J. (1982) 'Light organ and eyestalk compensation to body tilt in the luminescent midwater shrimp, *Sergestes similis*', *Journal of Experimental Biology*, 98(1), pp. 83–104.

Latz, M. I. (1995) 'Physiological mechanisms in the control of bioluminescent countershading in a midwater shrimp', *Marine & Freshwater Behaviour & Phy*, 26, pp. 207–18.

Lawrence, S. (1992) 'Sexual cannibalism in the praying mantid, *Mantis religiosa*: a field study', *Animal Behaviour*, 43(4), pp. 569–83.

Layberry, R. A., Hall, P. W., and Lafontaine, J. D. (1998) *The Butterflies of Canada*. Toronto: University of Toronto.

Le Poul, Y., Whibley, A., Chouteau, M., Prunier, F., Llaurens, V., and Joron, M. (2014) 'Evolution of dominance mechanisms at a butterfly mimicry supergene', *Nature Communications*, 5, 5644. doi:10.1038/ncomms6644

Leach, T. H., Williamson, C. E., Theodore, N., Fischer, J. M., and Olson, M. H. (2015) 'The role of ultraviolet radiation in the diel vertical migration of zooplankton: an experimental test of the transparency-regulator hypothesis', *Journal of Plankton Research*, 37(5), pp. 886–96.

Leal, M. (1999) 'Honest signalling during prey–predator interactions in the lizard *Anolis cristatellus*', *Animal Behaviour*, 58(3), pp. 521–6.

Leal, M., and Rodríguez-Robles, J. A. (1995) 'Antipredator responses of *Anolis cristatellus* (Sauria: Polychrotidae)', *Copeia*, 1995(1), pp. 155–61.

Leal, M., and Rodríguez-Robles, J. A. (1997) 'Signalling displays during predator–prey interactions in a Puerto Rican anole, *Anolis cristatellus*', *Animal Behaviour*, 54(5), pp. 1147–54.

Lederhouse, R. C., and Scriber, J. M. (1996) 'Intrasexual selection constrains the evolution of the dorsal color pattern of male black swallowtail butterflies, *Papilio polyxenes*', *Evolution*, 50(2), pp. 717–22.

Lee, T. J., Marples, N. M., and Speed, M. P. (2010) 'Can dietary conservatism explain the primary evolution of aposematism?', *Animal Behaviour*, 79(1), pp. 63–74.

Lee, W.-J., and Moss, C. F. (2016) 'Can the elongated hindwing tails of fluttering moths serve as false sonar targets to divert bat attacks?', *The Journal of the Acoustical Society of America*, 139(5), pp. 2579–88.

Leech, D. M., and Johnsen, S. (2006) 'Ultraviolet vision and foraging in juvenile bluegill (*Lepomis macrochirus*)', *Canadian Journal of Fisheries and Aquatic Sciences*, 63(10), pp. 2183–90.

Lehman, E. M., Jr, E. D., and Brodie Brodie III, E. D. (2004) 'No evidence for an endosymbiotic bacterial origin of tetrodotoxin in the newt, *Taricha granulosa*', *Toxicon*, 44(3), pp. 243–9.

Leimar, O., Enquist, M., and Sillén-Tullberg, B. (1986) 'Evolutionary stability of aposematic coloration and prey unprofitability: A theoretical analysis', *American Naturalist*, 128(4), pp. 469–90.

Levin, B.R. (1988) 'Frequency-dependent selection in bacterial populations', *Philosophical Transactions of the Royal Society of London Series B: Biological Sciences*, 319(1196), pp. 459–72.

Lev-Yadun, S. (2001) 'Aposematic (warning) coloration associated with thorns in higher plants', *Journal of Theoretical Biology*, 210(3), pp. 385–8.

Lev-Yadun, S. (2009) 'Aposematic (warning) coloration in plants', *Plant-Environment Interactions*. New York: Springer, pp. 167–202. Plant-Environment Interactions From Sensory Plant Biology to Active Plant Behavior Editors: Baluska, Frantisek (Ed.)

Lev-Yadun, S. (2013) 'Theoretical and functional complexity of white variegation of unripe fleshy fruits', *Plant Signaling & Behavior*, 8(10), pp. e25851.

Lev-Yadun, S. (2014a) 'Defensive masquerade by plants', *Biological Journal of the Linnean Society*, 113(4), pp. 1162–6.

Lev-Yadun, S. (2014b) 'Potential defence from herbivory by 'dazzle effects' and 'trickery coloration'of leaf variegation', *Biological Journal of the Linnean Society*, 111(3), pp. 692–7.

Lev-Yadun, S. (2015) 'The proposed anti-herbivory roles of white leaf variegation', *Progress in Botany*: Springer, pp. 241–69.

Lev-Yadun, S., Dafni, A., Flaishman, M. A., Inbar, M., Izhaki, I., Katzir, G., and Ne'eman, G. (2004) 'Plant coloration undermines herbivorous insect camouflage', *BioEssays*, 261126e113, pp. 1126–30.

Lev-Yadun, S., and Ne'eman, G. (2013) 'Bimodal colour pattern of individual *Pinus halepensis* Mill. seeds: a new type of crypsis', *Biological Journal of the Linnean Society*, 109(2), pp. 271–8.

Lewis, E. E., and Cane, J. H. (1990) 'Stridulation as a primary anti-predator defence of a beetle', *Animal Behaviour*, 40(5), pp. 1003–4.

Lichter-Marck, I. H., Wylde, M., Aaron, E., Oliver, J. C. and Singer, M. S. (2015) 'The struggle for safety: effectiveness of caterpillar defenses against bird predation', *Oikos*, 124(4), pp. 525–33.

Lin, J.-W., Chen, Y.-R., Wang, Y.-H., Hung, K.-C., and Lin, S.-M. (2017) 'Tail regeneration after autotomy revives survival: a case from a long-term monitored lizard population under avian predation'. *Proceedings of the Royal Society of London Series B: Biological Sciences*, 284(1847), 20162538. doi: 10.1098/rspb.2016.2538.

Lindgren, J., Sjövall, P., Carney, R. M., Uvdal, P., Gren, J. A., Dyke, G., Schultz, B. P., Shawkey, M. D., Barnes, K. R., and Polcyn, M. J. (2014) 'Skin pigmentation provides evidence of convergent melanism in extinct marine reptiles', *Nature*, 506, pp. 484–8.

Lindroth, C. H. (1971) 'Disappearance as a protective factor', *Entomol. Scand.*, 2, pp. 41–8.

Lindstedt, C., Huttunen, H., Kakko, M., and Mappes, J. (2011) 'Disentangling the evolution of weak warning signals: high detection risk and low production costs of chemical defences in gregarious pine sawfly larvae', *Evolutionary Ecology*, 25(5), pp. 1029–46.

Lindstedt, C., Lindstedt, Mappes, J., Päivinen, J., and Varama, M. (2006) 'Effects of group size and pine defence chemicals on Diprionid sawfly survival against ant predation', *Oecologia*, 150(3), pp. 519–26.

Lindstedt, C., Schroderus, E., Lindström, L., Mappes, T., and Mappes, J. (2016) 'Evolutionary constraints of warning signals: A genetic trade-off between the efficacy of larval and adult warning coloration can maintain variation in signal expression', *Evolution*, 70(11), pp. 2562–72.

Lindström, L., Alatalo, R. V., Lyytinen, A., and Mappes, J. (2004) 'The effect of alternative prey on the dynamics of imperfect Batesian and Mullerian mimicries', *Evolution*, 58(6), pp. 1294–302.

Lindström, L., Alatalo, R. V., Lyytinen, A. and Mappes, J. (2001a) 'Strong antiapostatic selection against novel rare aposematic prey', *Proceedings of the National Academy of Sciences*, 98(16), pp. 9181–4.

Lindström, L., Alatalo, R. V., and Mappes, J. (1997) 'Imperfect Batesian mimicry - The effects of the frequency and the distastefulness of the model', *Proceedings of the Royal Society of London Series B: Biological Sciences*, 264(1379), pp. 149–53.

Lindström, L., Alatalo, R. V., Mappes, J., Riipi, M., and Vertainen, L. (1999) 'Can aposematic signals evolve by gradual change?', *Nature*, 397(6716), pp. 249–51.

Lindström, L., Lyytinen, A., Mappes, J., and Ojala, K. (2006) 'Relative importance of taste and visual appearance for predator education in Müllerian mimicry', *Animal Behaviour*, 72, pp. 323–33.

Lindström, L., Rowe, C., and Guilford, T. (2001b) 'Pyrazine odour makes visually conspicuous prey aversive', *Proceedings of the Royal Society B: Biological Sciences*, 268(1463), pp. 159–62.

Linsley, E. G., Eisner, T., and Klots, A. B. (1961) 'Mimetic assemblages of sibling species of lycid beetles', *Evolution*, 15, pp. 15–29.

Liu, M.H., Blamires, S.J., Liao, C.P., and Tso, I.M. (2014) 'Evidence of bird dropping masquerading by a spider to avoid predators', *Scientific Reports*, 4, 5058.

Lloyd-Jones, T. J., and Luckhurst, L. (2002) 'Outline shape is a mediator of object recognition that is particularly important for living things', *Memory & Cognition*, 30(4), pp. 489–98.

Lobban, C. S., Hallam, S. J., Mukherjee, P., and Petrich, J. W. (2007) 'Photophysics and multifunctionality of Hypericin-like pigments in heterotrich ciliates: A phylogenetic perspective', *Photochemistry and Photobiology*, 83, pp. 1074–94.

Londoño, G. A., García, D. A., and Sánchez Martínez, M. A. (2015) 'Morphological and behavioral evidence of Batesian mimicry in nestlings of a lowland Amazonian bird', *The American Naturalist*, 185(1), pp. 135–41.

Long, C. (1993) 'Bivocal distraction nest-site display in the red squirrel, *Tamiasciurus hudsonicus*, with comments on outlier nesting and nesting behavior', *Canadian Field-Naturalist*, 107(1), pp. 104–6.

Lopanik, N. B. (2014) 'Chemical defensive symbioses in the marine environment', *Functional Ecology*, 28(2), pp. 328–40.

López-Palafox, T. G., Luis-Martínez, A., and Cordero, C. (2015) 'The movement of "false antennae" in butterflies with "false head" wing patterns', *Current Zoology*, 61(4), pp. 758–64.

Losey, G. S., Cronin, T. W., Goldsmith, T., Hyde, D., Marshall, N., and McFarland, W. (1999) 'The UV visual world of fishes: a review', *Journal of Fish Biology*, 54(5), pp. 921–43.

Lovell, P. G., Bloj, M., and Harris, J. M. (2012) 'Optimal integration of shading and binocular disparity for depth perception', *Journal of Vision*, 12, pp. 1.

Lovell, P. G., Ruxton, G. D., Langridge, K. V., and Spencer, K. A. (2013) 'Egg-laying substrate selection for optimal camouflage by quail', *Current Biology*, 23(3), pp. 260–4.

Lu, D., Willard, D., Patel, I. R., Kadwell, S., Overton, L., Kost, T., Luther, M., Chen, W., Woychik, R. P., Wilkison, W. O., and Cone, R. D. (1994) 'Agouti protein is an antagonist of the melanocyte-stimulating-hormone receptor', *Nature*, 371, pp. 799–802.

Luedeman, J. K., McMorris, F. R., and Warner, D. D. (1981) 'Predators encountering a model–mimic system with alternative prey', *American Naturalist*, 117(6), pp. 1040–8.

Lythgoe, J. N. (1979) *Ecology of vision*. Oxford: Clarendon Press.

Lyytinen, A., Brakefield, P. M., Lindström, L., and Mappes, J. (2004) 'Does prediction maintain eyespot plasticity in *Bicyclus anynana*?', *Proceedings of the Royal Society B:Biological Sciences*, 271(1536), pp. 279–83.

Lyytinen, A., Brakefield, P. M., and Mappes, J. (2003) 'Significance of butterfly eyespots as an anti-predator device in ground-based and aerial attacks', *Oikos*, 100(2), pp. 373–9.

Maan, M. E., and Cummings, M. E. (2011) 'Poison frog colors are honest signals of toxicity, particularly for bird predators', *The American Naturalist*, 179(1), pp. E1–E14.

MacDougall, A., and Dawkins, M. S. (1998) 'Predator discrimination error and the benefits of Müllerian mimicry', *Animal Behaviour*, 55, pp. 1281–8.

Machado, V., Araujo, A. M., Serrano, J., and Galian, J. (2004) 'Phylogenetic relationships and the evolution of mimicry in the *Chauliognathus* yellow-black species complex (Coleoptera: Cantharidae) inferred from mitochondrial COI sequences', *Genetics and Molecular Biology*, 27(1), pp. 55–60.

Mackintosh, N. J. (1976) 'Overshadowing and stimulus intensity', *Animal Learning & Behavior*, 4(2), pp. 186–92.

Maia, R. and White, T. (2018) 'Comparing colours using visual models', *Behavioral Ecology*, ary017. https://doi.org/10.1093/beheco/ary017

Majumdar, R., Sixt, M., and Parent, C. A. (2014) 'New paradigms in the establishment and maintenance of gradients during directed cell migration', *Current Opinion in Cell Biology*, 30, pp. 33–40.

Malcicka, M., Bezemer, T. M., Visser, B., Bloemberg, M., Snart, C. J. P., Hardy, I. C. W., and Harvey, J. A. (2015) 'Multi-trait mimicry of ants by a parasitoid wasp', *Scientific Reports*, 5, 8043. doi:10.1038/srep08043

Maldonado, H. (1970) 'The deimatic reaction in the praying mantis *Stagmatoptera biocellata*', *Zeitschrift für vergleichende Physiologie*, 68(1), pp. 60–71.

Mallet, J. (1999) 'Causes and consequences of a lack of coevolution in Müllerian mimicry', *Evolutionary Ecology*, 13(7–8), pp. 777–806.

Mallet, J. (2001) 'Mimicry: An interface between psychology and evolution', *Proceedings of the National Academy of Sciences of the United States of America*, 98(16), pp. 8928–30.

Mallet, J. (2010) 'Shift happens! Shifting balance and the evolution of diversity in warning colour and mimicry', *Ecological Entomology*, 35, pp. 90–104.

Mallet, J. (2014) 'Speciation: frog mimics prefer their own', *Current Biology*, 24, pp. R1094–6.

Mallet, J. (2015) 'New genomes clarify mimicry evolution', *Nature Genetics*, 47(4), pp. 306–7.

Mallet, J., Barton, N., Lamas, G., Santisteban, J., Muedas, M., and Eeley, H. (1990) 'Estimates of selection and gene flow from measures of cline width and linkage disequilibrium in *Heliconius* hybrid zones', *Genetics*, 124(4), pp. 921–36.

Mallet, J., and Barton, N. H. (1989) 'Strong natural selection in a warning-color hybrid zone', *Evolution*, 43(2), pp. 421–31.

Mallet, J., and Gilbert, L. E. (1995) 'Why are there so many mimicry rings: correlations between habitat, behavior and mimicry in *Heliconius* butterflies', *Biological Journal of the Linnean Society*, 55(2), pp. 159–80.

Mallet, J., and Joron, M. (1999) 'Evolution of diversity in warning color and mimicry: polymorphisms, shifting balance, and speciation', *Annual Review of Ecology and Systematics*, 30(1), pp. 201–33.

Mallet, J., and Singer, M. C. (1987) 'Individual selection, kin selection, and the shifting balance in the evolution of warning colours: the evidence from butterflies', *Biological Journal of the Linnean Society*, 32(4), pp. 337–50.

Manceau, M., Domingues, V. S., Linnen, C. R., Rosenblum, E. B., and Hoekstra, H. E. (2010) 'Convergence in pigmentation at multiple levels: mutations, genes and function', *Philosophical Transactions of the Royal Society B: Biological Sciences*, 365(1552), pp. 2439.

Manceau, M., Domingues, V. S., Mallarino, R., and Hoekstra, H. E. (2011) 'The developmental role of agouti in color pattern evolution', *Science*, 331, pp. 1062.

Mappes, J., Kokko, H., Ojala, K., and Lindström, L. (2014) 'Seasonal changes in predator community switch the direction of selection for prey defences', *Nature Communications*, 5. doi:10.1038/ncomms6016

Mappes, J., Marples, N., and Endler, J. A. (2005) 'The complex business of survival by aposematism', *Trends in Ecology & Evolution*, 20(11), pp. 598–603.

Marek, P., Papaj, D., Yeager, J., Molina, S., and Moore, W. (2011) 'Bioluminescent aposematism in millipedes', *Current Biology*, 21(18), pp. R680–1.

Marek, P. E., and Bond, J. E. (2009) 'A Müllerian mimicry ring in Appalachian millipedes', *Proceedings of the National Academy of Sciences of the United States of America*, 106(24), pp. 9755–60.

Marek, P. E., and Moore, W. (2015) 'Discovery of a glowing millipede in California and the gradual evolution of bioluminescence in Diplopoda', *Proceedings of the National Academy of Sciences*, 112(20), pp. 6419–24.

Marples, N., and Kelly, D. (1999) 'Neophobia and dietary conservatism: two distinct processes?', *Evolutionary Ecology*, 13(7), pp. 641–53.

Marples, N. M. (1993) 'Is the alkaloid in 2-spot ladybirds (*Adalia bipunctata*) a defence against ant predation?', *Chemoecology*, 4(1), pp. 29–32.

Marples, N. M., Brakefield, P. M., and Cowie, R. J. (1989) 'Differences between the 7-spot and 2-spot ladybird beetles (Coccinellidae) in their toxic effects on a bird predator', *Ecological Entomology*, 14(1), pp. 79–84.

Marples, N. M., Kelly, D. J., and Thomas, R. J. (2005) 'Perspective: the evolution of warning coloration is not paradoxical', *Evolution*, 59(5), pp. 933–40.

Marples, N. M., and Mappes, J. (2011) 'Can the dietary conservatism of predators compensate for positive frequency dependent selection against rare, conspicuous prey?', *Evolutionary Ecology*, 25(4), pp. 737–49.

Marples, N. M., Vanveelen, W., and Brakefield, P. M. (1994) 'The relative importance of color, taste and smell in the protection of an aposematic insect *Coccinella septempunctata*', *Animal Behaviour*, 48(4), pp. 967–74.

Marr, D. (1982) *Vision: a computational investigation into the human representation and processing of visual information.* San Francisco: W.H. Freeman.

Marr, D., and Hildreth, E. (1980) 'Theory of edge detection', *Proceedings of the Royal Society B: Biological Sciences*, 207, pp. 187–217.

Marshall, G. A. K. (1908) 'On diaposematism, with reference to some limitations of the Müllerian hypothesis of mimicry', *Transactions of the Entomological Society*, 1908, pp. 93–142.

Marshall, N. B. (1971) *Explorations in the life of fishes.* Cambridge, MA: Harvard University Press.

Marshall, N. J. (2000) 'Communication and camouflage with the same 'bright' colours in reef fishes', *Philosophical Transactions of the Royal Society of London B: Biological Sciences*, 355(1401), pp. 1243–8.

Mather, J. A., and Mather, D. L. (2004) 'Apparent movement in a visual display: the 'passing cloud' of *Octopus cyanea* (Mollusca: Cephalopoda)', *Journal of Zoology*, 263(1), pp. 89–94.

Mather, M. H., and Robertson, R. J. (1992) 'Honest advertisement in flight displays of bobolinks (*Dolichonyx oryzivorus*)', *The Auk*, 109(4), pp. 869–73.

Mathger, L. M., Barbosa, A., Miner, S., and Hanlon, R. T. (2006) 'Color blindness and contrast perception in cuttlefish (*Sepia officinalis*) determined by a visual sensorimotor assay', *Vision Res*, 46(11), pp. 1746–53.

Mathger, L. M., Chiao, C. C., Barbosa, A., Buresch, K. C., Kaye, S., and Hanlon, R. T. (2007) 'Disruptive coloration elicited on controlled natural substrates in cuttlefish, *Sepia officinalis*', *J Exp Biol*, 210(Pt 15), pp. 2657–66.

Matsumura, K., Fuchikawa, T., and Miyatake, T. (2017) 'Decoupling of behavioral trait correlation across life stages in two holometabolous insects', *Behavior Genetics*, 47(4), pp. 1–9.

Maunsell, J. H., and Van Essen, D. C. (1983) 'Functional properties of neurons in middle temporal visual area of the macaque monkey. I. Selectivity for stimulus direction, speed, and orientation', *Journal of Neurophysiology*, 49(5), pp. 1127–47.

McCall, A. C., and Fordyce, J. A. (2010) 'Can optimal defence theory be used to predict the distribution of plant chemical defences?', *Journal of Ecology*, 98(5), pp. 985–92.

McCosker, J. E. (1977) 'Fright posture of plesiopid fish *Calloplesiops altivelis*: example of Batesian mimicry', *Science*, 197(4301), pp. 400–1.

McCosker, J. E., and Rosenblatt, R. H. (1993) 'A revision of the snake eel genus *Myrichthys* (Anguilliformes: Ophichthidae) with the description of a new eastern Pacific species.', *Proceedings of the California Academy of Sciences*, 48, pp. 153–69.

McFall-Ngai, M., and Morin, J. G. (1991) 'Camouflage by disruptive illumination in leiognathids, a family of shallow-water, bioluminescent fishes', *Journal of Experimental Biology*, 156(1), pp. 119–37.

McFall-Ngai, M. J. (1990) 'Crypsis in the pelagic environment', *American Zoologist*, 30(1), pp. 175–88.

McGlothlin, J. W., Kobiela, M. E., Feldman, C. R., Castoe, T. A., Geffeney, S. L., Hanifin, C. T., Toledo, G., Vonk, F. J., Richardson, M. K., and Brodie, E. D. (2016) 'Historical contingency in a multigene family facilitates adaptive evolution of toxin resistance', *Current Biology*, 26(12), pp. 1616–21.

McIver, J. D., and Stonedahl, G. (1993) 'Myrmecomorphy: morphological and behavioral mimicry of ants', *Annual Review of Entomology*, 38, pp. 351–79.

McKaye, K. R. (1981) 'Field observation on death feigning: a unique hunting behavior by the predatory cichlid, *Haplochromis livingstoni*, of Lake Malawi', *Environmental Biology of Fishes*, 6(3), pp. 361–5.

McNamara, M. E., Briggs, D. E., Orr, P. J., Noh, H., and Cao, H. (2012) 'The original colours of fossil beetles'. *Proceedings of the Royal Society of London B: Biological Sciences*,279, 1114–21. doi:10.1098/rspb.2011.1677

McPhail, J. D. (1977) 'Possible function of caudal spot in characid fishes', *Canadian Journal of Zoology-Revue Canadienne De Zoologie*, 55(7), pp. 1063–6.

Meadows, D. W. (1993) 'Morphological variation in eyespots of the foureye butterflyfish (*Chaetodon capistratus*) - Implications for eyespot function', *Copeia*, 1993(1), pp. 235–40.

Meakin, C., and Qin, J. (2012) 'Growth, behaviour and colour changes of juvenile King George whiting (*Sillaginodes punctata*) mediated by light intensities', *New Zealand Journal of Marine and Freshwater Research*, 46, pp. 111–23.

Merilaita, S. (1998) 'Crypsis through disruptive coloration in an isopod', *Proceedings of the Royal Society of London B: Biological Sciences*, 265(1401), pp. 1059–64.

Merilaita, S., and Dimitrova, M. (2014) 'Accuracy of background matching and prey detection: predation by blue tits indicates intense selection for highly matching prey colour pattern', *Functional Ecology*, 28(5), pp. 1208–15.

Merilaita, S., and Kaitala, V. (2002) 'Community structure and the evolution of aposematic coloration', *Ecology Letters*, 5(4), pp. 495–501.

Merilaita, S., and Lind, J. (2005) 'Background-matching and disruptive coloration, and the evolution of cryptic coloration', *Proceedings of the Royal Society of London B: Biological Sciences*, 272(1563), pp. 665–70.

Merilaita, S., Lyytinen, A., and Mappes, J. (2001) 'Selection for cryptic coloration in a visually heterogeneous habitat', *Proceedings of the Royal Society of London B: Biological Sciences*, 268(1479), pp. 1925–9.

Merilaita, S., and Ruxton, G. D. (2007) 'Aposematic signals and the relationship between conspicuousness and distinctiveness', *Journal of Theoretical Biology*, 245(2), pp. 268–77.

Merilaita, S., Schaefer, H. M., and Dimitrova, M. (2013) 'What is camouflage through distractive markings?', *Behavioral Ecology*, 24(5), pp. e1271–2.

Merilaita, S., Scott-Samuel, N. E., and Cuthill, I. C. (2017) 'How camouflage works', *Philos Trans R Soc Lond B Biol Sci*, 372(1724). doi: 10.1098/rstb.2016.0341.

Merilaita, S., and Tullberg, B. S. (2005) 'Constrained camouflage facilitates the evolution of conspicuous warning coloration', *Evolution*, 59(1), pp. 38–45.

Merilaita, S., Tuomi, J., and Jormalainen, V. (1999) 'Optimization of cryptic coloration in heterogeneous habitats', *Biological Journal of the Linnean Society*, 67(2), pp. 151–61.

Merilaita, S., Vallin, A., Kodandaramaiah, U., Dimitrova, M., Ruuskanen, S., and Laaksonen, T. (2011) 'Number of eyespots and their intimidating effect on naïve predators in the peacock butterfly', *Behavioral Ecology*, 22(6), pp. 1326–31.

Merrill, R. M., Gompert, Z., Dembeck, L. M., Kronforst, M. R., McMillan, W. O., and Jiggins, C. D. (2011a) 'Mate preference across the speciation continuum in a clade of mimetic butterflies', *Evolution*, 65(5), pp. 1489–500.

Merrill, R. M., and Jiggins, C. D. (2009) 'Müllerian mimicry: sharing the load reduces the legwork', *Current Biology*, 19(16), pp. R687–9.

Merrill, R. M., Van Schooten, B., Scott, J. A., and Jiggins, C. D. (2011b) 'Pervasive genetic associations between traits causing reproductive isolation in *Heliconius* butterflies', *Proceedings of the Royal Society B: Biological Sciences*, 278(1705), pp. 511–18.

Michalis, C., Scott-Samuel, N. E., Gibson, D. P., and Cuthill, I. C. (2017) 'Optimal background matching camouflage', *Proceedings of the Royal Society B: Biological Sciences*, 284(1858), 20170709. doi: 10.1098/rspb.2017.0709

Midgley, J. J., White, J. D. M., Johnson, S. D., and Bronner, G. N. (2015) 'Faecal mimicry by seeds ensures dispersal by dung beetles', *Nature Plants*, 1, 15141. doi:10.1038/nplants.2015.141

Milinski, M., and Boltshauser, P. (1995) 'Boldness and predator deterrence: a critique of Godin & Davis'. *Proceedings of the Royal Society B: Biological Sciences*, 262(1363), pp. 103–5.

Millar, S. E., Miller, M. W., Stevens, M. E., and Barsh, G. S. (1995) 'Expression and transgenic studies of the mouse agouti gene provide insight into the mechanisms by which mammalian coat color patterns are generated', *Development*, 121, pp. 3223–32.

Miller, C. T., Beleza, S., Pollen, A. A., Schluter, D., Kittles, R. A., Shriver, M. D., and Kingsley, D. M. (2007) 'cis-Regulatory changes in Kit ligand expression and parallel evolution of pigmentation in sticklebacks and humans.', *Cell*, 131, pp. 1179–89.

Miller, M. C. (1974) 'Aeolid nudibranchs (Gastropoda-Opisthobranchia) of family Glaucidae from New-Zealand waters', *Zoological Journal of the Linnean Society*, 54(1), pp. 31-&.

Mills, M. G., and Patterson, L. B. (2009) 'Not just black and white: Pigment pattern development and evolution in vertebrates', *Seminars in Cell & Developmental Biology*, 20, pp. 72–81.

Milne, A. A. (1926) *Winnie the Pooh: the Complete Collection of Stories and Poems*. 1994 edn. London: Methuen Children's Books Limited.

Miyatake, T. (2001a) 'Diurnal periodicity of death-feigning in *Cylas formicarius* (Coleoptera: Brentidae)', *Journal of Insect Behavior*, 14(4), pp. 421–32.

Miyatake, T. (2001b) 'Effects of starvation on death-feigning in adults of *Cylas formicarius* (Coleoptera: Brentidae)', *Annals of the Entomological Society of America*, 94(4), pp. 612–16.

Miyatake, T., Nakayama, S., Nishi, Y., and Nakajima, S. (2009) 'Tonically immobilized selfish prey can survive by sacrificing others', *Proceedings of the Royal Society of London B: Biological Sciences*, 276(1668), pp. 2763–7. doi: 10.1098/rspb.2009.0558.

Miyatake, T., Okada, K., and Harano, T. (2008) 'Negative relationship between ambient temperature and death-feigning intensity in adult *Callosobruchus maculatus* and *Callosobruchus chinensis*', *Physiological Entomology*, 33(1), pp. 83–8.

Miyazawa, K., and Noguchi, T. (2001) 'Distribution and origin of tetrodotoxin', *Toxin Reviews*, 20(1), pp. 11–33.

Mobley, C. D. (1994) *Light and water: radiative transfer in natural waters*. Cambridge, MA: Academic Press.

Mochida, K. (2011) 'Combination of local selection pressures drives diversity in aposematic signals', *Evolutionary Ecology*, 25(5), pp. 1017.

Mohl, B., and Miller, L., A (1976) 'Ultrasonic clicks produced by the peacock butterfly: a possible bat-repellent mechanism', *Journal of Experimental Biology*, 64(3), pp. 639–44.

Moment, G. B. (1962) 'Reflexive selection: a possible answer to an old puzzle', *Science,* 136(3512), pp. 262–3.

Montgomery, M. E., and Nault, L. R. (1977) 'Comparative response of aphids to the alarm pheromone, (E)-ß-farnesene', *Entomologia Experimentalis et Applicata*, 22(3), pp. 236–42.

Moore, K. A., and Williams, D. D. (1990) 'Novel strategies in the complex defense repertoire of a stonefly (*Pteronarcys dorsata*) nymph', *Oikos*, 57, pp. 49–56.

Morgan, S. G., and Christy, J. H. (1996) 'Survival of marine larvae under the countervailing selective pressures of photodamage and predation', *Limnology and Oceanography*, 41(3), pp. 498–504.

Morris, R. L., and Reader, T. (2016) 'Do crab spiders perceive Batesian mimicry in hoverflies?', *Behavioral Ecology*, 27(3), pp. 920–31.

Mostler, G. (1935) 'Beobachtungen zur frage der wespen-mimikry.', *Z. Morphol. Oekol. Tierre*, 29, pp. 381–454.

Motychak, J. E., Brodie Jr, E. D., and Brodie III, E. D. (1999) 'Evolutionary response of predators to dangerous prey: preadaptation and the evolution of tetrodotoxin resistance in garter snakes', *Evolution*, 53(5), pp. 1528–35.

Mukherjee, R., and Kodandaramaiah, U. (2015) 'What makes eyespots intimidating – the importance of pairedness', *BMC Evolutionary Biology*, 15(1), pp. 34.

Müller, F. (1878) 'Ueber die vortheile der mimicry bei schmetterlingen', *Zoologischer Anzeiger*, 1, pp. 54–5.

Müller, F. (1879) 'Ituna and Thyridia: a remarkable case of mimicry in butterflies', *Transactions of the Entomological Society*, 1879, pp. xx–xxix.

Muntz, W. R. A. (1976) 'On yellow lenses in mesopelagic animals', *Journal of the Marine Biological Association of the United Kingdom*, 56, pp. 963–76.

Murali, G., and Kodandaramaiah, U. (2016) 'Deceived by stripes: conspicuous patterning on vital anterior body parts can redirect predatory strikes to expendable posterior organs', *Open Science*, 3(6), pp. 160057.

Murali, G., and Kodandaramaiah, U. (2017) 'Body size and evolution of motion dazzle coloration in lizards', *Behavioral Ecology*. doi: 10.1093/beheco/arx128

Murdoch, W. W. (1969) 'Switching in general predators: experiments on predator specificity and stability of prey populations', *Ecological monographs,* 39(4), pp. 335–54.

Murphy, S. M., Leahy, S. M., Williams, L. S., and Lill, J. T. (2009) 'Stinging spines protect slug caterpillars (Limacodidae) from multiple generalist predators', *Behavioral Ecology*, 21(1), pp. 153–60.

Murphy, T. G. (2006) 'Predator-elicited visual signal: why the turquoise-browed motmot wag-displays its racketed tail', *Behavioral Ecology*, 17(4), pp. 547–53.

Murphy, T. G. (2007) 'Dishonest 'preemptive' pursuit-deterrent signal? Why the turquoise-browed motmot wags its tail before feeding nestlings', *Animal Behaviour*, 73(6), pp. 965–70.

Murray, J. D. (2002) *Mathematical Biology II: Spatial Models and Biomedical Applications*. New York: Spring-Verlag.

Muscat, E., Rotenberg, E. L., and Machado, I. F. (2016) 'Death-feigning behaviour in an *Erythrolamprus miliaris* (Linnaeus, 1758)(Colubridae) water snake in Ubatuba, São Paulo, southeastern Brazil', *Herpetology Notes*, 9, pp. 95–7.

Nakayama, S., and Miyatake, T. (2009) 'Positive genetic correlations between life-history traits and death-feigning behavior in adzuki bean beetle (*Callosobruchus chinensis*)', *Evolutionary Ecology*, 23(5), pp. 711.

Nakayama, S., and Miyatake, T. (2010a) 'A behavioral syndrome in the adzuki bean beetle: genetic correlation among death feigning, activity, and mating behavior', *Ethology*, 116(2), pp. 108–12.

Nakayama, S., and Miyatake, T. (2010b) 'Genetic trade-off between abilities to avoid attack and to mate: a cost of tonic immobility', *Biology Letters*, 6(1), pp. 18–20.

Narayan, E. J., Cockrem, J. F., and Hero, J.-M. (2013) 'Sight of a predator induces a corticosterone stress response and generates fear in an amphibian', *PloS One*, 8(8), pp. e73564.

Nelson, X. J., and Jackson, R. R. (2009) 'Aggressive use of Batesian mimicry by an ant-like jumping spider', *Biology Letters*, 5(6), pp. 755–7.

Nicholson, A. J. (1927) 'Presidential Address. A new theory of mimicry in insects.', *Australian Zoologist*, 5, pp. 10–24.

Nijhout, H. F. (1991) *The Development and Evolution of Butterfly Wing Patterns*. Washington: Smithsonian Institute Press.

Nishikawa, H., Iga, M., Yamaguchi, J., Saito, K., Kataoka, H., Suzuki, Y., Sugano, S., and Fujiwara, H. (2013) 'Molecular basis of wing coloration in a Batesian mimic butterfly, *Papilio polytes*', *Scientific Reports*, 3, 3184. doi:10.1038/srep03184

Niu, Y., Chen, G., Peng, D.-L., Song, B., Yang, Y., Li, Z.-M., and Sun, H. (2014) 'Grey leaves in an alpine plant: a cryptic colouration to avoid attack?', *New Phytologist*, 203, pp. 953–63.

Niu, Y., Chen, Z., Stevens, M. and Sun, H. (2017). 'Divergence in cryptic leaf colour provides local camouflage in an alpine plant'. *Proceedings of the Royal Society of London Series B: Biological Sciences*, 284(1864), 20171654. doi: 10.1098/rspb.2017.1654

Nokelainen, O., Hegna, R. H., Reudler, J. H., Lindstedt, C., and Mappes, J. (2012) 'Trade-off between warning signal efficacy and mating success in the wood tiger moth'. *Proceedings of the Royal Society of London Series B: Biological Sciences*, 279(1727), 257–65. doi: 10.1098/rspb.2011.0880

Nokelainen, O., Valkonen, J., Lindstedt, C., and Mappes, J. (2014) 'Changes in predator community structure shifts the efficacy of two warning signals in Arctiid moths', *Journal of Animal Ecology*, 83(3), pp. 598–605.

Nonacs, P. (1985) 'Foraging in a dynamic mimicry complex', *American Naturalist*, 126(2), pp. 165–80.

Noonan, B. P., and Comeault, A. A. (2009) 'The role of predator selection on polymorphic aposematic poison frogs', *Biology Letters*, 5(1), pp. 51–4.

Noor, M.A., Parnell, R.S., and Grant, B.S. (2008) 'A reversible color polyphenism in American peppered moth (*Biston betularia cognataria*) caterpillars', *PloS One*, 3(9), e3142.

Norman, M. D., Finn, J., and Tregenza, T. (2001) 'Dynamic mimicry in an Indo-Malayan octopus', *Proceedings of the Royal Society of London Series B: Biological Sciences*, 268(1478), pp. 1755–8.

Norris, K. S., and Lowe, C. H. (1964) 'An analysis of background color-matching in amphibians and reptiles', *Ecology*, 45, pp. 565.

Novales Flamarique, I. N., and Browman, H. I. (2001) 'Foraging and prey-search behaviour of small juvenile rainbow trout (*Oncorhynchus mykiss*) under polarized light', *Journal of Experimental Biology*, 204(14), pp. 2415–22.

O'Brien, T. J., and Dunlap, W. P. (1975) 'Tonic immobility in the blue crab (*Callinectes sapidus*, Rathbun): Its relation to threat of predation', *Journal of Comparative and Physiological Psychology*, 89(1), pp. 86.

O'Brien, W. J., Kettle, D., and Riessen, H. (1979) 'Helmets and invisible armor: structures reducing predation from tactile and visual planktivores', *Ecology*, 60(2), pp. 287–94.

O'Donald, P., and Majerus, M. E. N. (1988) 'Frequency-dependent sexual selection', *Philosophical Transactions of the Royal Society of London Series B: Biological Sciences*, 319(1196), pp. 571–86.

O'Donald, P. and Pilecki, C. (1970) 'Polymorphic mimicry and natural selection', *Evolution*, 24(2), pp. 395–401.

O'Donald, P., and Pilecki, C. (1974) 'Polymorphic mimicry and natural selection - Reply', *Evolution*, 28(3), pp. 484–5.

O'Donnell, S. (1999) 'Dual mimicry in the dimorphic eusocial wasp *Mischocyttarus mastigophorus* Richards (Hymenoptera: Vespidae)', *Biological Journal of the Linnean Society*, 66(4), pp. 501–14.

Odén, K., Gunnarsson, S., Berg, C., and Algers, B. (2005) 'Effects of sex composition on fear measured as tonic immobility and vigilance behaviour in large flocks of laying hens', *Applied Animal Behaviour Science*, 95(1), pp. 89–102.

Ohno, T., and Miyatake, T. (2007) 'Drop or fly? Negative genetic correlation between death-feigning intensity and flying ability as alternative anti-predator strategies', *Proceedings of the Royal Society of London B: Biological Sciences*, 274(1609), pp. 555–60.

Ohsaki, N. (1995) 'Preferential predation of female butterflies and the evolution of Batesian mimicry', *Nature*, 378(6553), pp. 173–5.

Olofsson, M., Jakobsson, S., and Wiklund, C. (2012a) 'Auditory defence in the peacock butterfly (*Inachis io*) against mice (*Apodemus flavicollis* and *A. sylvaticus*)', *Behavioral Ecology and Sociobiology*, 66(2), pp. 209–15.

Olofsson, M., Jakobsson, S., and Wiklund, C. (2013) 'Bird attacks on a butterfly with marginal eyespots and the role of prey concealment against the background', *Biological Journal of the Linnean Society*, 109(2), pp. 290–7.

Olofsson, M., Løvlie, H., Tibblin, J., Jakobsson, S., and Wiklund, C. (2012b) 'Eyespot display in the peacock butterfly triggers antipredator behaviors in naïve adult fowl', *Behavioral Ecology*, 24(1), pp. 305–10.

Olofsson, M., Vallin, A., Jakobsson, S., and Wiklund, C. (2010) 'Marginal eyespots on butterfly wings deflect bird attacks under low light intensities with UV wavelengths', *PloS One*, 5(5), e10798. doi: 10.1371/journal.pone.0010798

Olofsson, M., Vallin, A., Jakobsson, S., and Wiklund, C. (2011) 'Winter predation on two species of hibernating butterflies: monitoring rodent attacks with infrared cameras', *Animal Behaviour*, 81(3), pp. 529–34.

Olsen, K. M., Kooyers, N. J., and Small, L. L. (2014) 'Adaptive gains through repeated gene loss: parallel evolution of cyanogenesis polymorphisms in the genus *Trifolium* (Fabaceae)', *Phil. Trans. R. Soc. B*, 369(1648), pp. 20130347.

Osorio, D., and Cuthill, I. C. (2013) 'Camouflage and perceptual organization in the animal kingdom', in Wagemans, J. (ed.), *The Oxford Handbook of Perceptual Organization*, Oxford: Oxford University Press, pp. 843–62.

Osorio, D., and Srinivasan, M. V. (1991) 'Camouflage by edge enhancement in animal coloration patterns and its implications for visual mechanisms.', *Proceedings of the Royal Society B: Biological Sciences*, 244, pp. 81–5.

Owen, D. F., and Whiteley, D. (1986) 'Reflexive selection: Moment's hypothesis resurrected', *Oikos*, 47, pp. 117–20.

Owen, D.F., and Whiteley, D. (1989) 'Evidence that reflexive polymorphisms are maintained by visual selection by predators', *Oikos*, 55(1), pp. 130–3.

Palmer, S. E. (1999). *Vision science: Photons to phenomenology*. Cambridge, MA: MIT Press.

Papageorgis, C. (1975) 'Mimicry in neotropical butterflies', *American Scientist*, 63, pp. 522–32.

Parker, A. (2003) *In the blink of an eye: how vision sparked the big bang of evolution*. New York: Basic Books.

Partridge, L. (1988) 'The rare-male effect: what is its evolutionary significance?', *Philosophical Transactions of the Royal Society of London Series B: Biological Sciences*, 319(1196), pp. 525–39.

Patel, H., Naik, V., and Tank, S. K. (2016) 'Death-feigning behavior in two species of *Lygosoma* (Squamata: Scincidae) from India', *Phyllomedusa: Journal of Herpetology*, 15(2), pp. 191–4.

Paul, N. D., and Gwynn-Jones, D. (2003) 'Ecological roles of solar UV radiation: towards an integrated approach', *Trends in Ecology & Evolution*, 18(1), pp. 48–55.

Paxton, C. G. M., Magurran, A. E., and Zschokke, S. (1994) 'Caudal eyespots on fish predators influence the inspection behaviour of Trinidadian guppies, *Poecilia reticulata*', *Journal of Fish Biology*, 44(1), pp. 175–7.

Pegram, K. V., and Rutowski, R. L. (2016) 'Effects of directionality, signal intensity, and short-wavelength components on iridescent warning signal efficacy', *Behavioral Ecology and Sociobiology*, 70(8), pp. 1331–43.

Pekár, S., Jarab, M., Fromhage, L., and Herberstein, M. E. (2011) 'Is the evolution of inaccurate mimicry a result of selection by a suite of predators? A case study using myrmecomorphic spiders', *American Naturalist*, 178(1), pp. 124–34.

Pekár, S., Petráková, L., Bulbert, M. W., Whiting, M. J., and Herberstein, M. E. (2017) 'The golden mimicry complex uses a wide spectrum of defence to deter a community of predators', *eLife*, 6, pp. e22089.

Penacchio, O., Cuthill, I. C., Lovell, P. G., Ruxton, G. D., and Harris, J. M. (2015a) 'Orientation to the sun by animals and its interaction with crypsis', *Functional Ecology*, 29(9), pp. 1165–77.

Penacchio, O., Harris, J. M., and Lovell, P. G. (2017) 'Establishing the behavioural limits for countershaded camouflage', *Scientific Reports*, 7(1), pp. 13672.

Penacchio, O., Lovell, P. G., Cuthill, I. C., Ruxton, G. D., and Harris, J. M. (2015b) 'Three-dimensional camouflage: Exploiting photons to conceal form', *American Naturalist*, 186(4), pp. 553–63.

Penacchio, O., Lovell, P. G., and Harris, J. M. (in review) 'Subtlety of countershading camouflage in different weather: countershading must be close to optimal to reduce visibility'.

Penacchio, O., Lovell, P. G., Sanghera, S., Cuthill, I., Ruxton, G., and Harris, J. (2015c) 'Countershading camouflage and the efficiency of visual search', *Perception*, 44(10), pp. 1240.

Penney, H. D., Hassall, C., Skevington, J. H., Abbott, K. R. and Sherratt, T. N. (2012) 'A comparative analysis of the evolution of imperfect mimicry', *Nature*, 483(7390), pp. 461–4.

Penney, H. D., Hassall, C., Skevington, J. H., Lamborn, B., and Sherratt, T. N. (2014) 'The relationship between morphological and behavioral mimicry in hover flies (Diptera: Syrphidae)', *American Naturalist*, 183(2), pp. 281–9.

Perrinet, L. U., and Bednar, J. A. (2015) 'Edge co-occurrences can account for rapid categorization of natural versus animal images', *Scientific Reports*, 5, pp. 11400.

Peterson, G. S., Johnson, L. B., Axler, R. P., and Diamond, S. A. (2002) 'Assessment of the risk of solar ultraviolet radiation to amphibians. II. In situ characterization of exposure in amphibian habitats', *Environmental Science & Technology*, 36(13), pp. 2859–65.

Peterson, M. A., Kimchi, R., and Reisberg, I. D. (2013) 'Perceptual organization in vision', *The Oxford handbook of cognitive psychology*, pp. 9–31. Edited by Daniel Reisberg. OUP.

Pfennig, D. W., Harcombe, W. R., and Pfennig, K. S. (2001) 'Frequency-dependent Batesian mimicry: Predators avoid look-alikes of venomous snakes only when the real thing is around', *Nature*, 410(6826), pp. 323-323.

Pfennig, D. W., Harper, G. R., Brumo, A. F., Harcombe, W. R., and Pfennig, K. S. (2007) 'Population differences in predation on Batesian mimics in allopatry with their model: selection against mimics is strongest when they are common', *Behavioral Ecology and Sociobiology*, 61(4), pp. 505–11.

Phillips, B. T., Gruber, D. F., Vasan, G., Roman, C. N., Pieribone, V. A., and Sparks, J. S. (2016) 'Observations of in situ deep-sea marine bioluminescence with a high-speed, high-resolution sCMOS camera', *Deep Sea Research Part I: Oceanographic Research Papers*, 111, pp. 102–9.

Phillips, G. A., How, M. J., Lange, J. E., Marshall, N. J., and Cheney, K. L. (2017) 'Disruptive colouration in reef fish: does matching the background reduce predation risk?', *Journal of Experimental Biology*, 220(11), pp.1962–74.

Piel, W. H., and Monteiro, A. (2011) 'Flies in the ointment make for convincing poop. ', *Yale Environmental News*, 16, pp. 13.

Pielowski, Z. (1959) 'Studies on the relationship: predator (goshawk)-prey (pigeon)', *Bull Acad Pol Sci Biol*, 7, pp. 401–3.

Pietrewicz, A. T., and Kamil, A. C. (1979) 'Search image formation in the blue jay (*Cyanocitta cristata*)', *Science*, 204(4399), pp. 1332–3.

Pilecki, C., and Odonald, P. (1971) 'Effects of predation on artificial mimetic polymorphisms with perfect and imperfect mimics at varying frequencies', *Evolution*, 25(2), pp. 365–70.

Pinheiro, C., Antezana, M., and Machado, L. (2014) 'Evidence for the deflective function of eyespots in wild *Junonia evarete* Cramer (Lepidoptera, Nymphalidae)', *Neotropical Entomology*, 43(1), pp. 39–47.

Pinheiro, C., Freitas, A., Campos, V., DeVries, P., and Penz, C. (2016) 'Both palatable and unpalatable butterflies use bright colors to signal difficulty of capture to predators', *Neotropical Entomology*, 45(2), pp. 107–13.

Pinheiro, C. E. G. (1996) 'Palatability and escaping ability in neotropical butterflies: Tests with wild kingbirds (*Tyrannus melancholicus*, Tyrannidae)', *Biological Journal of the Linnean Society*, 59(4), pp. 351–65.

Pinheiro, C. E. G. (2003) 'Does Müllerian mimicry work in nature? Experiments with butterflies and birds (Tyrannidae)', *Biotropica*, 35(3), pp. 356–64.

Pinheiro, C. E. G., and de Campos, V. C. (2013) 'Do rufous-tailed jacamars (*Galbula ruficauda*) play with aposematic butterflies?', *Ornitologia Neotropical*, 24(3), pp. 365–7.

Pinheiro, C. E. G., and Freitas, A. V. L. (2014) 'Some possible cases of escape mimicry in neotropical butterflies', *Neotropical Entomology*, 43(5), pp. 393–8.

Plaisted, K., and Mackintosh, N. (1995) 'Visual search for cryptic stimuli in pigeons: implications for the search image and search rate hypotheses', *Animal Behaviour*, 50(5), pp. 1219–32.

Platt, A. P., and Brower, L. P. (1968) 'Mimetic versus disruptive coloration in intergrading populations of *Limenitis arthemis* and *astyanax* butterflies', *Evolution*, 22(4), pp. 699–718.

Platt, A. P., Coppinger, R. P., and Brower, L. P. (1971) 'Demonstration of selective advantage of mimetic *Limenitis* butterflies presented to caged avian predators', *Evolution*, 25(4), pp. 692–701.

Plowright, R. C., and Owen, R. E. (1980) 'The evolutionary significance of bumble bee color patterns: a mimetic interpretation', *Evolution*, 34(4), pp. 622–37.

Polnaszek, T. J., Rubi, T. L., and Stephens, D. W. (2017) 'When it's good to signal badness: using objective measures of discriminability to test the value of being distinctive', *Animal Behaviour*, 129, pp. 113–25.

Poulton, E. B. (1888) 'Notes in 1887 upon lepidopterous larvae, etc., including a complete account of the life-history of the larvae of *Sphinx convolvuli* and *Aglia tau*', *Trans. Entomol. Soc. Lond*, 1888, pp. 515–606.

Poulton, E. B. (1890) *The colours of animals: their meaning and use, especially considered in the case of insects.* London: Kegan Paul, Trench, Trubner & Co. London.

Poulton, E. B. (1909) 'The value of colour in the struggle for life.', in Seward, A.C. (ed.) *Darwin and modern science; essays in commemoration of the centenary of the birth of Charles Darwin and of the fiftieth anniversary of the publication of the Origin of Species*: Cambridge University Press, pp. 207–27.

Poulton, E. B. (1924) '*Papilio dardanus*. The most interesting butterfly in the world', *Journal of the East African and Ugandan Natural History Society*, 20, pp. 4–22.

Powell, R. A. (1982) 'Evolution of black-tipped tails in weasels - Predator confusion', *American Naturalist*, 119(1), pp. 126–31.

Preißler, K., and Pröhl, H. (2017) 'The effects of background coloration and dark spots on the risk of predation in poison frog models', *Evolutionary Ecology*, 31(5), pp. 683–94.

Preston-Mafham, R., and Preston-Mafham, K. (1993) *The encyclopedia of land invertebrate behaviour.* Hong Kong: The MIT Press.

Priebe, N. J., and Ferster, D. (2006) 'Mechanisms underlying cross-orientation suppression in cat visual cortex', *Nature Neuroscience*, 9(4), pp. 552.

Prudic, K. L., and Oliver, J. C. (2008) 'Once a Batesian mimic, not always a Batesian mimic: mimic reverts back to ancestral phenotype when the model is absent', *Proceedings of the Royal Society B: Biological Sciences*, 275(1639), pp. 1125.

Prudic, K. L., Oliver, J. C., and Sperling, F. A. H. (2007) 'The signal environment is more important than diet or chemical specialization in the evolution of warning coloration', *Proceedings of the National Academy of Sciences of the United States of America*, 104(49), pp. 19381–6.

Prudic, K. L., Stoehr, A. M., Wasik, B. R. and Monteiro, A. (2015) 'Eyespots deflect predator attack increasing fitness and promoting the evolution of phenotypic plasticity'. *Proceedings of the Royal Society B: Biological Sciences,*, 282(1798). doi: 10.1098/rspb.2014.1531

Punnett, R. C. (1915) *Mimicry in Butterflies.* London: Cambridge University Press.

Purves, D., Paydarfar, J. A., and Andrews, T. J. (1996) 'The wagon wheel illusion in movies and reality', *Proceedings of the National Academy of Sciences*, 93(8), pp. 3693–7.

Putman, B. J., and Clark, R. W. (2014) 'The fear of unseen predators: ground squirrel tail flagging in the absence of snakes signals vigilance', *Behavioral Ecology*, 26(1), pp. 185–93.

Putman, B. J., Coss, R. G., and Clark, R. W. (2015) 'The ontogeny of antipredator behavior: age differences in California ground squirrels (*Otospermophilus beecheyi*) at multiple stages of rattlesnake encounters', *Behavioral Ecology and Sociobiology*, 69(9), pp. 1447–57.

Puurtinen, M., and Kaitala, V. (2006) 'Conditions for the spread of conspicuous warning signals: a numerical model with novel insights', *Evolution*, 60(11), pp. 2246–56.

Qadri, M. A. J., and Cook, R. G. (2015) 'Experimental divergences in the visual cognition of birds and mammals', *Comparative Cognition & Behavior Reviews*, 10, pp. 73–105.

Qadri, M. A. J., Romero, L. M., and Cook, R. G. (2014) 'Shape from shading in starlings (*Sturnus vulgaris*)', *Journal of Comparative Psychology (Washington, DC: 1983)*, 128, pp. 343.

Ramachandran, V. S. (1988) 'Perception of shape from shading'. *Nature*, 331, pp. 163–166.

Ranade, S. P., and Prakash, V. (2016) 'Nesting of lesser whistling-duck *Dendrocygna javanica* (Horsfield, 1821) (Aves: Anseriformes: Anatidae) and broken-wing distraction display at Kamrup District, Assam, India', *Journal of Threatened Taxa*, 8(5), pp. 8824–6.

Randall, J. E. (2005) 'A review of mimicry in marine fishes', *Zoological Studies*, 44, pp. 299–328.

Randler, C. (2006) 'Is tail wagging in white wagtails, *Motacilla alba*, an honest signal of vigilance?', *Animal Behaviour*, 71(5), pp. 1089–93.

Randler, C. (2007) 'Observational and experimental evidence for the function of tail flicking in Eurasian moorhen *Gallinula chloropus*', *Ethology*, 113(7), pp. 629–39.

Ratcliffe, J. M., and Nydam, M. L. (2008) 'Multimodal warning signals for a multiple predator world', *Nature*, 455(7209), pp. 96.

Regolin, L., and Vallortigara, G. (1995) 'Perception of partly occluded objects by young chicks', *Attention, Perception, & Psychophysics*, 57(7), pp. 971–6.

Reid, D. G. (1987) 'Natural selection for apostasy and crypsis acting on the shell colour polymorphism of a mangrove snail, Littoraria filosa (Sowerby)(Gastropoda: Littorinidae)', *Biological Journal of the Linnean Society*, 30(1), pp. 1–24.

Resetarits, E. J., and Raxworthy, C. J. (2016) 'Hidden in plain sight: How ventral line markings in chameleons may enhance camouflage', *Am Nat*, 187(2), pp. 262–73.

Rettenmeyer, C. W. (1970) 'Insect mimicry', *Annual Review of Entomology*, 15, pp. 43–74.

Reudler, J., Lindstedt, C., Pakkanen, H., Lehtinen, I., and Mappes, J. (2015) 'Costs and benefits of plant allelochemicals in herbivore diet in a multi enemy world', *Oecologia*, 179(4), pp. 1147–58.

Ridgway, M. S., and McPhail, J. D. (1987) 'Rival male effects on courtship behavior in the Enos Lake species pair of sticklebacks (*Gasterosteus*)', *Canadian Journal of Zoology-Revue Canadienne De Zoologie*, 65(8), pp. 1951–5.

Ries, L., and Mullen, S. P. (2008) 'A rare model limits the distribution of its more common mimic: a twist on frequency-dependent Batesian mimicry', *Evolution*, 62(7), pp. 1798–1803.

Riipi, M., Alatalo, R. V., Lindström, L., and Mappes, J. (2001) 'Multiple benefits of gregariousness cover detectability costs in aposematic aggregations', *Nature*, 413(6855), pp. 512–14.

Ritland, D. B., and Brower, L. P. (1991) 'The viceroy butterfly is not a Batesian mimic', *Nature*, 350(6318), pp. 497–8.

Rittenhouse, D. (1786) 'Explanation of an optical deception', *Transactions of the American Philosophical Society*, 2, pp. 37–42.

Robbins, R. K. (1981) 'The "false head" hypothesis: predation and wing pattern variation of lycaenid butterflies', *The American Naturalist*, 118(5), pp. 770–5.

Roberts, J. A., Taylor, P. W., and Uetz, G. W. (2006) 'Consequences of complex signaling: predator detection of multimodal cues', *Behavioral Ecology*, 18(1), pp. 236–40.

Robinson, M. H. (1969) 'The defensive behaviour of some orthopteroid insects from Panama', *Ecological Entomology*, 121(7), pp. 281–303.

Rocco, V., Barriga, J. P., Zagarese, H., and Lozada, M. (2002) 'How much does ultraviolet radiation contribute to the feeding performance of rainbow trout, *Oncorhynchus mykiss*, juveniles under natural illumination?', *Environmental Biology of Fishes*, 63(2), pp. 223–8.

Rock, I., and Brosgole, L. (1964) 'Grouping based on phenomenal proximity', *Journal of Experimental Psychology*, 67(6), pp. 531.

Rodriguez, J., Pitts, J. P., von Dohlen, C. D., and Wilson, J. S. (2014) 'Müllerian mimicry as a result of co-divergence between velvet ants and spider wasps', *PloS One*, 9, e112942.

Rogers, S. M., and Simpson, S. J. (2014) 'Thanatosis', *Current Biology*, 24(21), pp. R1031–3.

Rojas, B. (2017) 'Behavioural, ecological, and evolutionary aspects of diversity in frog colour patterns', *Biological Reviews*, 92(2), pp. 1059–80.

Rojas, B., Burdfield-Steel, E., Pakkanen, H., Suisto, K., Maczka, M., Schulz, S., and Mappes, J. (2017) 'How to fight multiple enemies: target-specific chemical defences in an aposematic moth', *Proceedings of the Royal Society B: Biological Sciences*, 284(1863). doi: 10.1098/rspb.2017.1424

Rojas, B., Devillechabrolle, J., and Endler, J. A. (2014a) 'Paradox lost: variable colour-pattern geometry is associated with differences in movement in aposematic frogs', *Biology Letters*, 10(6), pp. 20140193.

Rojas, B., and Endler, J. A. (2013) 'Sexual dimorphism and intra-populational colour pattern variation in the aposematic frog *Dendrobates tinctorius*', *Evolutionary Ecology*, 27(4), pp. 739–53.

Rojas, B., Rautiala, P., and Mappes, J. (2014b) 'Differential detectability of polymorphic warning signals under varying light environments', *Behavioural Processes*, 109, pp. 164–72.

Rojas, B., Valkonen, J., and Nokelainen, O. (2015) 'Aposematism', *Current Biology*, 25(9), pp. R350–1.

Romanes, G. J. (1883) *Mental evolution in animals*. New York: D. Appleton.

Rönkä, K., De Pasqual, C., Mappes, J., Gordon, S., and Rojas, B. (2018) 'Colour alone matters: no predator generalization among morphs of an aposematic moth', *Animal Behaviour*, 135, pp. 153–63.

Rose, T. A., Munn, A. J., Ramp, D., and Banks, P. B. (2006) 'Foot-thumping as an alarm signal in macropodoid marsupials: prevalence and hypotheses of function', *Mammal Review*, 36(4), pp. 281–98.

Rosser, N., Dasmahapatra, K. K., and Mallet, J. (2014) 'Stable *Heliconius* butterfly hybrid zones are correlated with a local rainfall peak at the edge of the Amazon basin', *Evolution*, 68(12), pp. 3470–84.

Rothschild, M. (1961) 'Defensive odours and Müllerian mimicry among insects', *Ecological Entomology*, 113(5), pp. 101–23.

Rothschild, M. (1963) 'Is the buff ermine (*Spilosoma lutea* (Huf.)) a mimic of the white ermine (*Spilosoma lubricipeda* (L.))?', *Proceedings of the Royal Entomological Society of London, Series A*, 38, pp. 159–64.

Rothschild, M. (1964) 'An extension of Dr Lincoln Brower's theory on bird predation and food specificity, together with some observations on bird memory in relation to aposematic colour patterns.', *Entomologist*, 97, pp. 73–8.

Rothschild, M. (1971) 'Speculations about mimicry with Henry Ford', in Creed, E.R. (ed.) *Ecological Genetics and Evolution*, Oxford: Blackwell, pp. 202–23.

Rothschild, M. (1975) 'Remarks on carotenoids in the evolution of signals', *Coevolution of animals and plants*, pp. 20–47. In: Gilbert LE, Raven PH (eds) Coevolution of animals and plants, University of Texas Press, Austin, pp 20–47.

Rowe, C., and Guilford, T. (1996) 'Hidden colour aversions in domestic chicks triggered by pyrazine odours of insect warning displays', *Nature*, 383(6600), pp. 520–2.

Rowe, C., and Halpin, C. (2013) 'Why are warning displays multimodal?', *Behavioral Ecology and Sociobiology*, 67(9), pp. 1425–39.

Rowe, C., Lindström, L., and Lyytinen, A. (2004) 'The importance of pattern similarity between Müllerian mimics in predator avoidance learning', *Proceedings of the Royal Society B: Biological Sciences*, 271(1537), pp. 407–13.

Rowe, M. P., and Owings, D. H. (1978) 'The meaning of the sound of rattling by rattlesnakes to California ground squirrels', *Behaviour*, 66(3), pp. 252–67.

Rowland, H. M. (2009) 'From Abbott Thayer to the present day: what have we learned about the function of countershading?', *Philosophical Transactions of the Royal Society B: Biological Sciences*, 364, pp. 519–27.

Rowland, H. M. (2011) 'The history, theory and evidence for a cryptic function of countershading', in Stevens, M. and Merilaita, S. (eds), *Animal Camouflage: Mechanisms and Function*, Cambridge: Cambridge University Press, pp. 53–72.

Rowland, H. M., Cuthill, I. C., Harvey, I. F., Speed, M. P., and Ruxton, G. D. (2008) 'Can't tell the caterpillars from the trees: countershading enhances survival in a woodland.', *Proceedings of the Royal Society B: Biological Sciences*, 275, pp. 2539–45.

Rowland, H. M., Hoogesteger, T., Ruxton, G. D., Speed, M. P., and Mappes, J. (2010a) 'A tale of 2 signals: signal mimicry between aposematic species enhances predator avoidance learning', *Behavioral Ecology*, 21(4), pp. 851–60.

Rowland, H. M., Ihalainen, E., Lindström, L., Mappes, J. and Speed, M. P. (2007a) 'Co-mimics have a mutualistic relationship despite unequal defences', *Nature*, 448(7149), pp. 64–7.

Rowland, H. M., Mappes, J., Ruxton, G. D., and Speed, M. P. (2010b) 'Mimicry between unequally defended prey can be parasitic: evidence for quasi-Batesian mimicry', *Ecology Letters*, 13(12), pp. 1494–502.

Rowland, H. M., Ruxton, G. D., and Skelhorn, J. (2013) 'Bitter taste enhances predatory biases against aggregations of prey with warning coloration', *Behavioral Ecology*, 24(4), pp. 942–8.

Rowland, H. M., Speed, M. P., Ruxton, G. D., Edmunds, M., Stevens, M., and Harvey, I. F. (2007b) 'Countershading enhances cryptic protection: an experiment with wild birds and artificial prey', *Animal Behaviour*, 74, pp. 1249–58.

Rowland, H. M., Wiley, E., Ruxton, G. D., Mappes, J., and Speed, M. P. (2010c) 'When more is less: the fitness consequences of predators attacking more unpalatable prey when more are presented', *Biology Letters*, 6(6), pp. 732–5.

Rudh, A., Breed, M. F., and Qvarnström, A. (2012) 'Does aggression and explorative behaviour decrease with lost warning coloration?', *Biological Journal of the Linnean Society*, 108(1), pp. 116–26.

Rundus, A. S., Owings, D. H., Joshi, S. S., Chinn, E., and Giannini, N. (2007) 'Ground squirrels use an infrared signal to deter rattlesnake predation', *Proceedings of the National Academy of Sciences*, 104(36), pp. 14372–6.

Ruxton, G. D. (2009) 'Non-visual crypsis: a review of the empirical evidence for camouflage to senses other than vision', *Philosophical Transactions of the Royal Society of London B: Biological Sciences*, 364(1516), pp. 549–57.

Ruxton, G. D., Franks, D. W., Balogh, A. C. V., and Leimar, O. (2008) 'Evolutionary implications of the form of predator generalization for aposematic signals and mimicry in prey', *Evolution*, 62(11), pp. 2913–21.

Ruxton, G. D., and Kennedy, M. W. (2006) 'Peppers and poisons: the evolutionary ecology of bad taste', *Journal of Animal Ecology*, 75(5), pp. 1224–6.

Ruxton, G. D., and Sherratt, T. N. (2006) 'Aggregation, defence and warning signals: the evolutionary relationship', *Proceedings of the Royal Society of London B: Biological Sciences*, 273(1600), pp. 2417–24.

Ruxton, G. D., Sherratt, T. N., and Speed, M. P. (2004a) *Avoiding attack: the evolutionary ecology of crypsis, warning signals, and mimicry.* Oxford: Oxford University Press.

Ruxton, G. D., Speed, M., and Sherratt, T. N. (2004b) 'Evasive mimicry: when (if ever) could mimicry based on difficulty of capture evolve?', *Proceedings of the Royal Society B: Biological Sciences*, 271(1553), pp. 2135–42.

Ruxton, G. D., Speed, M. P., and Broom, M. (2009) 'Identifying the ecological conditions that select for intermediate levels of aposematic signalling', *Evolutionary Ecology*, 23(4), pp. 491–501.

Ruxton, G. D., Speed, M. P. and Kelly, D. J. (2004c) 'What, if anything, is the adaptive function of countershading?', *Animal Behaviour,* 68, pp. 445–51.

Sagonas, K., Karambotsi, N., Bletsa, A., Reppa, A., Pafilis, P., and Valakos, E. D. (2017) 'Tail regeneration affects the digestive performance of a Mediterranean lizard', *The Science of Nature*, 104(3–4), pp. 22.

Sanchez Paniagua, K., and Abarca, J. G. (2016) 'Thanatosis in four poorly known toads of the genus *Incilius* (Amphibia: Anura) from the highlands of Costa Rica', *Mesoamerican Herpetology*, 3(1), pp. 135–40.

Santana, S. E., Dial, T. O., Eiting, T. P., and Alfaro, M. E. (2011) 'Roosting ecology and the evolution of pelage markings in bats', *PLoS One*, 6, pp. e25845.

Santer, R. D. (2013) 'Motion dazzle: a locust's eye view', *Biology Letters*, 9(6), pp. 20130811.

Santos, J. C., Baquero, M., Barrio-Amorós, C., Coloma, L. A., Erdtmann, L. K., Lima, A. P., and Cannatella, D. C. (2014) 'Aposematism increases acoustic diversification and speciation in poison frogs', *Proceedings of the Royal Society of London B: Biological Sciences*, 281(1796). doi: 10.1098/rspb.2014.1761

Santos, J. C., and Cannatella, D. C. (2011) 'Phenotypic integration emerges from aposematism and scale in poison frogs', *Proceedings of the National Academy of Sciences*, 108(15), pp. 6175–80.

Santos, M. B. d., Oliveira, M. C. L. M. d., Verrastro, L., and Tozetti, A. M. (2010) 'Playing dead to stay alive: death-feigning in *Liolaemus occipitalis* (Squamata: Liolaemidae)', *Biota Neotropica*, 10(4), pp. 361–4.

Saporito, R. A., Spande, T. F., Garraffo, H. M., and Donnelly, M. A. (2009) 'Arthropod alkaloids in poison frogs: a review of the "dietary hypothesis"', *Heterocycles*, 79, pp. 277–97.

Sasaki, A., Kawaguchi, I., and Yoshimori, A. (2002) 'Spatial mosaic and interfacial dynamics in a Müllerian mimicry system', *Theoretical Population Biology*, 61(1), pp. 49–71.

Satoh, S., Takahashi, T., Tada, S., Tanaka, H., and Kohda, M. (2017) 'Parental females of a nest-brooding cichlid improve and benefit from the protective value of young masquerading as snails', *Animal Behaviour*, 124, pp. 75–82.

Savage, J. M., and Slowinski, J. B. (1992) 'The colouration of the venomous coral snakes (family Elapidae) and their mimics (families Aniliidae and Colubridae)', *Biological Journal of the Linnean Society*, 45, pp. 235–54.

Savage, W. K., and Mullen, S. P. (2009) 'A single origin of Batesian mimicry among hybridizing populations of admiral butterflies (*Limenitis arthemis*) rejects an evolutionary reversion to the ancestral phenotype', *Proceedings of the Royal Society B: Biological Sciences*, 276(1667), pp. 2557–65.

Saxena, S. (1957) 'An experimental study of thanatosis in *Armadillidium vulgare* (Latreille)', *Journal of the Zoological Society of India*, 9, pp. 192–9.

Schaefer, H. M., and Ruxton, G. D. (2009) 'Deception in plants: mimicry or perceptual exploitation?', *Trends in Ecology & Evolution*, 24(12), pp. 676–85.

Schaefer, H. M., and Ruxton, G. D. (2011). *Plant–animal communication*. Oxford: Oxford University Press.

Schaefer, H. M., and Stobbe, N. (2006) 'Disruptive coloration provides camouflage independent of background matching', *Proc Biol Sci*, 273(1600), pp. 2427–32.

Schlenoff, D. H. (1985) 'The startle responses of blue jays to *Catocala* (Lepidoptera: Noctuidae) prey models', *Animal Behaviour*, 33(4), pp. 1057–67.

Schmidt, J. O. (2004) 'Venom and the good life in tarantula hawks (Hymenoptera: Pompilidae): how to eat, not be eaten, and live long', *J Kan Ent Soc*, 77, pp. 402–13.

Schmidt, R. S. (1958) 'Behavioural evidence on the evolution of Batesian mimicry', *Animal Behaviour*, 6, pp. 129–38.

Schmidt, R. S. (1960) 'Predator behaviour and the perfection of incipient mimetic resemblances', *Behaviour*, 16, pp. 149–58.

Schmied, H., Lambertz, M., and Geissler, P. (2013) 'New case of true mimicry in cockroaches (Blattodea)', *Entomological Science*, 16(1), pp. 119–21.

Schuler, W., and Roper, T. J. (1992) 'Responses to warning coloration in avian predators', *Advances in the Study of Behavior*, 21, pp. 111–46.

Scott-Brown, K. C., and Heeley, D. W. (2001) 'The effect of the spatial arrangement of target lines on perceived speed', *Vision Research*, 41(13), pp. 1669–82.

Scott-Samuel, N. E., Baddeley, R., Palmer, C. E., and Cuthill, I. C. (2011) 'Dazzle camouflage affects speed perception', *PloS One*, 6, pp. e20233.

Seapy, R., and Young, R. (1986) 'Concealment in epipelagic pterotracheid heteropods (Gastropoda) and cranchiid squids (Cephalopoda)', *Journal of Zoology*, 210(1), pp. 137–47.

Seehausen, M., and Van Alphen, J. J. M. (1999) 'Evolution of colour patterns in East African cichlid fish', *J Evol Biol*, 12(3), 514–34.

Sermanet, P., Eigen, D., Zhang, X., Mathieu, M., Fergus, R., and LeCun, Y. (2013) 'Overfeat: Integrated recognition, localization and detection using convolutional networks', *arXiv preprint arXiv:1312.6229*.

Servedio, M. R. (2000) 'The effects of predator learning, forgetting, and recognition errors on the evolution of warning coloration', *Evolution*, 54(3), pp. 751–63.

Seymoure, B. M., and Aiello, A. (2015) 'Keeping the band together: evidence for false boundary disruptive coloration in a butterfly', *J Evol Biol*, 28(9), pp. 1618–24.

Shashar, N., Hagan, R., Boal, J. G., and Hanlon, R. T. (2000) 'Cuttlefish use polarization sensitivity in predation on silvery fish', *Vision Research*, 40(1), pp. 71–5.

Shashar, N., Hanlon, R. T., and Petz, A. de M. (1998) 'Polarization vision helps detect transparent prey', *Nature*, 393(6682), pp. 222–3.

Sheppard, P. M. (1958) *Natural selection and heredity*. New York: Harper Torchbooks.

Sheppard, P. M., Turner, J. R. G., Brown, K. S., Benson, W. W., and Singer, M. C. (1985) 'Genetics and the evolution of Muellerian mimicry in *Heliconius* butterflies', *Philosophical Transactions of the Royal Society of London Series B: Biological Sciences*, 308(1137), pp. 433–610. doi: 10.1098/rstb.1985.0066

Sherratt, T. N. (2002a) 'The coevolution of warning signals', *Proceedings of the Royal Society of London B: Biological Sciences*, 269(1492), pp. 741–6.

Sherratt, T. N. (2002b) 'The evolution of imperfect mimicry', *Behavioral Ecology*, 13(6), pp. 821–6.

Sherratt, T. N. (2003) 'State-dependent risk-taking in systems with defended prey', *Oikos*, 103, pp. 93–100.

Sherratt, T. N. (2006) 'Spatial mosaic formation through frequency-dependent selection in Müllerian mimicry complexes', *Journal of Theoretical Biology*, 240(2), pp. 165–74.

Sherratt, T. N. (2011) 'The optimal sampling strategy for unfamiliar prey', *Evolution*, 65(7), pp. 2014–25.

Sherratt, T. N., and Beatty, C. D. (2003) 'The evolution of warning signals as reliable indicators of prey defense', *American Naturalist*, 162(4), pp. 377–89.

Sherratt, T. N., and Harvey, I. F. (1993) 'Frequency-dependent food selection by arthropods: a review', *Biological Journal of the Linnean Society*, 48(2), pp. 167–86.

Sherratt, T. N. and Macdougall, A. D. (1995) 'Some population consequences of variation in preference among individual predators', *Biological Journal of the Linnean Society*, 55(2), pp. 93–107.

Sherratt, T. N., and Peet-Pare, C. A. (2017) 'The perfection of mimicry: an information approach', *Philosophical Transactions of the Royal Society B: Biological Sciences*, 372(1724). doi: 10.1098/rstb.2016.0340.

Sherratt, T. N., Pollitt, D., and Wilkinson, D. M. (2007) 'The evolution of crypsis in replicating populations of web-based prey', *Oikos*, 116(3), pp. 449–60.

Sherratt, T. N., Rashed, A., and Beatty, C. D. (2004a) 'The evolution of locomotory behavior in profitable and unprofitable simulated prey', *Oecologia*, 138(1), pp. 143–50.

Sherratt, T. N., Roberts, G., and Kassen, R. (2009) 'Evolutionary stable investment in products that confer both an individual benefit and a public good', *Front. Biosci*, 14, pp. 340–7.

Sherratt, T. N., Speed, M. P., and Ruxton, G. D. (2004b) 'Natural selection on unpalatable species imposed by state-dependent foraging behaviour', *Journal of Theoretical Biology*, 228(2), pp. 217–26.

Sherratt, T. N., Whissell, E., Webster, R., and Kikuchi, D. W. (2015) 'Hierarchical overshadowing of stimuli and its role in mimicry evolution', *Animal Behaviour*, 108, pp. 73–9.

Sherratt, T. N., Wilkinson, D. M., and Bain, R. S. (2005) 'Explaining *Dioscorides*' "double difference": why are some mushrooms poisonous, and do they signal their unprofitability?', *The American Naturalist*, 166(6), pp. 767–75.

Shettleworth, S. J. (2010) *Cognition, Evolution and Behaviour*. 2nd edn. New York: Oxford University Press.

Shine, R. (1990) 'Function and evolution of the frill of the frillneck lizard, *Chlamydosaurus kingii* (Sauria: Agamidae)', *Biological Journal of the Linnean Society*, 40(1), pp. 11–20.

Shirai, L. T., Saenko, S. V., Keller, R. A., Jeronimo, M. A., Brakefield, P. M., Descimon, H., Wahlberg, N., and Beldade, P. (2012) 'Evolutionary history of the recruitment of conserved developmental genes in association to the formation and diversification of a novel trait', *BMC Evolutionary Biology*, 12, pp. 21.

Shreeve, T. G., Dennis, R. L., and Wakeham-Dawson, A. (2000) 'Phylogenetic, habitat, and behavioural aspects of possum behaviour in European lepidoptera', *Journal of Research on the Lepidoptera*, 39, pp. 80–5.

Silberglied, R. E., Aiello, A., and Windsor, D. M. (1980) 'Disruptive coloration in butterflies: lack of support in *Anartia fatima*', *Science*, 209(4456), pp. 617–19.

Sillén-Tullberg, B. (1985) 'Higher survival of an aposematic than of a cryptic form of a distasteful bug', *Oecologia*, 67(3), pp. 411–15.

Sillén-Tullberg, B. (1988) 'Evolution of gregariousness in aposematic butterfly larvae: a phylogenetic analysis', *Evolution*, 42, pp. 293–305.

Sillén-Tullberg, B. (1990) 'Do predators avoid groups of aposematic prey? An experimental test', *Animal Behaviour*, 40(5), pp. 856–60.

Sillén-Tullberg, B., and Leimar, O. (1988) 'The evolution of gregariousness in distasteful insects as a defense against predators', *The American Naturalist*, 132(5), pp. 723–34.

Simoncelli, E. P., and Heeger, D. J. (1998) 'A model of neuronal responses in visual area MT', *Vision Research*, 38(5), pp. 743–61.

Skelhorn, J. (2015) 'Masquerade', *Current Biology*, 25(15), pp. R643–4.

Skelhorn, J., Griksaitis, D., and Rowe, C. (2008) 'Colour biases are more than a question of taste', *Animal Behaviour*, 75, pp. 827–35.

Skelhorn, J., Halpin, C. G., and Rowe, C. (2016a) 'Learning about aposematic prey', *Behavioral Ecology*, 27(4), pp. 955–64.

Skelhorn, J., Holmes, G. G., Hossie, T. J., and Sherratt, T. N. (2016b) 'Multicomponent deceptive signals reduce the speed at which predators learn that prey are profitable', *Behavioral Ecology*, 27(1), pp. 141–7.

Skelhorn, J., Holmes, G. G., and Rowe, C. (2016c) 'Deimatic or aposematic?', *Animal Behaviour*, 113, pp. e1–e3.

Skelhorn, J., and Rowe, C. (2005) 'Frequency-dependent taste-rejection by avian predation may select for defence chemical polymorphisms in aposematic prey', *Biology Letters*, 1(4), pp. 500–3.

Skelhorn, J., and Rowe, C. (2006) 'Prey palatability influences predator learning and memory', *Animal Behaviour*, 71, pp. 1111–18.

Skelhorn, J., and Rowe, C. (2007a) 'Automimic frequency influences the foraging decisions of avian predators on aposematic prey', *Animal Behaviour*, 74(5), pp. 1563–72.

Skelhorn, J., and Rowe, C. (2007b) 'Predators' toxin burdens influence their strategic decisions to eat toxic prey', *Current Biology*, 17(17), pp. 1479–83.

Skelhorn, J., and Rowe, C. (2009) 'Distastefulness as an antipredator defence strategy', *Animal Behaviour*, 78(3), pp. 761–6.

Skelhorn, J., and Rowe, C. (2010) 'Birds learn to use distastefulness as a signal of toxicity', *Proceedings of the Royal Society B: Biological Sciences*, 277(1688), pp. 1729–34.

Skelhorn, J. and Rowe, C. (2016) 'Cognition and the evolution of camouflage'. *Proc. R. Soc. B*: The Royal Society, 20152890. Proc Biol Sci. 2016 Feb 24;283(1825):20152890. doi: 10.1098/rspb.2015.2890.

Skelhorn, J., Rowland, H. M., Delf, J., Speed, M. P., and Ruxton, G. D. (2011) 'Density-dependent predation influences the evolution and behavior of masquerading prey', *Proceedings of the National Academy of Sciences of the United States of America*, 108(16), pp. 6532–6.

Skelhorn, J., Rowland, H. M., and Ruxton, G. D. (2010a) 'The evolution and ecology of masquerade', *Biological Journal of the Linnean Society*, 99(1), pp. 1–8.

Skelhorn, J., Rowland, H. M., Speed, M. P., and Ruxton, G. D. (2010b) 'Masquerade: camouflage without crypsis', *Science*, 327(5961), pp. 51.

Skelhorn, J., and Ruxton, G. D. (2011) 'Mimicking multiple models: polyphenetic masqueraders gain additional benefits from crypsis', *Behavioral Ecology*, 22(1), pp. 60–5.

Skelhorn, J., and Ruxton, G. D. (2013) 'Size-dependent microhabitat selection by masquerading prey', *Behavioral Ecology*, 24(1), pp. 89–97.

Skelhorn, J., and Ruxton, G. D. (2014) 'Viewing distance affects how the presence of inedible models influence the benefit of masquerade', *Evolutionary Ecology*, 28(3), pp. 441–55.

Skelly, D. K. (1994) 'Activity level and the susceptibility of anuran larvae to predation', *Animal Behaviour*, 47(2), pp. 465–8.

Smith, K. E., Halpin, C. G., and Rowe, C. (2014) 'Body size matters for aposematic prey during predator aversion learning', *Behavioural Processes*, 109, pp. 173–9.

Smith, S. M. (1975) 'Innate recognition of coral snake pattern by a possible avian predator', *Science*, 187, pp. 759–60.

Smith, S. M. (1977) 'Coral-snake pattern recognition and stimulus generalisation by naive great kiskadees (Aves: Tyrannidae)', *Nature*, 265, pp. 535–6.

Smithers, S. P., Wilson, A., and Stevens, M. (2017) 'Rock pool gobies change their body pattern in response to background features', *Biological Journal of the Linnean Society*, 121(1), pp. 109–21.

Smithwick, F. M., Nicholls, R., Cuthill, I. C., and Vinther, J. (2017) 'Countershading and stripes in the theropod dinosaur Sinosauropteryx reveal heterogeneous habitats in the early Cretaceous Jehol Biota', *Current Biology*. 2017 Nov 6;27(21):3337-3343.e2. doi: 10.1016/j.cub.2017.09.032. Epub 2017 Oct 26.

Sonerud, G. A. (1988) 'To distract display or not: Grouse hens and foxes', *Oikos*, 51(2), pp. 233–7.

Sourakov, A. (2013) 'Two heads are better than one: false head allows *Calycopis cecrops* (Lycaenidae) to escape predation by a Jumping Spider, *Phidippus pulcherrimus* (Salticidae)', *Journal of Natural History*, 47(15–16), pp. 1047–54.

Sovrano, V. A., and Bisazza, A. (2008) 'Recognition of partly occluded objects by fish', *Animal Cognition*, 11(1), pp. 161–6.

Speed, M. P. (1993) 'Muellerian mimicry and the psychology of predation', *Animal Behaviour*, 45(3), pp. 571–80.

Speed, M. P. (2001) 'Can receiver psychology explain the evolution of aposematism?', *Animal Behaviour*, 61(1), pp. 205–16.

Speed, M. P., Alderson, N. J., Hardman, C., and Ruxton, G. D. (2000) 'Testing Mullerian mimicry: an experiment with wild birds', *Proceedings of the Royal Society of London Series B: Biological Sciences*, 267(1444), pp. 725–31.

Speed, M. P., Brockhurst, M. A., and Ruxton, G. D. (2010) 'The dual benefits of aposematism: predator avoidance and enhanced resource collection', *Evolution*, 64(6), pp. 1622–33.

Speed, M. P., Fenton, A., Jones, M. G., Ruxton, G. D., and Brockhurst, M. A. (2015) 'Coevolution can explain defensive secondary metabolite diversity in plants', *New Phytologist*, 208(4), pp. 1251–63.

Speed, M. P., and Franks, D. W. (2014) 'Antagonistic evolution in an aposematic predator–prey signaling system', *Evolution*, 68(10), pp. 2996–3007.

Speed, M. P., Kelly, D. J., Davidson, A., and Ruxton, G. D. (2004) 'Countershading enhances crypsis with some bird species but not others', *Behavioral Ecology*, 16, pp. 327–34.

Speed, M. P., and Ruxton, G. D. (2005a) 'Warning displays in spiny animals: one (more) evolutionary route to aposematism', *Evolution*, 59(12), pp. 2499–508.

Speed, M. P., and Ruxton, G. D. (2005b) 'Aposematism: what should our starting point be?', *Proceedings of the Royal Society of London B: Biological Sciences*, 272(1561), pp. 431–8.

Speed, M. P., Ruxton, G. D., and Broom, M. (2006) 'Automimicry and the evolution of discrete prey defences', *Biological Journal of the Linnean Society*, 87(3), pp. 393–402.

Speed, M. P., Ruxton, G. D., Mappes, J., and Sherratt, T. N. (2012) 'Why are defensive toxins so variable? An evolutionary perspective', *Biological Reviews*, 87(4), pp. 874–84.

Speed, M. P., and Turner, J. R. G. (1999) 'Learning and memory in mimicry: II. Do we understand the mimicry spectrum?', *Biological Journal of the Linnean Society*, 67(3), pp. 281–312.

Srygley, R. B. (1999) 'Incorporating motion into investigations of mimicry', *Evolutionary Ecology*, 13(7), pp. 691–708.

Srygley, R. B. (2004) 'The aerodynamic costs of warning signals in palatable mimetic butterflies and their distasteful models', *Proceedings of the Royal Society B: Biological Sciences*, 271(1539), pp. 589–94.

Srygley, R. B., and Chai, P. (1990) 'Predation and the elevation of thoracic temperature in brightly colored neotropical butterflies', *American Naturalist*, 135(6), pp. 766–87.

Staddon, J. E. R., and Gendron, R. P. (1983) 'Optimal detection of cryptic prey may lead to predator switching', *The American Naturalist*, 122(6), pp. 843–8.

Stamp, N. (2003) 'Out of the quagmire of plant defense hypotheses', *The Quarterly Review of Biology*, 78(1), pp. 23–55.

Stamp, N. E., and Wilkens, R. T. (1993) 'On the cryptic side of life: being unapparent to enemies and the consequences for foraging and growth of caterpillars', in Stamp, N. E., and Casey, T. M. (eds), *Caterpillars: Ecological and Evolutionary Contraints on Foraging*, New York: Routledge, Chapman and Hall, pp. 283–330.

Stang, A. T., and McRae, S. B. (2009) 'Why some rails have white tails: the evolution of white undertail plumage and anti-predator signaling', *Evolutionary Ecology*, 23(6), pp. 943–61.

Stankowich, T. (2008) 'Tail-flicking, tail-flagging, and tail position in ungulates with special reference to black-tailed deer', *Ethology*, 114(9), pp. 875–85.

Stankowich, T., Caro, T., and Cox, M. (2011) 'Bold coloration and the evolution of aposematism in terrestrial carnivores', *Evolution*, 65(11), pp. 3090–9.

Stankowich, T., and Coss, R. G. (2008) Alarm walking in Columbian black-tailed deer: its characterization and possible antipredatory signaling functions, *Journal of Mammalogy*, 89(3), pp. 636–45. https://doi.org/10.1644/07-MAMM-A-203R.1

Stankowich, T., Haverkamp, P. J., and Caro, T. (2014) 'Ecological drivers of antipredator defenses in carnivores', *Evolution*, 68(5), pp. 1415–25.

Starostová, Z., Gvoždík, L., and Kratochvíl, L. (2017) 'An energetic perspective on tissue regeneration: The costs of tail autotomy in growing geckos', *Comparative Biochemistry and Physiology Part A: Molecular & Integrative Physiology*, 206, pp. 82–6.

Staudinger, M. D., Hanlon, R. T., and Juanes, F. (2011) 'Primary and secondary defences of squid to cruising and ambush fish predators: variable tactics and their survival value', *Animal Behaviour*, 81(3), pp. 585–94.

Stauffer, J. R., Half, E. A., and Seltzer, R. (1999) 'Hunting strategies of a Lake Malawi cichlid with reverse countershading', *Copeia*, 1999, pp. 1108.

Steiner, C. C., Weber, J. N., and Hoekstra, H. E. (2007) 'Adaptive variation in beach mice produced by two inter-acting pigmentation genes', *PLoS Biology*, 5, pp. e219.

Sternburg, J. G., Waldbauer, G. P., and Jeffords, M. R. (1977) 'Batesian mimicry - selective advantage of color pattern', *Science*, 195(4279), pp. 681–3.

Stevens, M. (2005) 'The role of eyespots as anti-predator mechanisms, principally demonstrated in the Lepidoptera', *Biological Reviews*, 80(4), pp. 573–88.

Stevens, M. (2007) 'Predator perception and the interrela-tion between different forms of protective coloration.', *Proceedings of the Royal Society B: Biological Sciences*, 274, pp. 1457–64.

Stevens, M. (2013) *Sensory Ecology, Behaviour and Evolution.* Oxford, England: Oxford University Press.

Stevens, M. (2016) 'Color change, phenotypic plasticity, and camouflage', *Frontiers in Ecology and Evolution*, 4, pp. 51.

Stevens, M., and Cuthill, I. C. (2006) 'Disruptive color-ation, crypsis and edge detection in early visual pro-cessing.', *Proceedings of the Royal Society B: Biological Sciences*, 273, pp. 2141–7.

Stevens, M., Cuthill, I. C., Windsor, A. M., and Walker, H. J. (2006) 'Disruptive contrast in animal camouflage', *Proc Biol Sci*, 273(1600), pp. 2433–8.

Stevens, M., Hardman, C. J., and Stubbins, C. L. (2008a) 'Conspicuousness, not eye mimicry, makes "eyespots" effective antipredator signals', *Behavioral Ecology*, 19(3), pp. 525–31.

Stevens, M., Marshall, K. L., Troscianko, J., Finlay, S., Burnand, D., and Chadwick, S. L. (2012) 'Revealed by conspicuousness: distractive markings reduce camou-flage', *Behavioral Ecology*, 24(1), pp. 213–22.

Stevens, M., and Merilaita, S. (2009a) 'Animal camouflage: current issues and new perspectives', *Philosophical Transactions of the Royal Society B: Biological Sciences*, 364(1516), pp. 423–7.

Stevens, M., and Merilaita, S. (2009b) 'Defining disruptive coloration and distinguishing its functions', *Philosophical Transactions of the Royal Society B: Biological Sciences*, 364, pp. 481–8.

Stevens, M., and Merilaita, S. (2011) *Animal camouflage: mechanisms and function.* Cambridge: Cambridge Uni-versity Press.

Stevens, M., and Ruxton, G. D. (2012) 'Linking the evolu-tion and form of warning coloration in nature', *Proceedings of the Royal Society B: Biological Sciences*, 279(1728), pp. 417–26.

Stevens, M., and Ruxton, G. D. (2014) 'Do animal eyespots really mimic eyes?', *Current Zoology*, 60(1), pp. 26–36.

Stevens, M., Searle, W. T., Seymour, J. E., Marshall, K. L., and Ruxton, G. D. (2011) 'Motion dazzle and camouflage as distinct anti-predator defenses', *BMC Biol*, 9, pp. 81.

Stevens, M., Winney, I. S., Cantor, A., and Graham, J. (2009) 'Outline and surface disruption in animal camou-flage', *Proc Biol Sci*, 276(1657), pp. 781–6.

Stevens, M., Yule, D. H., and Ruxton, G. D. (2008b) 'Dazzle coloration and prey movement', *Proceedings of the Royal Society of London B: Biological Sciences*, 275(1651), pp. 2639–43.

Stobbe, N., and Schaefer, H. M. (2008) 'Enhancement of chromatic contrast increases predation risk for striped butterflies', *Proc Biol Sci*, 275(1642), pp. 1535–41.

Stoddard, M. C., Kupan, K., Eyster, H. N., Rojas-Abreu, W., Cruz-Lopez, M., Serrano-Meneses, M. A., and Kupper, C. (2016) 'Camouflage and clutch survival in plovers and terns', *Sci Rep*, 6, pp. 32059.

Stone, L. S., and Thompson, P. (1992) 'Human speed perception is contrast dependent', *Vision Research*, 32, pp. 1535–49.

Stoner, C. J., Bininda-Emonds, O. R. P., and Caro, T. M. (2003) 'The adaptive significance of coloration in lago-morphs', *Biological Journal of the Linnean Society*, 79, pp. 309–28.

Straube, N., Li, C., Claes, J. M., Corrigan, S., and Naylor, G. J. P. (2015) 'Molecular phylogeny of Squaliformes and first occurrence of bioluminescence in sharks', *BMC Evolutionary Biology*, 15, pp. 162.

Strauss, S. Y., Rudgers, J. A., Lau, J. A., and Irwin, R. E. (2002) 'Direct and ecological costs of resistance to her-bivory', *Trends in Ecology & Evolution*, 17(6), pp. 278–85.

Stroud, J. T., and Losos, J. B. (2016) 'Ecological opportunity and adaptive radiation', *Annual Review of Ecology, Evolution, and Systematics*, 47, pp. 507–32.

Stuart-Fox, D., and Moussalli, A. (2009) 'Camouflage, communication and thermoregulation: lessons from colour changing organisms', *Philosophical Transactions of the Royal Society of London B: Biological Sciences*, 364(1516), pp. 463–70.

Stuckert, A. M. M., Saporito, R. A., Venegas, P. J., and Summers, K. (2014a) 'Alkaloid defenses of co-mimics in a putative Müllerian mimetic radiation', *BMC Evolutionary Biology*, 14, 76. https://doi.org/10.1186/1471-2148-14-76

Stuckert, A. M. M., Venegas, P. J., and Summers, K. (2014b) 'Experimental evidence for predator learning and Müllerian mimicry in Peruvian poison frogs (*Ranitomeya*, Dendrobatidae)', *Evolutionary Ecology*, 28(3), pp. 413–26.

Su, S., Lim, M., and Kunte, K. (2015) 'Prey from the eyes of predators: colour discriminability of aposematic and mimetic butterflies from an avian visual perspective', *Evolution*, 69, pp. 2985–94.

Summers, K., and Clough, M. E. (2001) 'The evolution of coloration and toxicity in the poison frog family (Dendrobatidae)', *Proceedings of the National Academy of Sciences*, 98(11), pp. 6227–32.

Summers, K., Speed, M., Blount, J., and Stuckert, A. (2015) 'Are aposematic signals honest? A review', *Journal of Evolutionary Biology*, 28(9), pp. 1583–99.

Sun, J., and Perona, P. (1998) 'Where is the sun?', *Nature Neuroscience*, 1, pp. 183–4.

Sun, P., Chubb, C., and Sperling, G. (2015) 'Two mechanisms that determine the Barber-Pole Illusion', *Vision Research*, 111, pp. 43–54.

Suzuki, K., Ikebuchi, M., and Okanoya, K. (2013) 'The impact of domestication on fearfulness: a comparison of tonic immobility reactions in wild and domesticated finches', *Behavioural Processes*, 100, pp. 58–63.

Suzuki, T. K., Tomita, S., and Sezutsu, H. (2014) 'Gradual and contingent evolutionary emergence of leaf mimicry in butterfly wing patterns', *BMC Evolutionary Biology*, 14(1), pp. 229.

Svennungsen, T. O., and Holen, Ø. H. (2007) 'The evolutionary stability of automimicry', *Proceedings of the Royal Society of London B: Biological Sciences*, 274(1621), pp. 2055–62.

Sword, G. A., Simpson, S. J., El Hadi, O. T. M., and Wilps, H. (2000) 'Density-dependent aposematism in the desert locust', *Proceedings of the Royal Society of London B: Biological Sciences*, 267(1438), pp. 63–8.

Symula, R., Schulte, R., and Summers, K. (2001) 'Molecular phylogenetic evidence for a mimetic radiation in Peruvian poison frogs supports a Müllerian mimicry hypothesis', *Proceedings of the Royal Society of London Series B: Biological Sciences*, 268(1484), pp. 2415–21.

Tan, K., Wang, Z., Li, H., Yang, S., Hu, Z., Kastberger, G., and Oldroyd, B. P. (2012) 'An 'I see you' prey–predator signal between the Asian honeybee, *Apis cerana*, and the hornet, *Vespa velutina*', *Animal Behaviour*, 83(4), pp. 879–82.

Tankus, A., and Yeshurun, Y. (2009) 'Computer vision, camouflage breaking and countershading', *Philosophical Transactions of the Royal Society B: Biological Sciences*, 364, pp. 529–36.

Taylor, C. H., Reader, T., and Gilbert, F. (2016) 'Why many Batesian mimics are inaccurate: evidence from hoverfly colour patterns', *Proceedings of the Royal Society B: Biological Sciences*, 283(1842). doi:10.1098/rspb.2016.1585

Telemeco, R. S., Baird, T. A., and Shine, R. (2011) 'Tail waving in a lizard (*Bassiana duperreyi*) functions to deflect attacks rather than as a pursuit-deterrent signal', *Animal Behaviour*, 82(2), pp. 369–75.

Terhune, E. C. (1977) 'Components of a visual stimulus used by scrubjays to discriminate a Batesian model', *The American Naturalist*, 111, pp. 435–51.

Thayer, A. H. (1896) 'The law which underlies protective coloration', *The Auk*, 13, pp. 124–9.

Thayer, C. H. (1909) *Concealing-coloration in the animal kingdom*. New York: Macmillan.

Théry, M., and Casas, J. (2009) 'The multiple disguises of spiders: web colour and decorations, body colour and movement', *Philosophical Transactions of the Royal Society B: Biological Sciences*, 364(1516), pp. 471–80.

Théry, M., and Gomez, D. (2010) 'Insect colours and visual appearance in the eyes of their predators', in Simpson, S., and Casas, J. (eds), *Advances in Insect Physiology* 38, *Insect Integument and Colour*, pp. 267–353. https://doi.org/10.1016/S0065-2806(10)38001-5 Thomas, R., Marples, N., Cuthill, I., Takahashi, M., and Gibson, E. (2003) 'Dietary conservatism may facilitate the initial evolution of aposematism', *Oikos*, 101(3), pp. 458–66.

Thompson, P. (1982) 'Perceived rate of movement depends on contrast', *Vision Research*, 22, pp. 377–80.

Thorogood, R., Kokko, H., and Mappes, J. (2018) 'Social transmission of avoidance among predators facilitates the spread of novel prey', *Nature Ecology & Evolution*, 2, pp. 254–61.

Tinbergen, L. (1960) *The dynamics of insect and bird populations in pine woods.* Brill Archive.

Tobler, M. (2005) 'Feigning death in the Central American cichlid *Parachromis friedrichsthalii*', *Journal of Fish Biology,* 66(3), pp. 877–81.

Todd, P.A., Phua, H., and Toh, K.B. (2015) 'Interactions between background matching and disruptive colouration: Experiments using human predators and virtual crabs', *Current Zoology*, 61(4), pp. 718–28.

Toh, K. B., and Todd, P. (2017) 'Camouflage that is spot on! Optimization of spot size in prey-background matching', *Evolutionary Ecology*, 31(4), pp. 447–61.

Tojo, S. (1991) 'Variation in phase polymorphism in the common cutworm, *Spodoptera litura* (Lepidoptera: Noctuidae)', *Applied Entomology and Zoology*, 26(4), pp. 571–8.

Toledo, L. F., Sazima, I., and Haddad, C. F. (2010) 'Is it all death feigning? Case in anurans', *Journal of Natural History*, 44(31–32), pp. 1979–88.

Tomonaga, M. (1998) 'Perception of shape from shading in chimpanzees (*Pan troglodytes*) and humans (*Homo sapiens*)', *Animal Cognition*, 1, pp. 25–35.

Torres-Campos, I., Abram, P. K., Guerra-Grenier, E., Boivin, G., and Brodeur, J. (2016) 'A scenario for the evolution of selective egg coloration: the roles of enemy-free space, camouflage, thermoregulation and pigment limitation', *Royal Society Open Science*, 3, pp. 150711.

Touchon, J. C., and Warkentin, K. M. (2008) 'Fish and dragonfly nymph predators induce opposite shifts in color and morphology of tadpoles', *Oikos*, 117(4), pp. 634–40.

Trigo, J. R. (2011) 'Effects of pyrrolizidine alkaloids through different trophic levels', *Phytochemistry Reviews*, 10, pp. 83–98.

Trnka, A., Trnka, M., and Grim, T. (2015) 'Do rufous common cuckoo females indeed mimic a predator? An experimental test', *Biological Journal of the Linnean Society*, 116(1), pp. 134–43.

Troscianko, J., Lown, A. E., Hughes, A. E., and Stevens, M. (2013) 'Defeating crypsis: detection and learning of camouflage strategies', *PLoS One*, 8(9), pp. e73733.

Troscianko, J., Skelhorn, J., and Stevens, M. (2017) 'Quantifying camouflage: how to predict detectability from appearance', *BMC Evol Biol*, 17(1), p. 7.

Troscianko, J., Wilson-Aggarwal, J., Stevens, M., and Spottiswoode, C. N. (2016) 'Camouflage predicts survival in ground-nesting birds', *Scientific Reports*, 6, 19966. doi:10.1038/srep19966

Troscianko, T. S., Benton, C. P., Lovell, P. G., Tolhurst, D. J., and Pizlo, Z. (2009) 'Camouflage and visual perception.', *Philosophical Transactions of the Royal Society B: Biological Sciences*, 364, pp. 449–61.

Tse, P. U., and Hsieh, P. J. (2006) 'The infinite regress illusion reveals faulty integration of local and global motion signals', *Vision Research*, 46(22), pp. 3881–5.

Tseng, L., and Tso, I. M. (2009) 'A risky defence by a spider using conspicuous decoys resembling itself in appearance', *Animal Behaviour*, 78(2), pp. 425–31.

Tsuda, A., Hiroaki, S., and Hirose, T. (1998) 'Effect of gut content on the vulnerability of copepods to visual predation', *Limnology and Oceanography*, 43(8), pp. 1944–7.

Tu, Z., Chen, X., Yuille, A. L., and Zhu, S.-C. (2005) 'Image parsing: Unifying segmentation, detection, and recognition', *International Journal of Computer Vision*, 63(2), pp. 113–40.

Tucker, S., Hipfner, J. M., and Trudel, M. (2016) 'Size- and condition-dependent predation: a seabird disproportionately targets substandard individual juvenile salmon', *Ecology*, 97(2), pp. 461–71.

Tullberg, B. S., and Hunter, A. F. (1996) 'Evolution of larval gregariousness in relation to repellent defences and warning coloration in tree-feeding Macrolepidoptera: a phylogenetic analysis based on independent contrasts', *Biological Journal of the Linnean Society*, 57(3), pp. 253–76.

Tullberg, B. S., Merilaita, S., and Wiklund, C. (2005) 'Aposematism and crypsis combined as a result of distance dependence: functional versatility of the colour pattern in the swallowtail butterfly larva', *Proceedings of the Royal Society of London B: Biological Sciences*, 272, pp. 1315–21.

Turing, A. M. (1952) 'The chemical basis of morphogenesis', *Philosophical Transactions of the Royal Society B: Biological Sciences*, 237, pp. 37–72.

Turner, E. R. A. (1961) 'Survival values of different methods of camouflage as shown in a model population', *Proceedings of the Zoological Society of London*, 136, pp. 273–84.

Turner, J. R., Kearney, E. P., and Exton, L. S. (1984) 'Mimicry and the Monte Carlo predator: the palatability spectrum,

and the origins of mimicry', *Biological Journal of the Linnean Society*, 23(2–3), pp. 247–68.

Turner, J. R. G. (1971) 'Studies of Müllerian mimicry and its evolution in burnet moths and heliconid butterflies', in Creed, E.R. (ed.) *Ecological Genetics and Evolution*. Oxford: Blackwell.

Turner, J. R. G. (1978) 'Why male butterflies are non-mimetic: natural selection, sexual selection, group selection, modification and sieving', *Biological Journal of the Linnean Society*, 10(4), pp. 385–432.

Turner, J. R. G. (1984) 'Mimicry: the palatability spectrum and its consequences', in Vane-Wright, R.I. & Ackery, P. (eds.) *The Biology of Butterflies*. Princeton: Princeton University Press.

Turner, J. R. G. (2000) 'Mimicry.', *Encyclopedia of Life Sciences*. London: Nature Publishing Group.

Twomey, E., Vestergaard, J. S., and Summers, K. (2014) 'Reproductive isolation related to mimetic divergence in the poison frog *Ranitomeya imitator*', *Nature Communications*, 5, pp. 4749.

Twomey, E., Vestergaard, J. S., Venegas, P. J., and Summers, K. (2016) 'Mimetic divergence and the speciation continuum in the mimic poison frog *Ranitomeya imitator*', *American Naturalist*, 187(2), pp. 205–24.

Twomey, E., Yeager, J., Brown, J. L., Morales, V., Cummings, M., and Summers, K. (2013) 'Phenotypic and genetic divergence among poison frog populations in a mimetic radiation', *PloS One*, 8(2). https://doi.org/10.1371/journal.pone.0055443

Tyrie, E. K., Hanlon, R. T., Siemann, L. A., and Uyarra, M. C. (2015) 'Coral reef flounders, *Bothus lunatus*, choose substrates on which they can achieve camouflage with their limited body pattern repertoire', *Biological Journal of the Linnean Society*, 114(3), pp. 629–38.

Uiblein, F., and Nielsen, J. G. (2005) 'Ocellus variation and possible functions in the genus *Neobythites* (Teleostei: Ophidiidae)', *Ichthyological Research*, 52(4), pp. 364–72.

Umbers, K. D., De Bona, S., White, T. E., Lehtonen, J., Mappes, J., and Endler, J. A. (2017) 'Deimatism: a neglected component of antipredator defence', *Biology Letters*, 13(4), pp. 20160936.

Umbers, K. D., and Mappes, J. (2016) 'Towards a tractable working hypothesis for deimatic displays', *Animal Behaviour*, 113, pp. e5-e7.

Umbers, K. D. L., and Mappes, J. (2015) 'Postattack deimatic display in the mountain katydid, *Acripeza reticulata*', *Animal Behaviour*, 100, pp. 68–73.

Utne-Palm, A. (1999) 'The effect of prey mobility, prey contrast, turbidity and spectral composition on the reaction distance of *Gobiusculus flavescens* to its planktonic prey', *Journal of Fish Biology*, 54(6), pp. 1244–58.

Valkonen, J., Niskanen, M., Björklund, M., and Mappes, J. (2011) 'Disruption or aposematism? Significance of dorsal zigzag pattern of European vipers', *Evolutionary Ecology*, 25(5), pp. 1047–63.

Valkonen, J. K., Nokelainen, O., Niskanen, M., Kilpimaa, J., Björklund, M., and Mappes, J. (2012) 'Variation in predator species abundance can cause variable selection pressure on warning signaling prey', *Ecology and Evolution*, 2(8), pp. 1971–6.

Vallin, A., Dimitrova, M., Kodandaramaiah, U., and Merilaita, S. (2011) 'Deflective effect and the effect of prey detectability on anti-predator function of eyespots', *Behavioral Ecology and Sociobiology*, 65(8), pp. 1629–36.

Vallin, A., Jakobsson, S., Lind, J., and Wiklund, C. (2005) 'Prey survival by predator intimidation: an experimental study of peacock butterfly defence against blue tits', *Proceedings of the Royal Society of London B: Biological Sciences*, 272(1569), pp. 1203–7.

Vallin, A., Jakobsson, S., and Wiklund, C. (2007) ' "An eye for an eye?" — on the generality of the intimidating quality of eyespots in a butterfly and a hawkmoth', *Behavioral Ecology and Sociobiology*, 61(9), pp. 1419–24.

Van Alphen, J. J. M. (1999) 'Evolution of colour patterns in East African cichlid fish', *Journal of Evolutionary Biology*, 12(3), pp. 514–34.

Van Buskirk, J., And Anderwald, P., Lupold, S., Reinhardt, L., and Schuler, H. (2003) 'The lure effect, tadpole tail shape, and the target of dragonfly strikes', *Journal of Herpetology*, 37(2), pp. 420–4.

Van Buskirk, J., Aschwanden, J., Buckelmuller, I., Reolon, S., and Ruttiman, S. (2004) 'Bold tail coloration protects tadpoles from dragonfly strikes', *Copeia*, (3), pp. 599–602.

Van Santen, J. P., and Sperling, G. (1985) 'Elaborated reichardt detectors', *JOSA A*, 2(2), pp. 300–21.

van Someren, V. G. L., and Jackson, T. H. E. (1959) 'Some comments on the adaptive resemblance amongst African Lepidoptera (Rhopalocera)', *Journal of the Lepidopterist's Society*, 13, pp. 121–50.

Van Veen, J., Sommeijer, M. J., and Monge, I. A. (1999) 'Behavioural development and abdomen inflation of gynes and newly mated queens of *Melipona beecheii* (Apidae, Meliponinae)', *Insectes Sociaux*, 46(4), pp. 361–5.

Vanin, S. A., and Guerra, T. J. (2012) 'A remarkable new species of flesh-fly mimicking weevil (Coleoptera: Curculionidae: Conoderinae) from Southeastern Brazil', *Zootaxa*, 3413, pp. 55–6.

VanRullen, R., Reddy, L., and Koch, C. (2005) 'Attention-driven discrete sampling of motion perception', *Proceedings of the National Academy of Sciences of the United States of America*, 102(14), pp. 5291–6.

Vaughan, F. A. (1983) 'Startle responses of blue jays to visual stimuli presented during feeding', *Animal Behaviour*, 31(2), pp. 385–96.

Vega-Redondo, F., and Hasson, O. (1993) 'A game-theoretic model of predator–prey signaling', *Journal of Theoretical Biology*, 162(3), pp. 309–19.

Vencl, F. V., Ottens, K., Dixon, M. M., Candler, S., Bernal, X. E., Estrada, C., and Page, R. A. (2016) 'Pyrazine emission by a tropical firefly: An example of chemical aposematism?', *Biotropica*, 48(5), pp. 645–55.

Vermaas, W. F. J., Timlin, J. A., Jones, H. D. T., Sinclair, M. B., Nieman, L. T., Hamad, S. W., Melgaard, D. K., and Haaland, D. M. (2008) 'In vivo hyperspectral confocal fluorescence imaging to determine pigment localization and distribution in cyanobacterial cells', *Proceedings of the National Academy of Sciences*, 105, pp. 4050–5.

Vester, H. I., Folkow, L. P., and Blix, A. (2004) 'Click sounds produced by cod (*Gadus morhua*)', *The Journal of the Acoustical Society of America*, 115(2), pp. 914–19.

Vestheim, H., and Kaartvedt, S. (2006) 'Plasticity in coloration as an antipredator strategy among zooplankton', *Limnology and Oceanography*, 51(4), pp. 1931–4.

Vidal-Cordero, J. M., Moreno-Rueda, G., López-Orta, A., Marfil-Daza, C., Ros-Santaella, J. L., and Ortiz-Sánchez, F. J. (2012) 'Brighter-colored paper wasps (*Polistes dominula*) have larger poison glands', *Frontiers in Zoology*, 9(1), pp. 20.

Vincent, B. T., Baddeley, R. J., Troscianko, T., and Gilchrist, I. D. (2005) 'Is the early visual system optimised to be energy efficient?', *Network: Computation in Neural Systems*, 16(2–3), pp. 175–90.

Vinther, J., Nicholls, R., Lautenschlager, S., Pittman, M., Kaye, T. G., Rayfield, E., Mayr, G., and Cuthill, I. C. (2016) '3D camouflage in an ornithischian dinosaur', *Current Biology*, 26, pp. 2456–62.

Vinyard, G. L., and O'Brien, W. J. (1975) 'Dorsal light response as an index of prey preference in bluegill (*Lepomis macrochirus*)', *Journal of the Fisheries Board of Canada*, 32(10), pp. 1860–3.

Vlieger, L., and Brakefield, P. M. (2007) 'The deflection hypothesis: eyespots on the margins of butterfly wings do not influence predation by lizards', *Biological Journal of the Linnean Society*, 92(4), pp. 661–7.

Vogel, H. H. (1950) 'Observations on social behavior in turkey vultures', *The Auk*, 67(2), pp. 210–16.

von Helversen, B., Schooler, L. J., and Czienskowski, U. (2013) 'Are stripes beneficial? Dazzle camouflage influences perceived speed and hit rates', *PLoS One*, 8(4), pp. e61173.

Vondrick, C., Khosla, A., Malisiewicz, T. and Torralba, A. (2013) 'HOGgles: Visualizing object detection features'. *Proceedings of the IEEE International Conference on Computer Vision*, 1–8. IEEE.

Vrieling, H., Duhl, D. M., Millar, S. E., Miller, K. A., and Barsh, G. S. (1994) 'Differences in dorsal and ventral pigmentation result from regional expression of the mouse agouti gene', *Proceedings of the National Academy of Sciences*, 91, pp. 5667.

Wagemans, J., Elder, J. H., Kubovy, M., Palmer, S. E., Peterson, M. A., Singh, M., and von der Heydt, R. (2012) 'A century of Gestalt psychology in visual perception: I. Perceptual grouping and figure–ground organization', *Psychological Bulletin*, 138(6), pp. 1172.

Waldbauer, G. P. (1988) 'Aposematism and Batesian mimicry - measuring mimetic advantage in natural habitats', *Evolutionary Biology*, 22, pp. 227–59.

Waldbauer, G. P. (1996) *Insects Through the Seasons*. Cambridge, Massachusetts: Harvard University Press.

Waldbauer, G. P., and Laberge, W. E. (1985) 'Phenological relationships of wasps, bumblebees, their mimics and insectivorous birds in Northern Michigan', *Ecological Entomology*, 10(1), pp. 99–110.

Waldbauer, G. P., and Sheldon, J. K. (1971) 'Phenological relationships of some aculeate Hymenoptera, their dipteran mimics, and insectivorous birds', *Evolution*, 25(2), pp. 371–82.

Waldbauer, G. P., and Sternburg, J. G. (1975) 'Saturniid moths as mimics - alternative interpretation of attempts to demonstrate mimetic advantage in nature', *Evolution*, 29(4), pp. 650–8.

Waldbauer, G. P., and Sternburg, J. G. (1987) 'Experimental field demonstration that two aposematic butterfly color patterns do not confer protection against birds in Northern Michigan', *American Midland Naturalist*, 118(1), pp. 145–52.

Waldbauer, G. P., Sternburg, J. G., and Maier, C. T. (1977) 'Phenological relationships of wasps, bumblebees, their mimics, and insectivorous birds in an Illinois sand area', *Ecology*, 58(3), pp. 583–91.

Waldron, S. J., Endler, J. A., Valkonen, J. K., Honma, A., Dobler, S., and Mappes, J. (2017) 'Experimental evidence suggests that specular reflectance and glossy appearance help amplify warning signals', *Scientific Reports*, 7(1), pp. 257.

Wallace, A. R. (1865) 'On the phenomena of variation and geographical distribution as illustrated by the Papilionidae of the Malayan region', *Transactions of the Linnean Society of London*, 25, pp. 1–71.

Wallace, A. R. (1867) [Untitled] *Proceedings of the Entomological Society of London*, March 4th, pp. IXXX–IXXXi.

Wallace, A. R. (1877) 'The colors of animals and plants', *The American Naturalist*, 11, pp. 641–62.

Wallace, A. R. (1882) 'Dr Fritz Müller on some difficult cases of mimicry', *Nature*, 26, pp. 86–7.

Wallace, A. R. (1889) *Darwinism: An exposition of the theory of natural selection with some of its applications*. London: MacMillan & Co.

Wallach, H. (1935) 'Ueber visuell wahrgenommene Bewegungsrichtung', *Psychologishe Forschung*, 20, pp. 325–80.

Walther, B. A. (2002) 'Vertical stratification and use of vegetation and light habitats by neotropical forest birds', *Journal of Ornithology*, 143, pp. 64–81.

Wang, I. J. (2011) 'Inversely related aposematic traits: reduced conspicuousness evolves with increased toxicity in a polymorphic poison-dart frog', *Evolution*, 65(6), pp. 1637–49.

Wang, I. J., and Shaffer, H. B. (2008) 'Rapid color evolution in an aposematic species: a phylogenetic analysis of color variation in the strikingly polymorphic strawberry poison-dart frog', *Evolution*, 62(11), pp. 2742–59.

Warrant, E. J., and Locket, N. A. (2004) 'Vision in the deep sea', *Biological Reviews*, 79, pp. 671–712.

Weber, M. G., and Agrawal, A. A. (2014) 'Defense mutualisms enhance plant diversification', *Proceedings of the National Academy of Sciences*, 111(46), pp. 16442–7.

Webster, R. J. (2015) 'Does disruptive camouflage conceal edges and features?', *Current Zoology*, 61(4), pp. 708–17.

Webster, R. J., Callahan, A., Godin, J.-G. J., and Sherratt, T. N. (2009) 'Behaviourally mediated crypsis in two nocturnal moths with contrasting appearance', *Philosophical Transactions of the Royal Society of London B: Biological Sciences*, 364(1516), pp. 503–10.

Webster, R. J., Godin, J.-G. J., and Sherratt, T. N. (2015) 'The role of body shape and edge characteristics on the concealment afforded by potentially disruptive marking', *Animal Behaviour*, 104, pp. 197–202.

Webster, R. J., Hassall, C., Herdman, C. M., Godin, J. G., and Sherratt, T. N. (2013) 'Disruptive camouflage impairs object recognition', *Biol Lett*, 9(6), pp. 20130501.

Wehrle-Haller, B. (2003) 'The role of Kit-ligand in melanocyte development and epidermal homeostasis', *Pigment Cell & Melanoma Research*, 16, pp. 287–96.

Weldon, P. J. (2013) 'Chemical aposematism', *Chemoecology*, 23(4), pp. 201–2.

Weldon, P. J. (2017) 'Poison frogs, defensive alkaloids, and sleepless mice: critique of a toxicity bioassay', *Chemoecology*, 27(4), pp. 123–6.

Wertheimer, M. (1923) 'A brief introduction to gestalt, identifying key theories and principles', *Psychol Forsch*, 4, pp. 301–50.

West, D. A. (1994) 'Unimodal Batesian polymorphism in the Neotropical swallowtail butterfly *Eurytides lysithous* (Hbn)', *Biological Journal of the Linnean Society*, 52(3), pp. 197–224.

White, S. L. and Gowan, C. (2014) 'Social learning enhances search image acquisition in foraging brook trout', *Environmental biology of fishes*, 97(5), pp. 523–8.

Whitman, D. W., Blum, M. S., and Alsop, D. W. (1990) 'Allomones: chemicals for defense', *Insect Defenses: adaptive mechanisms and strategies of prey and predators*, pp. 289–351. David L. Evans and Justin O. Schmidt [eds.] State University of New York, Albany, 1990.

Whoriskey, F. G. (1991) 'Stickleback distraction displays: Sexual or foraging deception against egg cannibalism', *Animal Behaviour*, 41, pp. 989–95.

Whoriskey, F. G., and Fitzgerald, G. J. (1985) 'Sex, cannibalism and sticklebacks', *Behavioral Ecology and Sociobiology*, 18(1), pp. 15–18.

Widder, E. A. (1999) 'Bioluminescence', in Editors: Archer, S., Djamgoz, M.B., Loew, E., Partridge, J.C., Vallerga, S. (Eds.) *Adaptive mechanisms in the ecology of vision*: Dordrecht: Springer, pp. 555–81.

Wiklund, C., and Järvi, T. (1982) 'Survival of distasteful insects after being attacked by naive birds: a reappraisal of the theory of aposematic coloration evolving through individual selection', *Evolution*, 36(5), pp. 998–1002.

Wiklund, C., Vallin, A., Friberg, M., and Jakobsson, S. (2008) 'Rodent predation on hibernating peacock and small tortoiseshell butterflies', *Behavioral Ecology and Sociobiology*, 62(3), pp. 379–89.

Wilkening, D. A. (2000) 'A simple model for calculating ballistic missile defense effectiveness', *Science & Global Security*, 8(2), pp. 183–215.

Wilkinson, D. M., and Sherratt, T., N. (2008) 'The art of concealment', *Biologist*, 55, pp. 10.

Williams, B. L., Hanifin, C. T., Brodie Jr, E. D., and Brodie III, E. D. (2010) 'Tetrodotoxin affects survival probability of rough-skinned newts (*Taricha granulosa*) faced with TTX-resistant garter snake predators (*Thamnophis sirtalis*)', *Chemoecology*, 20(4), pp. 285–90.

Williams, D. (2001) *Naval Camouflage 1914–1945: a complete visual reference.* London: Chatham Publishing.

Williams, P. (2007) 'The distribution of bumblebee colour patterns worldwide: possible significance for thermoregulation, crypsis, and warning mimicry', *Biological Journal of the Linnean Society*, 92(1), pp. 97–118.

Williams, P. H. (1998) 'An annotated checklist of bumble bees with an analysis of patterns of description (Hymenoptera: Apidae, Bombini)', *Bulletin of the Natural History Museum, London (Entomology)*, 67, pp. 79–152.

Willink, B., Brenes-Mora, E., Bolaños, F., and Pröhl, H. (2013) 'Not everything is black and white: color and behavioral variation reveal a continuum between cryptic and aposematic strategies in a polymorphic poison frog', *Evolution*, 67(10), pp. 2783–94.

Willmott, K. R., and Mallet, J. (2004) 'Correlations between adult mimicry and larval host plants in ithomiine butterflies', *Proc. R. Soc. Lond. B. (Suppl.) Biology Letters*, 271, pp. S266–9.

Wilson, J. S., Jahner, J. P., Forister, M. L., Sheehan, E. S., Williams, K. A. and Pitts, J. P. (2015) 'North American velvet ants form one of the world's largest known Müllerian mimicry complexes', *Current Biology*, 25(16), pp. R704–R706.

Wilson, J. S., Jahner, J. P., Williams, K. A., and Forister, M. L. (2013) 'Ecological and evolutionary processes drive the origin and maintenance of imperfect mimicry', *PloS One*, 8(4). https://doi.org/10.1371/journal.pone.0061610

Wilson, J. S., Williams, K. A., Forister, M. L., von Dohlen, C. D., and Pitts, J. P. (2012) 'Repeated evolution in overlapping mimicry rings among North American velvet ants', *Nature Communications*, 3, 1272. doi:10.1038/ncomms2275

Wilson, R. P., Griffiths, I. W., Mills, M. G., Carbone, C., Wilson, J. W., and Scantlebury, D. M. (2015) 'Mass enhances speed but diminishes turn capacity in terrestrial pursuit predators', *Elife*, 4, pp. e06487.

Winemiller, K. O. (1990) 'Caudal eyespots as deterrents against fin predation in the neotropical cichlid *Astronotus ocellatus*', *Copeia*, (3), pp. 665–73.

Winters, A. E., Stevens, M., Mitchell, C., Blomberg, S. P., and Blount, J. D. (2014) 'Maternal effects and warning signal honesty in eggs and offspring of an aposematic ladybird beetle', *Functional Ecology*, 28(5), pp. 1187–96.

Woodland, D., Jaafar, Z., and Knight, M.-L. (1980) 'The "pursuit deterrent" function of alarm signals', *The American Naturalist*, 115(5), pp. 748–53.

Wourms, M. K., and Wasserman, F. E. (1985) 'Butterfly wing markings are more advantageous during handling than during the initial strike of an avian predator', *Evolution*, 39(4), pp. 845–51.

Wright, D. I., and O'Brien, W. J. (1982) 'Differential location of *Chaoborus* larvae and Daphnia by fish: the importance of motion and visible size', *American Midland Naturalist*, 108(1), pp. 68–73.

Wright, J. J. (2011) 'Conservative coevolution of Müllerian mimicry in a group of rift lake catfish', *Evolution*, 65(2), pp. 395–407.

Wright, S. (1932) 'The roles of mutation, inbreeding, crossbreeding, and selection in evolution', Proceedings of the Sixth Annual Congress of Genetics, 1, pp. 356–66. Reprinted in *Sewall Wright, Evolution: Selected Papers*, William B. Provine (ed.). Chicago: University of Chicago Press, pp. 161–177.

Wu, G.-M., Boivin, G., Brodeur, J., Giraldeau, L.-A., and Outreman, Y. (2010) 'Altruistic defence behaviours in aphids', *BMC Evolutionary Biology*, 10(1), pp. 19.

Wuerger, S., Shapley, R., and Rubin, N. (1996) '"On the visually perceived direction of motion" by Hans Wallach: 60 years later', *Perception*, 25(11), pp. 1317–67.

Wüster, W., Allum, C. S., Bjargardóttir, I. B., Bailey, K. L., Dawson, K. J., Guenioui, J., Lewis, J., McGurk, J., Moore, A. G., and Niskanen, M. (2004) 'Do aposematism and Batesian mimicry require bright colours? A test, using European viper markings', *Proceedings of the Royal Society of London B: Biological Sciences*, 271(1556), pp. 2495–9.

Xiao, F., and Cuthill, I. C. (2016) 'Background complexity and the detectability of camouflaged targets by birds and humans'. *Proceedings of the Royal Society of London B:*

Biological Sciences, 283(1838), 20161527. doi: 10.1098/rspb.2016.1527

Yachi, S., and Higashi, M. (1998) 'The evolution of warning signals', *Nature*, 394(6696), pp. 882–4.

Yack, J. E., and Fullard, J. H. (2000) 'Ultrasonic hearing in nocturnal butterflies', *Nature*, 403, pp. 265–6.

Yeager, J., Brown, J. L., Morales, V., Cummings, M., and Summers, K. (2012) 'Testing for selection on color and pattern in a mimetic radiation', *Current Zoology*, 58(4), pp. 668–76.

Yokoo, R., Hood, R. D., and Savage, D. F. (2015) 'Live-cell imaging of cyanobacteria', *Photosynthesis Research*, 126, pp. 33–46.

York, J. R., and Baird, T. A. (2016) 'Juvenile collared lizards adjust tail display frequency in response to variable predatory threat', *Ethology*, 122(1), pp. 37–44.

Yosef, R., and Whitman, D. W. (1992) 'Predator exaptations and defensive adaptations in evolutionary balance: no defence is perfect', *Evolutionary Ecology*, 6(6), pp. 527–36.

Young, B. A., Solomon, J., and Abishahin, G. (1999) 'How many ways can a snake growl? The morphology of sound production in *Ptyas mucosus* and its potential mimicry of *Ophiophagus*', *Herpetological Journal*, 9(3), pp. 89–94.

Young, R., and Arnold, J. (1982) 'The functional morphology of a ventral photophore from the mesopelagic squid, *Abralia trigonura*', *Malacologia*, 23(1), pp. 135–63.

Young, R. E. (1983) 'Oceanic bioluminescence: an overview of general functions', *Bulletin of Marine Science*, 33(4), pp. 829–45.

Young, R. E., and Mencher, F. M. (1980) 'Bioluminescence in mesopelagic squid: diel color change during counter-illumination', *Science*, 208, pp. 1286–8.

Young, R. E., and Roper, C. F. (1976) 'Bioluminescent countershading in midwater animals: evidence from living squid', *Science*, 191, pp. 1046–8.

Young, R. E., and Roper, F. E. (1977) 'Intensity regulation of bioluminescence during countershading in living midwater animals', *Fishery Bulletin*, 75, pp. 239–52.

Zangerl, A. R., and Berenbaum, M. R. (2005) 'Increase in toxicity of an invasive weed after reassociation with its coevolved herbivore', *Proceedings of the National Academy of Sciences of the United States of America*, 102(43), pp. 15529–32.

Zaret, T. M. (1972) 'Predators, invisible prey, and the nature of polymorphism in the Cladocera (Class Crustacea)', *Limnology and Oceanography*, 17(2), pp. 171–84.

Zaret, T. M., and Kerfoot, W. C. (1975) 'Fish predation on *Bosmina longirostris*: Body-size selection versus visibility selection', *Ecology*, 56(1), pp. 232–7.

Zrzavy, J., and Nedved, O. (1999) 'Evolution of mimicry in the New World *Dysdercus* (Hemiptera: Pyrrhocoridae)', *Journal of Evolutionary Biology*, 12(5), pp. 956–69.

Zverev, V., Kozlov, M. V., and Zvereva, E. L. (2017) 'Variation in defensive chemistry within a polyphagous Baikal population of *Chrysomela lapponica* (Coleoptera: Chrysomelidae): potential benefits in a multi-enemy world', *Population Ecology*, 59(4), pp. 329–41.

Zvereva, E. L., and Kozlov, M. V. (2016) 'The costs and effectiveness of chemical defenses in herbivorous insects: a meta-analysis', *Ecological Monographs*, 86(1), pp. 107–24.

Zylinski, S., Darmaillacq, A.-S., and Shashar, N. (2012) 'Visual interpolation for contour completion by the European cuttlefish (*Sepia officinalis*) and its use in dynamic camouflage', *Proceedings of the Royal Society of London B: Biological Sciences*, 279(1737), pp. 2386–90.

Zylinski, S., and Johnsen, S. (2011) 'Mesopelagic cephalopods switch between transparency and pigmentation to optimize camouflage in the deep', *Current Biology*, 21(22), pp. 1937–41.

Zylinski, S., Osorio, D., and Johnsen, S. (2016) 'Cuttlefish see shape from shading, fine-tuning coloration in response to pictorial depth cues and directional illumination', *Proceedings of the Royal Society B: Biological Sciences*, 283, pp. 20160062.

Zylinski, S., Osorio, D., and Shohet, A. J. (2009a) 'Cuttlefish camouflage: context-dependent body pattern use during motion', *Proc Biol Sci*, 276(1675), pp. 3963–9.

Zylinski, S., Osorio, D., and Shohet, A. J. (2009b) 'Perception of edges and visual texture in the camouflage of the common cuttlefish, *Sepia officinalis*', *Philos Trans R Soc Lond B Biol Sci*, 364(1516), pp. 439–48.

Index